T0211823

Economies, Institutions and Territories

Presenting multidisciplinary and global insights, this book explores the nexus between economies, institutions, and territories and how global phenomena have local consequences.

It examines how original and innovative economic related processes embed themselves in societies at the local level; how boundaries between the state and the market are placed under stress by unexpected changes. It explores whether new types of elites and forms of social inequalities are emerging as a result of institutional and economic changes, and whether peripheral areas are experiencing insidious forms of economic and institutional lock-in. Presenting empirical cases and useful analytical and conceptual tools, the book makes current economic and territorial phenomena more understandable.

This is an important read for students and scholars in the fields of geography, sociology, political sciences, anthropology, economics, regional science, and international relations. It is also a valuable resource for policymakers, well-educated lay readers and economic, political and international relations journalists.

Luca Storti, Associate Professor in Economic Sociology, University of Torino, Italy, and Research Fellow of the Ralph Bunche Institute for International Studies in the Graduate Center (City University of New York). His research interests involve the topic of organized crime groups at boundaries between legal and illegal markets, and the relationships between Institutions and the Economy.

Giulia Urso, Tenured Assistant Professor of Economic Geography, Gran Sasso Science Institute, Social Sciences, L'Aquila, Italy. Her research mostly focusses on slow-burning pressures of peripheral areas and their resilience to different kind of shocks.

Neil Reid, Professor, Department of Geography and Planning, University of Toledo, USA. He is an economic and urban geographer. Much of his current research examines the growth of the American craft brewing industry, with a focus on the role of craft breweries in neighborhood revitalization.

The Dynamics of Economic Space

This series aims to play a leading international role in the development, promulgation and dissemination of new ideas in economic geography. It has as its goal the development of a strong analytical perspective on the processes, problems, and policies associated with the dynamics of local and regional economies as they are incorporated into the globalizing world economy. In recognition of the increasing complexity of the world economy, the Commission's interests include: industrial production; business, professional and financial services, and the broader service economy including e-business; corporations, corporate power, enterprise and entrepreneurship; the changing world of work and intensifying economic interconnectedness.

Rural-Urban Linkages for Sustainable Development
Edited by Armin Kratzer and Jutta Kister

Beyond Free Market
Social Inclusion and Globalization
Edited by Fayyaz Baqir and Sanni Yaya

Culture, Creativity and Economy
Collaborative practices, value creation and spaces of creativity
Edited by Brian J. Hracs, Taylor Brydges, Tina Haisch, Atle Hauge, Johan Jansson and Jenny Sjöholm

Social Protection and Informal Workers in Sub-Saharan Africa
Lived Realities and Associational Experiences from Tanzania and Kenya
Edited by Lone Riisgaard, Winnie Mitullah and Nina Torm

Economies, Institutions and Territories
Dissecting Nexuses in a Changing World
Edited by Luca Storti, Giulia Urso and Neil Reid

Excessive Inequality and Socio-Economic Progress
Ona Gražina Rakauskienė, Dalia Streimikiene and Lina Volodzkienė

Economies, Institutions and Territories

Dissecting Nexuses in a Changing World

Edited by Luca Storti, Giulia Urso and Neil Reid

Routledge
Taylor & Francis Group

LONDON AND NEW YORK

First published 2023
by Routledge
4 Park Square, Milton Park, Abingdon, Oxon OX14 4RN

and by Routledge
605 Third Avenue, New York, NY 10158

Routledge is an imprint of the Taylor & Francis Group, an informa business

© 2023 selection and editorial matter, Luca Storti, Giulia Urso and Neil Reid; individual chapters, the contributors

The right of Luca Storti, Giulia Urso and Neil Reid to be identified as the authors of the editorial material, and of the authors for their individual chapters, has been asserted in accordance with sections 77 and 78 of the Copyright, Designs and Patents Act 1988.

British Library Cataloguing-in-Publication Data
A catalogue record for this book is available from the British Library

Library of Congress Cataloging-in-Publication Data
A catalog record has been requested for this book

ISBN: 978-1-032-04233-6 (hbk)
ISBN: 978-1-032-04234-3 (pbk)
ISBN: 978-1-003-19104-9 (ebk)

DOI: 10.4324/9781003191049

Typeset in Times New Roman
by SPi Technologies India Pvt Ltd (Straive)

Contents

PART II

Coordination between State and Market: Emerging Problems 89

PART III

Social Inequalities, Displacement and Conflicts between Social Groups 165

Figures

Tables

Contributors

Aykut Aniç has a Bachelor's Degree in Business Administration and has a Master of Accounting. He is currently a PhD Candidate at Ankara University studying Economic Geography. He works for UNDP Turkey as a project coordinator on a development project. Previously, he has worked for DG Regional Development and DG Development Agencies mainly on preparing regional development programs, conducting spatial analysis especially on regional disparities. He coordinated the KAYS/DAMIS project, which was evaluated as a public sector innovation by OECD. He is keen on the regional development field as well as the regional ecosystem and institutional context, and is also interested in skill and employment-related research.

Neil Argent is Professor in Human Geography and Head of the Department of Geography and Planning at the University of New England (UNE), Australia. His research focusses on the geography of rural economic, demographic and social change in developed world nations. He is an Associate Editor of the international *Journal of Rural Studies* and an editorial board member of the *Journal of Rural Studies* and the *Journal of Community and Rural Development*.

Laura Birou, PhD, serves as the Alico Chair of Operations Management at Florida Gulf Coast University. She has worked in the supply chain management field since 1983. A highly sought-after public speaker and educator, she has over a 30-year career in higher education and SCM Talent Development. Laura is known for innovative educational approaches in the areas of Purchasing, Operations, Supply Chain Management, Knowledge Management, Diversity, International Business, Strategy, and Ethics.

Catherine Campbell, PhD, MPH, is Assistant Professor and Extension Specialist of Community Food Systems in the Department of Family, Youth and Community Sciences at University of Florida's Institute of Food and Agricultural Sciences (UF/IFAS). She conducts social science research on food systems to support community health, sustainability, equity, and resilience. Her research focuses on understanding the behavior, motivation, and decision-making of food systems stakeholders—including producers, consumers, and local governments—with a special focus on urban food systems.

Christa D. Court, PhD, is an Assistant Professor of Regional Economics in the Food and Resource Economics Department at the University of Florida and directs the UF/IFAS Economic Impact Analysis Program. She earned her BS in Economics from Middle Tennessee State University and her MA and PhD in Economics from West Virginia University. She also serves as Extension Specialist within UF/IFAS Extension with research and extension programs focused on regional economic analysis and disaster impact analysis. Her current research focuses on the rapid assessment of agricultural losses and damages after natural and human-induced disaster events.

Diana Dakhlallah is an Assistant Professor of Organizational Behavior at the Desautels Faculty of Management, McGill University. Her research falls at the intersection of economic sociology, political economy, and organizational theory, in the contexts of public-sector organizations and healthcare markets.

Zachary DeGroot is a graduate of UCLA.

Elizabeth C. Delmelle is an Associate Professor of Geography at the University of North Carolina at Charlotte. Her research and teaching interests include geographic information science and urban and transportation geography. She is especially interested in examining how and why neighborhoods change and in discovering new datasets and methods that can further knowledge of these processes. She also serves as Associate Editor for the Journal of Transport Geography.

Francesco Duina is Professor of Sociology at Bates College (USA). He has held appointments at Harvard University, Copenhagen Business School, and the University of British Columbia (where he was Professor and Head of Sociology during 2013–2015). His work lies at the intersections of economic and political sociology, international political economy, and cultural sociology. His articles have appeared in journals such as the *Journal of European Public Policy*, *Review of International Political Economy*, *Regulation & Governance*, and *Economy and Society*. He has authored six books, including *The Social Construction of Free Trade: The EU, NAFTA, and Mercosur* and *Broke and Patriotic: Why Poor Americans Love Their Country*. He recently edited *States and Nations, Power and Civility: Hallsian Perspectives* and is a regular contributor to media outlets such as *The Guardian* and *Los Angeles Times*.

Rebecca Jean Emigh is Professor of Sociology at UCLA. She authored numerous books, articles, and chapters on comparative and historical sociology, focusing on long-term processes of social change. *How Societies and States Count* (with Dylan Riley and Patricia Ahmed) won honorable mention from the ASA CHS section, and *The Racialization of Legal Categories in the First US Census* (with Dylan Riley and Patricia Ahmed) won the SSHA 2015 Founders Prize. She was chair of the Comparative/ Historical Section of the American Sociological Association and is a co-editor of *Social Science History*.

Liz Felter, PhD, is a Regional Specialized Extension Agent in central Florida. She is located at the University of Florida, IFAS Mid-Florida Research & Education Center in Apopka, FL. She received her doctorate in Agricultural Leadership, and Communications from the University of Florida in 2013. Using qualitative research methods, specifically focus groups, her dissertation focused on behavior change that would increase water conservation practices for homeowners with automated irrigation systems. Her Extension programs address food systems from the agricultural production discipline.

Sebastian Galindo, PhD, is a Research Associate Professor in the Department of Agricultural Education and Communication at the University of Florida. His research focuses on the use of mixed methodologies for the evaluation of research, teaching, and extension/outreach initiatives. He leads UF's Strategic Collaborative for Engagement and Evaluation (SCENE) and serves as Director for the Evaluation Program of the Southeastern Coastal Center for Agricultural Health and Safety and as Technical Monitoring and Evaluation Supervisor for the Feed the Future Innovation Lab for Livestock Systems. Galindo also teaches graduate level courses on Program Evaluation, Qualitative Research, Mixed Methods, Statistical Thinking, and Methodology of Planned Change.

Jennifer Hagen is a Family Consumer Sciences Agent with UF/IFAS Extension Lee County. She holds a BA in Sociology from Eckerd College and an MS in Environmental Education from Lesley University. Her program areas include, food systems, food safety and regulations, and community resource development, specializing in land use and zoning. Jenn is a Leadership in Energy and Environmental Design (LEED) Accredited Professional with experience in green building practices and sustainable design. She is also a Certified ServSafe Instructor and Proctor for the ServSafe Food Protection Manager Certification. She is passionate about building effective relationships with stakeholders to work collectively on issues at the intersection of agriculture, food, and community health.

Brittany Hall-Scharf, is a Marine Agent with UF/IFAS Extension and the Florida Sea Grant Program. Scharf oversees coastal and marine programs addressing fisheries, habitats, water quality, and sustainable economic resource activities. She holds a BS in Biological Science and a Coastal and Marine Science Resources Certificate from Florida State University, and an MS in Biological Oceanography from the University of South Florida's College of Marine Science.

Greg Halseth is a Professor in the Geography Program at the University of Northern British Columbia, where he is also the Canada Research Chair in Rural and Small-Town Studies and the founder and Co-Director of UNBC's Community Development Institute. His research examines rural and small-town community development, and community strategies for coping with social and economic change, all with a focus upon northern B.C.'s resource-based towns.

Lisa Hickey is a Sustainable Agriculture and Food Systems Extension Agent for UF/IFAS Extension Manatee County. She has worked at the Manatee County Extension Office since March 2007. Currently in her role as an educator, Lisa works with the commercial fruit and vegetable producers. She offers diagnostic assistance, commodity resources, and pesticide safety training to assist the producers. Her outreach focuses on positive environmental impacts through sustainable farming.

Molly Jameson is the Sustainable Agriculture and Community Food Systems Agent with UF/IFAS Extension Leon County. Her program areas include community food systems, small farms education, and 4-H youth agriculture education. Within these programs, she teaches about composting, vegetable production and gardening techniques, and hands-on soil and plant science concepts for youth. Molly has an MS degree in Soil Science from North Carolina State University and a BS degree in Horticulture, Organic Crop Production from University of Florida.

Elias T. Kirche, PhD, is an Associate Professor at Florida Gulf Coast University where he currently teaches courses in Supply Chain Management, Transportation Management, and Business Analytics. He earned his PhD from the University of Houston. His research interests include applications of technology and decision support systems in supply chain management and logistics. He worked as a construction manager in the USA and in Brazil and has provided consulting services for process improvement and procurement of IT for companies in Latin America.

Ken Cai Kowalski is a PhD student in the Sociology Department of the University of North Carolina at Chapel Hill. He is interested in workers' experiences and perceptions of their labor, public trust in political institutions and expertise, and cultural justifications for social stratification. His research has been published in *Sociological Perspectives*, *Socius*, and *New Media & Society*. Currently, he is preparing to undertake dissertation research examining the relations between discursive sources of legitimation for economic inequality, political identities, and popular beliefs about the past, present, and futures of capitalism in the US context.

Alexandre Magnier, PhD, is an Assistant Professor of Agribusiness at Florida Gulf Coast University and is affiliated with the Center for Agribusiness. Before coming to Florida, he was a Program Director at the Economics and Management Center of the University of Missouri, where his work focused on the economics of biotechnologies. He obtained a PhD in Agricultural and Applied Economics from the University of Missouri. His work has been published in a variety of peer-reviewed journals and book chapters.

Michelle Marinello is a graduate student in sociology at UCLA. Their research interests are in political and economic sociology, including topics such as regional economic policies and economic integration. Prior to studying sociology, they studied mathematics at Carleton College.

Sean Markey (PhD, MCIP, RPP) is a professor and certified planner with the School of Resource and Environmental Management at Simon Fraser University. His research concerns issues of local and regional economic development, rural and small-town development, and sustainable planning and infrastructure.

Andrea Marpillero-Colomina is a spatial policy scholar. She researches the intersections of infrastructure, policy, and place. Her passion is figuring out how cities can work and feel better for people, by advancing equity, supporting anti-racist practices, honoring history, constructing sustainable infrastructure, and creating healthy and beautiful public space. In 2020–2021, she was the Robert David Lion Gardiner Fellow at the New York Historical Society. She holds a PhD in Public and Urban Policy from The New School, an MS in Urban Planning from Columbia University, and a BA from Sarah Lawrence College. She lives and loves in Brooklyn, NY.

Fiona Haslam-McKenzie is a Winthrop Professorial Research Fellow within the School of Social Science at the University of Western Australia. She has expertise in regional economic development with extensive global experience in population and socio-economic change, housing, and analysis of remote, regional and urban socio-economic indicators. She has published widely and undertaken work for the corporate and small business sectors both nationally and in Western Australia. She has conducted work for all three tiers of government, and since 2015 has been director of the Centre for Regional Development at The University of Western Australia. She is Regional Economic Development program lead on the Co-operative Research Centre for Transformation in Mining Economies. She also currently serves as a board member of the Western Australian Environmental Protection Authority.

Isabelle Nilsson is an Associate Professor of Geography and a core faculty in the Public Policy Ph.D. program at the University of North Carolina at Charlotte. Her research is focused on housing, transportation, and local economic development issues. More specifically, how firms and households make location choices and how it affects social and economic outcomes in urban settings. She also serves on the editorial boards of

Computers, Environment and Urban Systems and Regional Science Policy and Practice.

David Outerbridge is the Sustainable Food Systems Agent for UF/IFAS Extension Lee County. He holds a BA in Anthropology from SUNY Purchase and an MSc in Geography from Durham University. His research has consisted of ethnography of traditional cultural food production in the South Pacific and multivariate analysis of society hazard interaction. His work in Extension focuses on localized food systems, production, natural resources, and community resource development.

Alexandrea J. Ravenelle is an Assistant Professor in the Sociology Department of the University of North Carolina at Chapel Hill. Her research focuses on the lived experience of workers in the gig economy, and the impact of the coronavirus pandemic on gig and precarious workers. She is the author of *Hustle and Gig: Struggling and Surviving in the Sharing Economy*, which has been translated into Korean, Spanish, and Traditional Chinese. Her research has been published in *Regions, Economy and Society*, *Sociological Perspectives*, *Journal of Managerial Psychology*, *Socius*, and *New Media & Society*.

Neil Reid, Professor, Department of Geography and Planning, University of Toledo, USA. He is an economic and urban geographer. Much of his current research examines the growth pf the American craft brewing industry, with a focus on the role of craft breweries in neighborhood revitalization.

Fritz Roka, PhD, is Director of the Florida Gulf Coast University Center of Agribusiness. He works in the areas of farm management and sustainable agriculture. Prior to joining FGCU in August 2018, he worked as an Agricultural Economist at the University of Florida's Southwest Florida Research and Education Center (Immokalee). After earning BS and MS degrees in Agricultural Economics from the University of Maryland, he Piled it Higher and Deeper at North Carolina State University studying the economics of swine manure.

Laura Ryser is the Research Manager for Rural and Small-Town Studies at the University of Northern British Columbia. Her research focuses on rural economic and social restructuring processes, shared service and infrastructure arrangements in small municipalities, resource revenue sharing arrangements, municipal fiscal and jurisdictional reforms, and local government entrepreneurialism.

Luca Storti, PhD, is an Associate Professor of Economic Sociology at the University of Torino and Research Fellow of the Ralph Bunche Institute for International Studies in the Graduate Center at The City University of New York. His main research interests and publications involve the topic of international organized crime groups at boundaries between legal and illegal markets, and the relationships between institutions and the economy.

Vassilis Tselios is an Associate Professor of Regional Analysis and Policy at Panteion University of Social and Political Sciences (Athens, Greece). He holds a PhD at the LSE. He was an Assistant and Associate Professor of Regional Economics at the University of Thessaly, a Visiting Fulbright Scholar at the University of Chicago Harris School of Public Policy, a Lecturer in Economic Geography at the University of Southampton and at Newcastle University, a Researcher at the LSE and at the University of Groningen, and a Consultant at the World Bank. His current research concerns regional development and policy, institutions and decentralization.

Dimitrios Tsiotas is an engineer and mathematician and holds a PhD in Regional Science and Network Science from the University of Thessaly (Volos, Greece). He is a Tenured Assistant Professor at the Department of Regional and Economic Development, Agricultural University of Athens (Greece). He was also an Adjunct Lecturer at the University of Thessaly (at the Departments of Regional Planning and Development, and of Business Administration) and an Academic Staff at the School of Social Sciences, Hellenic Open University (Patra, Greece). His research interests include regional science and economics, transportation and spatial economics, complex network analysis, and quantitative methods in spatial planning.

Giulia Urso, Tenured Assistant Professor of Economic Geography, Gran Sasso Science Institute, Social Sciences, L'Aquila, Italy. Her research mostly focusses on slow-burning pressures of peripheral areas and their resilience to different kind of shocks.

Haoying Wang is an Assistant Professor of management at New Mexico Tech (USA). His research specializes in agricultural and environmental economics, sustainability management, and regional economics and policy. He holds a PhD degree in applied economics and operations research from Penn State. His current research focuses on climate change, water resources and irrigation, precision agriculture, remote sensing applications in natural resources management, and indigenous economies in the US Southwest. He teaches financial economics, decision theory, risk management, and internet economics. He has built rich research experiences working with minority students, local stakeholders in the US Southwest, and government agencies.

Nuri Yavan is Associate Professor at the Department of Geography at Ankara University, Turkey. His research interest and publication focus on three areas: Economic Geography, particularly industrial and FDI location, Regional Development and Policy, specifically innovation, incentives and institutions, and History and Philosophy of Geography, especially critical analysis of Turkish Geography. He has published 8 books, 21 book chapters, 32 research/project reports and 30 articles in academic journals such as *Applied Geography* and *European Planning Studies*. He has presented more than 130 papers at various national and international conferences, and also served as a consultant to many public and private organizations. He is the editor of the Dynamics of Economic Space Book Series as well as the Economic Geography Book Series. He also currently serves as a Chair of the IGU Commission on the Dynamics of Economic Spaces.

Şükrü Yılmaz has Bachelor's Degree in Political Science and Public Administration at Ankara University in Turkey and Master Degree in Regional Development and Spatial Planning at Newcastle University in the UK. He wrote a master thesis on the smart specialization strategies of Turkish regions. He is now a PhD student in Economic Geography at Kiel University in Germany. His research interests focus on regional innovation and development policy, new path development, the roles of institutions and agencies on regional development.

Hermione Xiaoqing Zhou is an undergraduate student at Bates College (USA). Her interests focus on environmental sociology, environmental justice, international political economy, and political sociology. She has published articles in *Environmental Management* and *Territory, Politics, Governance.*

Xueguang Zhou is a Professor of Sociology, the Kwoh-Ting Li Professor in Economic Development, and a senior fellow at Freeman Spogli Institute for International Studies, Stanford University (USA). His book *The Logic of Governance in China: An Organizational Approach* summarizes his decade-long research on the institutional logic of governance in China. Professor Zhou's current research examines patterns of personnel flow in the Chinese bureaucracy and the historical origins of the bureaucratic state in China.

Ling Zhu is an Assistant Professor of Sociology at Chinese University of Hong Kong. Her overarching research interests consist of two substantive topics: (1) state governance in authoritarian regimes – how the bureaucratic organizations are structured and the bureaucrats are incentivized to maintain a delicate balance between political control (which requires power centralization) and regional development (which requires power decentralization), and (2) mechanisms of reproducing economic inequality, gender segregation, and family advantages/disadvantages in China and in the United States.

Introduction

How do original and innovative economic related processes embed themselves in societies at the local level? How are the boundaries between the state and the market put under stress by unexpected changes? Are new types of elites and new forms of social inequalities emerging as a result of institutional and economic changes? Are some peripheral areas experiencing insidious forms of economic and/or institutional lock-in?

The present book deals with such intriguing questions by looking at the nexuses between economies, institutions, and territories. To this aim, the book starts with three broad definitional assumptions for the dimensions in question. Economy is conceived as a process of interactions and exchanges between actors to allocate resources. Institutions are conceived in both their formal (i.e., rules, laws, and organizations) and informal or tacit dimensions (i.e., individual habits, group routines, and social norms and values). Lastly, territories are interpreted as arenas that are socially built and constantly negotiated, where context-specific processes unfold. The chapters collected in the book are divided into four sections:

- **Integration between the Economy and Society: Innovation, Tensions and Dilemmas**. The chapters of this section mainly deal with topics such as the emerging and institutionalization of new economic processes from below involving territorial and social resources; new combinations between market and reciprocity implying forms of collective actions to enhance common goods and positive externalities (environmental issues); arising conflicts in shaping, using, converting private/public spaces in both urban and extra-urban contexts.
- **Coordination between State and Market: Emerging Problems**. In this section are chapters concerning the emerging issues related to regulating and producing public goods that make territories attractive (i.e., infrastructures, public transportation); the establishing of supra-national institutions governing economic processes across individual states, and economic institutions (i.e., monetary unions) that interconnect otherwise independent states and territorial areas.
- **Social inequalities, displacement and conflicts between social groups**. This section has chapters on the spatial restructuring within urban areas that

have social effects that are difficult to predict (i.e., market-driven gentrification); the multifaceted relationships between local and central elite groups, the global competition concerning new types of international labor divisions, and the institutionalization of informal activities.
- **The challenges of peripherality**. The chapters of this last section deal with the topics related to peripheral areas experiencing either an economic, cognitive, political, and/or institutional lock-in; the lack of voice and/or limited agency of rural/peripheral policymakers, the prospects of places left behind by globalization; the role of institutions in the development of non-core areas, institutional inertia, and policy and governance structures that can enable change.

Within each section of the book, contributions vary in terms of geographical and thematic focuses, and relate to case studies from around the world. Also, the book includes contributions that are empirically driven and based on robust theoretical insights. It mixes different methods (i.e., qualitative tools such as interviews or ethnography, and quantitative ones using secondary data) and disciplines (i.e., geography, sociology, and economic sociology). In addition, some chapters entail comparative research paying attention to the dynamics of change over time and between different territories in relation to economies–institutions–territories. These are relevant and original features of the book. At the same time, the book has a robust unitary structure, grounded in an overarching approach. Briefly speaking, the topics explored in each chapter are not only relevant in themselves: they are also "analytical gateways" for looking at the puzzling and multifaceted links between the different dimensions of society. The unifying approach and an overview of the analysis framing the relationships between economy-institutions-territories will be presented in the first chapter of the book, which is theoretical in character. A concluding chapter, summarizing and synthesizing the insights from the contributors, will close the book.

Lastly, several insights of the book concerning how economies and institutions are structured at the local level can be of interest to policymakers, economic development planners, and entrepreneurs.

Luca Storti, Giulia Urso and Neil Reid

Acknowledgements

The editors wish to extend a collective thanks to all the authors of the chapters, Faye Leerink and the team of editors at Routledge for their support and enthusiasm about this book project.

1 Economies, Territories, Institutions

Analytical Fragments of a Complicated Relationship

Luca Storti, Giulia Urso, and Neil Reid

1.1 Introduction

From early 2020, the COVID-19 pandemic has shaken the economies world-wide. However, even before it, economic processes had long been challenged by the difficult transition towards post-industrial economies, the unexpected effects of the growing globalization, and instability concerning finance. Formal institutions are struggling to face challenges stemming from uncertainties in several spheres (e.g., pollution, climate change, reconciling economic growth with social inclusion, recessionary disturbances, etc.). At the same time, these challenges have an impact on routines, interaction patterns and cognitive orientations, thus affecting informal institutions. Territories, for their part, have experienced increasing pressures on their "identities". After a naive early phase of analyses of globalization – in which the death of distance (Cairncross, 1997) or a flat world (Friedman, 2005) were postulated and the relevance of territories diminished – an actual spatial turn has begun (Warf and Arias, 2009).

Given these relevant processes, the three broad spheres of economy, institutions, and territories, have been intensely scrutinized within the academic community. However, the cogs and wheels through which they are interconnected need to be further investigated. Against this background, this book aims to disentangle the behaviours of different kinds of social actors who take (economic) decisions in territorial settings and within institutional environments. In other words, the present co-edited book will deal with the puzzling nexuses between economies, institutions, and territories. Accounting for these nexuses will allow us to gain further insights and knowledge about the mutual interconnections between macro-spheres of the society and the singular features of each of the three dimensions.

Assuming a similar perspective to that we will follow in this book, several scholars have pointed out how institutions are an important ingredient for economic growth both at the local and national/supra-national level. Still, within territories, socio-economic differences are growing, so that we need to shed a better light on the pattern of governance shaping economic processes and outcomes. A relevant stream of literature has been analysing how social networks affect economic phenomena (i.e., emergence of new entrepreneurship,

DOI: 10.4324/9781003191049-1

processes of job-matching, etc.), but shocking events such as severe and sudden economic crises or global health issues can break old social networks and bring new ones into existence (Granovetter, 2017; Urso et al., 2021). These phenomena are raising new empirical questions. In addition, the attempt to expand and clarify how routines and conventions, which become established and spread locally, can help some economic processes or, conversely, hinder others, remains highly relevant, particularly for its broad practical implications (Trabalzi, 2007).

In sum, we can claim that economic processes take shape at the territorial level, and they have a significant role in affecting places and institutions. At the same time, places and institutions are deeply linked, and they impact economic dynamics as well. These mutual relationships recall the *quantum entanglement*, a physical phenomenon occurring when particles share spatial proximity and interact so that each feature of a particle cannot be analysed independently from the features of the others (Brody, 2020). In line with this approach, the chapters of the book include analyses and research underscoring how economies, institutions, and territories are interconnected at multiple levels and numerous ways. This heterogeneity has been increasing during recent years due to a combination of multiple factors such as globalization, financialization, de-industrialization, and the emergence of the above-mentioned types of crises. To understand the multifaceted connections between economies, territories, and institutions, attention should be given to transnational phenomena, local processes, new links between formal and informal spheres, and micro-macro dynamics through a comprehensive approach. From this perspective, all the chapters – covering a variety of empirical topics – explore this "new world", how it works, and its impacts. This is the main reason why such an edited book is timely.

Consistent with these premises, the book will make from the start three broad definitional assumptions for the dimensions in question. Economy will be conceived as an instituted process of interactions and exchanges between actors to allocate resources. Institutions will be conceived in both their formal (i.e., rules, laws, and organizations) and informal or tacit dimensions (i.e., individual habits, group routines and social norms and values). Lastly, territories will be interpreted as arenas which are socially built and constantly negotiated, where context-specific processes and tensions unfold.

The remaining of this chapter is organized as follows. First, we better clarify the three macro spheres upon which the book relies and try to lay out a framework to analyse their mutual interconnections. Then, we illustrate the four dimensions in which we aim to explore the interlinkages between economy, institutions, and territories, which correspond to the sections of the book. Lastly, we will briefly synthesize the contents of the chapters.

1.2 Analytical Toolkits and Framework

A good starting point is to illustrate the basic definition of the concept of economy arising from neoclassical approaches in economics. According to them, the economy is the allocation of scarce resources carried out by individuals. These individuals are conceived as strongly standardized, i.e., atomistic, provided with Olympian rationality and have comparable or similar utility functions (*homo economicus*). The outcome of actors' economic choices consists of stable equilibria as long as no exogenous shocks occur (Trigilia, 2002).

Social scientists have always been puzzled by these models, which are undoubtedly elegant and parsimonious but pay little attention to concrete empirical processes.

The first reaction we might have is to turn to an audience of economists and tell them that "the world is different and far more complicated than their models". However, this is not the optimal response. An easy way out for neoclassical economists would be to state that they know the real world is much more complicated than their models predict. We were mistaken to have treated their scientific models realistically. This would be, indeed, a huge misunderstanding for anyone applying scientific methods. Scientific models are not a reflection of reality and should primarily be evaluated with respect to their internal consistency (Boudon, 1986). Neoclassical economists would be right.

Let's try, thus, to move the debate to a strictly analytical level. Some coordinates of the main assumptions of economics that we have recalled may be helpful for our purposes. But it is at best a "zero model". First, this standard definition applies to a narrow range of phenomena. To use Polanyi's well-known dichotomy, we have evoked the formal meaning of the concept of economics. In this volume, instead, we will adopt the substantive understanding of the concept of economy (Polanyi, 2002). Thus, the economy is a broad sphere of activities, processes, and choices that involves exchanging resources, goods, and services of various kinds. These exchanges occur between social actors driven by heterogeneous mixes of motivations, cognitive and value orientations.

Moreover, the final and general outcome is not the simple sum of individual choices. Individuals are not isolated atoms; they interact. Interactions taking place within a structure of interdependence (e.g., social networks in which actors are embedded) affect the course of individual actions. There is, therefore, an aggregate effect: the whole is greater than the mere sum of the parts, viz., individual actions. Moreover, this aggregate effect is not a proper equilibrium but a kind of "spontaneous order" (Hayek, 1969), that is, the combination of self-interested individuals who are not intentionally trying to create a general and collective outcome. In other words, we share the idea that the economy shall be conceived as a sphere internal to human activities and the social fabric (Granovetter, 2017). This way of understanding the economy allows us to observe it in relation to its intersections with institutions and to focus on how it unfolds within and across territories. Before delving into the intersections, however, let's dwell on the other individual dimensions the analysis is based on, i.e., institutions and territories.

Defining institutions is not an easy task. Not surprisingly, "alternative definitions [of this concept] abound" according to "different intellectual traditions [...] and academic disciplines" (Duina, 2011: 3; Lecours, 2005).

Institutions are norms and conventions that "take on a rule-like status in social thought and action" (Douglas, 1986: 46–48, quoted in Powell and DiMaggio, 1991: 9), and "that cannot be changed easily or instantaneously" (Granovetter, 2017: 36. See also Mahoney and Thelen, 2009: 4). In other terms, they are: "sets of persistent patterns defining how some specified collection of social actions are and should be carried out" (Granovetter, 2017: 136). Thanks to their rule-like status, institutions facilitate decision-making in the presence of complex environments (Dequech, 2001). Thus, they reduce complexity (Beckert, 1996).

According to this perspective, the concept of institutions is both restrictive – they should be norms, conventions having a rule-like status – and encompassing (Granovetter, 2017; Powell and DiMaggio, 1991). Indeed, we "find institutions everywhere, from handshakes to marriages to strategic-planning departments" (Powell and DiMaggio, 1991: 9). Moreover, we "view behaviors as potentially institutionalizable over a wide territorial range, from understandings within a single family to myths of rationality and progress in the world system" (Granovetter, 2017: 9). It is almost impossible to think of social (inter)actions as something separate from institutions.

This restrictive and encompassing way of conceptualizing institutions have several corollaries that can be synthetized in two main groups.

First, though many would agree that we could think of each institution based on its unique features, institutions should also be understood with respect to their mutual interconnections. In this vein, we can view how institutions are intertwined with each other thereby giving rise to 'institutional environments' (Powell and DiMaggio, 1983, 1991) that have a certain degree of internal stability and consistency. Boundaries and internal features of institutional environments are not easy to define. We can even assume that it is not useful to attempt defining the perimeter of institutional environments in a strict way, but we can surely imagine narrowing down such an abstract concept to territorial areas or specific kinds of activities in which we recognize a consistent and stable institutional life.

Second, the concept of institutional environment is consistent with the idea that institutions are far more than "definite constellations of activity oriented to a particular kind of outcome in a well-defined sphere such as the economy or the polity" (Granovetter, 2017: 137). Institutions are both a "normative guide to behavior in defined spheres", and factors deeply shaping "individuals' cognition about the choices and frameworks they operate in" (Ibid.: 138). For the sake of empirical investigations, we can postulate distinctions between "families of institutions" such as political institutions, legal institutions, economic institutions, etc. (Ibid.: 138). Still, we do not have to reify the boundaries between each of these families since institutions provide a transversal framework in which one can make sense of broad expected events, curses of actions, and legitimated ways of doing things. Without assuming that institutions are always consistent with each other, they can be imagined as unitary frames connecting all the major aspects of social life, economic, political, cultural, ideological, religious, and others (Granovetter, 2017).

The effort to contextualize institutions in space is consistent with the attempt to introduce the third macro dimension underlying the analysis, namely, territory. The concept of territory is one of the founding concepts of the spatialized analysis of social phenomena. It is often treated as a synonym of space and place. These concepts have some overlapping areas, but they are not interchangeable (Tuan, 1977). They also have their mutual uniqueness (Duarte, 2017). In this volume, we have decided to take on the idea of territory because it implies a multilevel and dynamic process of construction. A territory is the product of interrelationships and is continuously built up through interactions involving different individual and collective actors. An incisive definition of the concept of territory is proposed by Duarte (2017: 44).

> [...] territory is a portion of space whose elements are endowed with values that reflect the culture of a person or group. In the case of the construction of place, the signifying process is centripetal, with the values coming from individuals and groups and enveloping entities and flows. A person or group projects their values on a portion of space; they see themselves reflected on it – and turn this portion of space into a place. In the case of territory the process of attributing values is centrifugal; it is a way of marking these elements with values, with the intention that any other person, entity or action that is present or occurs within this same portion of space is guided by, or even subject to the values imposed on the space.

Hence, territories are a portion of space to which social actors attribute particular social meanings, shared representations, and values. Therefore, territories are characterized by internal homogeneity since the aforementioned components tend to assume a certain uniformity. At the same time, territories are also affected by conflicts and tensions, as the different social groups perceive, value, and construct spaces in different ways. From these complex mixes of pressures towards uniformity and conflicts arise clashes or syntheses that are hard to predict. Moreover, territories are simultaneously the object of social construction processes and autonomous activators of social construction dynamics. In fact, the social meanings coagulating in territories, once they have been constructed, exert autonomous conditioning on social actors. In other words, values, social meanings, traditions, and routines with which the territories are impregnated have normative characteristics, assume a rule-like status, thereby shaping and conditioning behaviours.

We can now reconstruct the deep interconnections between the three macro dimensions under consideration as a sort of matrix.

First, it is important to notice that the concept of territory echoes the idea of institutions that we proposed above. The wide-ranging definition of institution that we have assumed, on the one hand, and the idea of territory as context activating normative pressure, on the other, tend to converge. As we have argued, institutional environments and families of institutions have persistent underpinning aspects, but they unfold in (partially) different ways throughout spatially and temporally situated contexts (Urso et al., 2021). Hence, territories are gateways to observe institutions in their concretely "being there", *sensu* Heidegger (1962). On the other hand, territories are portions of institutionalized space in as much as they are marked by relatively shared meanings, values, and routines. While restating that institutions and territories have different meanings and relate to different phenomena, they tend to merge from an analytical standpoint. Institutions and territories, in fact, are best observed and understood based on their multiple and various intersections.

As posited elsewhere (Storti, 2018), the interdependencies between economy and institutions can be synthetically schematised in the light of the main perspectives of analysis and research of Economic Sociology (see Table 1.1).[1] The PE considers how formal institutions, or in other words those manifested as systems and organizations, give shape to the economy through substantive regulation. The NIS, on the other hand, highlights how both formal and informal institutions guide actors' economic goals, the means judged to be legitimate in reaching them, and the ways considered appropriate to achieve them. Lastly, the NES, for its part, proposes a structure and relational vision of institutions (Swedberg, 2007): they are the crystallization of typical structural configurations (as regards morphology, the nature of the nodes, the type

Table 1.1 Interconnections between economy and institutions according to Economic Sociology

Perspectives of Economic Sociology	Intersection between the economy and institutions	Unit and level of analysis	Economic outcomes
PE	Regulatory and political embeddedness	Institutional collective actors. Processes of regulating and negotiating interests.	Models of capitalism. Opportunity structure and structure of substantive and formal incentives.
NES	Structural embeddedness	Individual and collective economic actors (entrepreneurs, enterprises). Relational networks and ties.	Formation and circulation of trust. Processes of opportunism and/or cooperation.
NI	Cultural and normative embeddedness	Cluster of norms regulating economic life. Population of individual or collective economic actors.	Formation of organizational fields. Processes of isomorphism.

(Adapted from Storti, 2018).

and density of ties, etc.) which influence the normative constraints and bonds of trust, opportunism, and cooperation between actors, thus giving rise to different attitudes to economic action (Burt, 2000; Granovetter, 2017). Accordingly, the NES perspective holds that it is important to examine networks' specific "topology", since the structure of social interaction is crucial to explaining emerging economic effects (Hedström, 2005; Stark, 2009).

In sum, Economic Sociology highlights that an "intimate and complex relationship" does exist between institutions and economic processes (Brinton and Nee, 1998 in Duina, 2011: 19). Institutions and the economy are deeply and mutually complementary. The empirical analysis must focus on the forms, the modalities, the intensity of this interpenetration. Still, the two macro dimensions – economy and institutions – are not conceivable if one pretends them to be independent. Institutions are the context or environment in which economic life takes place (Duina, 2011). This brings the territorial dimension back into play. Beyond the analytical aspects, in fact, if institutions are the environment in which the economic game unfolds, it is then useful to observe how this happens within spatially delimited areas.

Hence, we now turn to look at the recurring interconnections that occur between the economy and territories (see Table 1.2). First, territories are characterized by concentrations of routines, beliefs, customs, and value orientations that influence economic processes, thereby giving rise to typical economic structures at the local level (i.e., clusters, industrial districts). Second, territories 'host' (and produce) formal institutions that provide services and infrastructural endowments supporting certain economic processes (i.e., the 'local collective competitions goods' sustaining local development, see Crouch et al., 2001). Third, territories are "formed by the enactment, the dynamic articulations of entities (natural and man-made, material and immaterial) and flows" (Duarte, 2017: 19). These components constitute the backbone of economic processes, dynamics, and productions of goods and services.

In short, we have provided an essential definition of three macro dimensions underpinning socio-economic and geographical processes: economy,

Table 1.2 Interconnections between economy and territories

Territorial features	*Possible economic outcomes*
Territories as *arenas* of routines, beliefs, customs, and value orientations.	Economic structures located at the local level (i.e., clusters, industrial districts).
Territories as localization of formal institutions providing services and infrastructural endowments supporting economic processes.	'Collective competitions goods' sustaining local development.
Territories as articulations of entities (natural and man-made, material and immaterial) and flows.	Economic processes, dynamics, and productions of goods and services structured around the geographical and physical configuration of territories.

institutions, and territories. We have stated that these concepts are highly abstract. They must be narrowed down to be treated empirically. We have also highlighted the reciprocal interconnections of these three dimensions, which – if empirically observed – prove to be marked by a dense web of relationships. We have already applied a metaphor coming from the idea of quantum entanglement. We can extend the analogy with physics by recalling the quantum theory (Sciarrone, 2021: 25; see also Brody, 2020; Rovelli, 2019). According to the latter, we do not observe entities with predefined characteristics but entities that have characteristics in relation to other entities and that we grasp when they come into interaction (Sciarrone, 2021: 25).[2] This approach can be applied to economy–institutions–territories as well. It is a matter of fact that these three dimensions have their own distinct identity. Still, we observe their characteristics, structures, and properties when they interact with each other. Thus, the analytical posture of the observer and the spatio-temporal context under consideration become essential. To further articulate and specify these multiple interconnections, we break them down into four major thematic areas, each corresponding to a section of the book: Integration between the Economy and Society: Innovation, Tensions, and Dilemmas; Coordination between State and Market: Emerging Problems; Social Inequalities, Displacement, and Conflicts between Social Groups; and The Challenge of Peripherality. In the following, we highlight in depth the main contents of each section.

1.3 Integration between the Economy and Society: Innovation, Tensions, and Dilemmas

Innovation is a multifaceted social phenomenon that comes into tension with territories and institutional structures.

There is undoubtedly a naive way of approaching the phenomenon of innovation, which highlights individual factors concerning the creation of new goods and services. However, this is more about inventions than innovations. Yet, the romantic perspective pointing exclusively to individual creative talent does not help in delivering strong explanations even in relation to inventions. The moment of creation is undoubtedly exceptional, and deeply affected by the genius of individuals, but it is also the alchemy of a long process of making, exchanging information, and accessing resources.

Explanations pointing to individual factors are especially weak with respect to innovation in that innovations are, by definition, inventions becoming actual and concrete social phenomena (Ramella, 2019). In order to become established and spread – we could say in order to institutionalize themselves – innovations should be embedded in social structures.

In a well-known paper, Uzzi (1997) explains the devices whereby embeddedness in social networks shapes economic outcomes. Embeddedness is seen as a presupposition: all economic phenomena are embedded, and what makes the difference at the empirical level is the different degree of embeddedness with respect to the structural features of the concept (i.e., the crucial relevance of social networks in shaping economic outcomes). Consequently, concrete cases lie along a continuum, with over-embeddedness (when economic processes are firmly rooted in social ties) at one pole, and under-embeddedness (when market transactions prevail, and there is little involvement of relational mechanisms) at the other (Uzzi, 1996). Many studies have shown that innovation is supported by a favourable mix of strong and weak ties (Ramella, 2019), thereby finding itself in the middle between the over- and the under-embeddedness poles. Strong ties convey trust, a sense of belonging, and support stability to undertake processes that break established routines. An excess of strong ties, however, encourages conformist and routine behaviours. Therefore, weak ties are also needed. They provide, in fact, heterogeneous information and broad vision, which helps in not overlooking the signals and opportunities arising outside of deep-but-narrow social networks.

Consistently, Ramella (2019) argues that to understand innovation processes, it is necessary to look at the actors of innovation, the relationships that exist between them, and the sectoral and territorial contexts in which they operate.

As stated earlier, empirical evidence shows that heterogeneous ties are functional for innovation. A correlation between heterogeneity of ties, on the one hand, and that of nodes, on the other, has to be found. Therefore, given a social network characterized by different types of ties, we are also likely to find different kinds of actors. Both factors support innovation. As has been shown, the most ground-breaking innovations often occur in those territorial areas where complex networks are established (Ferrary and Granovetter, 2009), linking together heterogeneous actors (i.e., entrepreneurs, scientists, investors, public decision-makers). Moreover, within the complex networks that interconnect heterogeneous actors, conventions often circulate that support breakthrough changes, socially legitimize the rupture of established patterns, and morally sanction excess conformism and adherence to choices previously made by others. These environments, thus, embrace a certain level of social differences.

In short, innovation – more so if it is economic in character – takes root in social and territorial environments in which helping aspects are institutionalized (i.e., heterogeneous networks, orientation towards change, and adequate support by local politics). Given these premises, we can identify two typical modes of innovative social action, which are not necessarily antithetical but can coexist in specific institutional and territorial environments. These two modes of innovative social action have been depicted by Ron Burt and David Stark (Burt, 1995; Stark, 2009).

The first mode concerns the brokerage, i.e., those who exploit the absence of social ties and connections between two social actors, individual and/or collective ones. By bringing together otherwise separate actors, brokers facilitate the flow of innovative ideas and information (Burt, 1995).

In contrast, the second mode relates predominantly to the generation of, rather than access to, innovative ideas (Stark, 2009: 27). In network analytic terms, this suggests that innovative action "occurs at the overlap of cohesive structures where different communities (defined by their cohesive ties) intersect without dissolving their distinctive network identities" (Stark, 2009: 27). Brokerage consists of an "information flow at the gap" while the second perspective exalts the "creative friction at the overlap" (Ibid.: 28).

Both views share the idea that innovation does not grow in social and institutional environments that are too homogeneous and irenic – if any. Instead, innovation requires disconnections, holes and gaps, friction, and overlap between different social actors.

These very last remarks allow us to point out some aspects regarding the outcomes that innovations might produce. Innovations do not realize a collective or general interest as such. Instead, innovations can facilitate and support some social groups and marginalize others. Thus, innovations also generate asymmetries of power and resources and new conflicts between social groups. This happens mostly in the case of contrast between private interests and public goods.

Think, for example, of the regeneration of certain urban or suburban territorial areas. These regeneration processes imply innovative ways of organizing and building spaces. Social groups controlling a massive amount of economic resources can take control of some territorial areas, making them exclusive and inaccessible (i.e., the dynamics of extreme gentrification involving the rich and super-rich) (Butler and Lees, 2006; Lees, 2003; Storti and Dagnes, 2021).

Such puzzling and conflictual processes do not occur only in the presence of macroscopic social asymmetries and wealthy social groups. Instead, they can also happen when new ways of using or appropriating spaces emerge from below as a spontaneous form of social and collective innovation. One example among others is that of the so-called *Social Streets* (Morelli, 2019). The purpose of Social Streets is to promote good neighbourliness and social interaction, share needs, exchange knowledge, and carry out collective projects of common interest. Social Streets are remarkable experiences and experiments, proving high social effervescence. Still, regardless of the intentions of the social innovators creating and joining them, they can also have puzzling social effects. In fact, they generate new asymmetries and social disparities between territories. For several reasons, Social Streets cannot emerge anywhere. The areas where they do not arise can experience new disadvantages compared to those areas hosting this type of collective social innovation. It has been noted that such experiences emerge in middle-class and lower-middle-class areas. The most impoverished areas are less likely to generate collective social innovations. As a result, the social distance between the deprived areas and the others, both the rich and those characterized by medium income levels, increases. Also, such experiences provide localized territories with public goods that could – or perhaps should – be provided by political institutions.

In sum, innovation practices of different kinds lie at the intersection of economy and society and are a strategic 'point of view' for analysing territorially located institutional environments. Through these topics, several chapters of the book will reconstruct some dynamics mainly concerning economic resources and producing conflicts and tensions between different social groups. In addition, it will be highlighted how global dynamics having disruptive effects on territories (i.e., climate change or pandemic shocks) have facilitated the reorganization of economic sectors and stimulated innovative practices of resistance by local communities.

1.4 Coordination between State and Market: Emerging Problems

It is impossible to explain the existence of markets without acknowledging the role of institutions. When conceptualizing the multi-faceted intersections among institutions, the economy, and territories, as Bathelt and Glückler (2014: 342–343) point out from a geographical standpoint, "a natural starting point is to think about the incentives and constraints that impact economic action and interaction in spatial perspective". This means to primarily acknowledge their structuring role of economic spaces – first through the regulation of the markets – in terms of the kinds of incentives for action they generate and how these incentives change (Rodríguez-Pose and Storper, 2006). Reflecting on this aspect, i.e., on the incentives for individuals to do some things and not do others, is extremely important in that this determines the basic ways they can participate in the economy (Rodríguez-Pose and Storper, 2006). Influencing perceptions and preferences of economic agents and, thus, the interaction patterns in market transactions (Bathelt and Glückler, 2011), institutions produce and reproduce the geography of market-making (Bathelt and Glückler, 2014). From this perspective (see Martin, 2000; Wood, 2011), the economy is therefore interpreted as a "sociocultural process, founded on contested norms and values to which institutions give expression and which shape the incentives for investment and enterprise" (Tomaney, 2014: 133). This is consistent with the endeavour of stressing the interpenetration between the economy and institutions (see Section 1.2).

While it appears obvious that "first-nature" geography – the physical and natural conditions that shape spaces and differentiate them in the first place – influences economic growth via natural resource endowments; centrality vs remoteness and hence accessibility, logistics, and transport costs; knowledge and technology diffusion, which, among other elements, affect productivity, human capital accumulation and opportunities for innovation (Ketterer and Rodríguez-Pose, 2018), this is not automatically clear in the case of institutions. A number of scholars (Acemoglu et al., 2001; Rodrik, 2000; Rodrik et al., 2004, among others) maintain that institutional aspects, and especially the quality of local institutions, which is reflected in a society's formal and informal rules and norms, are as crucial as – or even more crucial than – underlying factors explaining differentials in economic development.

What are then the channels through which institutions operate in producing economic spaces? Broadly speaking, it is rather intuitive to see how a territory's political and legal setting, and more specifically its strength in contract enforcement and in the protection of property rights, affects its government credibility and ability to reduce uncertainty, thereby reducing the monetary costs of economic activity (Ketterer and Rodríguez-Pose, 2018). These conditions are referred to collectively as a territory's business climate (Motoyama and Hui, 2015). Farole et al. (2010) explain how both community-level and society-level institutions reduce uncertainty about the likely behaviour of others, in that way facilitating commitments and possible investment. More specifically, "community does this largely through interpersonal trust; society by enabling cooperation through transparency of rules and confidence in their enforcement" (Farole et al., 2010: 62). In their attempt to conceptualize how contextually mediated institutions influence micro-economic action, forming economic agents' expectations, the authors indicate three main mechanisms through which they do it: "first through their impact on the level of economic exchange and the transactions costs of that exchange; then by the way in which they influence the rate of technological change; and finally, by their indirect effects through socio-political channels" (Ibid.: 62). Institutions affect the efficiency (and the relative costs) of economic transactions as well as the dynamic of search for, cost of finding, and social access to resources. Consequently, given that innovation is highly dependent on searching, finding and matching resources, and on the capacity of places to absorb new knowledge and learn, institutions also impact technological progress. Besides influencing productivity, through political processes and the underlying values informing them, they guide the distribution of the outputs of economic growth across society, the resolution of common pool resources and public goods provision. By implication, institutions not only influence the levels of economic development but also its quality, because of the distributional consequences of governmental/political choices (Farole et al., 2010), which allocate resources and power to some groups (Acemoglu et al., 2005) or places instead of others and formally determine the structure of participation, which in turn determine distributional outcomes and shape the political institutions that emerge (Farole et al., 2010). In the United States, for instance, inter-state competition for investment, results in state governments providing large multinational corporations with incentive packages (tax breaks, worker training, infrastructure, etc,) as an enticement to locate within their territorial boundaries (Hanson, 2021). Appropriate distributional arrangements encourage participation both in the economy and the social/civic life, by providing incentives for those who are not the first beneficiaries of economic gains or new technologies. The function of institutions as moderators of inequalities is thus central (see also Section 1.5). As well explained by Rodríguez-Pose and Storper (2006: 8), in fact, this "improves overall investments in the creation of skill, raises the incentives to participate fully in the formal economy and to become an entrepreneur (hence, participation rates and levels), and improves the willingness to pay taxes and to invest". Institutions are then called to give voice to agents and groups who otherwise would remain unheard, first and foremost by markets (see Section 1.6 for an account of how this issue intertwines with peripherality). This is extremely relevant to avoid incomplete representation and hence disparities due to misalignment of privileges. In fact, by aggregating preferences (hopefully on a base as large as possible), enabling the matching of individuals' interests (and mitigating the conflicts arising from it), and formulating strategies to fully satisfy them, institutions fix the priorities which enter the political agenda, also in the economic field.

With groups and places constantly competing for power, also through markets, there will always be winners and losers, with some of them extracting more rents and wealth than others (Farole et al., 2010) and struggling to keep the status quo.

As basic regulators of human interaction and economic activity, it is now easier to grasp how, through the above processes, institutions operate, and therefore produce spatial patterning and differentiation of development, both within and between places.

Groups and sub-national spaces compete for power at the same time internally (within competition) and externally at different geographical scales (between competition). Competition, through the selective forces of markets, produces spatial unevenness in the access to the gains of economic growth. In Tomaney's interpretation, "regional disparities reflect the rational response of agents to market signals indicating where the best returns on skills and capital are obtained" (Tomaney, 2014: 132). Institutions, fostering competitiveness over cohesion, may magnify this process. For instance, a policy paradigm emphasizing agglomeration economies, hence privileging a city-centric approach, focussing on "people" instead of "places" has for decades further exacerbated spatial imbalances in Europe and beyond (see also the previous section in the chapter). Institutions in this sense may alter the environment within which future economic decisions are made: it pushes some firms out of the market, encourages others to enter, and reshuffles the relative efficiency of competing agents. In other words, they may act as selection environments equal to markets, privileging certain techniques, economic sectors, organizational forms, groups (through distributional systems) over others and pushing places towards a constant imperative to innovate and increase attractiveness.

The same applies to sectors within an economic system. As shown by Christopherson (2002) in her work on national investment regimes, the dominant rules that shape and constrain investment time horizons and decisions produce the societal conditions for nationally divergent competitive advantages, favouring some organizational forms and practices over others, hence promoting particular sectoral strengths within the U.S. economy. As summarized by Gertler (2010), the institutional architectures through which the state operates define ground rules, produce incentives, and transmit signals to economic actors, as well as redistributing income between individuals, social classes, and regions.

By favouring, as stated, some economic and social outcomes over others (which is by no means a neutral process), an important corollary is that institutions also select the values of a society: "the structure of the decision-making process determines whose beliefs matter" (Tomaney, 2014: 135). This means that, besides establishing the technical conditions for growth, they are concerned with the production of social and political values (Tomaney, 2014), and as strongly stressed by Pike et al. (2007), are the arena where the dilemma of what kind of local and regional development and for whom comes to be debated. Speaking of the quality of growth that is pursued by governments, it should be also noticed that acceptable distributional arrangements help economies to adjust successfully to the inevitable shocks and setbacks of the development process (Rodrik, 1999). These arguments are amplified under conditions of austerity or crises, when choices about the allocation of resources become even more salient and where institutions, including local and regional governments, are more likely to be transmission mechanisms (Tomaney, 2014) for the priorities of capital, markets, and elites (Streek and Schäfer, 2013; Tomaney et al., 2010).

Given the worldwide threat of the COVID-19 pandemic, and its socio-economic effects, it is worth stressing that markets are highly sensitive to shocks and institutions can be key in reassuring them and absorbing the disturbance engendered by the shock. High-quality institutions are of paramount importance because, especially in uncertain periods, countries deemed to have stronger and trustworthy institutions may attract investments more than those with weak ones, thereby fuelling self-reinforcing mechanisms of divergence. They also contribute to "adaptive efficiency" of places and societies, that is their ability to adjust to shocks in a world characterized by ubiquitous uncertainty and ergodicity and under conditions of bounded rationality (North, 1990, 1991, 2005). Going beyond their view as instruments responding to the exigencies of market pressures, during such times, institutions are more than ever called to exert their function as mechanisms for the mitigation of collective action problems in the face of the challenge of development (Tomaney, 2014). The distinction operated by Acemoglu and Robinson (2012) in both the economy and polity between inclusive vs extractive institutions may be enlightening here. Inclusive institutions distribute power broadly, ensuring the proceeds of growth are allocated equitably. Extractive institutions limit political rights and redistribute resources to elites that have few incentives to invest and innovate. In the face of a shocking event, being in presence of the latter might have a destructive impact on the economy and the social fabric, widening inequalities to economically and socially unsustainable levels.

1.5 Social Inequalities, Displacement, and Conflicts between Social Groups

In recent years, comparative studies have shown that many countries have been struggling to combine economic development with social cohesion and a reduction in the level of social inequalities.

During the golden age of Fordism, which ended in the early 1980s, most Western countries had managed to create a model of inclusive growth, i.e., good economic performance alongside a containment of inequality and a good chance of relative social mobility (Trigilia, 2020). In the current phase of a highly globalized and financial economy, inequalities are growing, both across and within countries (Crouch, 2019; Piketty, 2014; Storti and Dagnes, 2021).

These trends have also given rise to processes of increasing territorial differences. The current global scenario is much more contradictory and chaotic than the one prevailing in the last decades of the twentieth century (Trigilia, 2020). Just think, as an example, of some territorial cleavages that can be historically identified in some countries.

In Italy, the so-called North–South divide is once again growing after a few decades in which it diminished (Viesti, 2021). Above all, the internal variance within the areas of Southern Italy has increased (Busso and Storti, 2014). In an overall difficult moment, some territories of the South manage to push back the crisis, while others are marked by growing social and economic difficulties.

The differences between macro-areas within the U.K. have started again to make headlines. The South of the country, basically southeast England and the London metropolitan area, has increased its central position in international financial activities and service economies. Other industrial areas of the country, which are located elsewhere in England and Scotland, are marked by a more stagnant economy and rising social inequalities. These issues are compounded by some of the political tensions regarding Brexit (Keating, 2020), which could reinvigorate the demands for independence in some countries of the United Kingdom, such as Scotland and Northern Ireland.

In the United States, the debate on the existence of an area of the South, which coincides historically with the so-called Solid South (Feldman, 2013, 2015), marked by deep-rooted racial and socio-economic inequalities, by a traditional and agricultural economy, and by a conservative political orientation, has become topical again. The East–West coasts have been following a different path. They are characterized by a more innovative economy and a cosmopolitan political milieu, even if inequalities have also been growing. In addition, social mobility has been dramatically decreasing in East–West coasts areas during recent decades. Within these macro-territorial distinctions in the United States, there is also the so-called Rust Belt, whose "Shrinking Cities" are searching for new pathways towards economic prosperity (Shetty and Reid, 2014), and vast swathes of the central part of the United States (both urban and rural areas), known as "fly-over country", feel ignored by East and West coast elites (Harkins, 2016; Kendzior, 2015). Whether in Europe or North America, scholars have developed a new lexicon of descriptors when writing about such places, referring to them as "forgotten places" (Kendzior, 2015), "places that don't matter" (Rodríguez-Pose, 2018), and "places left behind" (Rickardsson et al., 2021).

Beyond the specific cases – Italy, U.K., and the U.S. as examples of territorial fractures persisting over time – we can identify four trends that have consolidated in recent years and generated puzzles within the economy–institutions–territories nexuses. First, the growth of territorial differences, giving rise to new marginal inner areas (see also Section 1.6). Second, the increase of social inequalities, poverty rates, and "social distances" between classes, and the decrease of the "relative social mobility". As has been shown by some studies, this set of variables is positively correlated with the emergence of political movements labelled as populist and with racist and discriminatory behaviours (Gidron and Hall, 2020; see next section). Third, the deanthropization and depopulation happening also in some areas of rich countries: young people with high human capital resources tend to emigrate internally, and some areas are likely to look like a 'social desert' in the long run. Fourth, the new challenges posed by climate change (i.e., warming and changing rainfall patterns) can further weaken (already) low-intensity agricultural activities. These four trends are common across both prosperous and impoverished areas of the world. Of course, they emerge within the territories differently, based on local resources, the filter exerted by formal and informal institutions, and the path-dependency trajectories that different societies and local economies follow.

For these reasons, and in the presence of a highly competitive and unstable macroeconomic scenario, it is hard for institutions to modify (or correct) the territories' path where impoverishment and loss of competitiveness are triggered. Moreover, even in territories characterized by economic vivacity and innovation, social inequalities often tend to grow, generating a loss of collective well-being. This has been happening, for instance, in Silicon Valley (Cooper, 2008).

Given such a fluid situation, which threatens even consolidated economic and social positions, it is relevant to observe the social groups that compete within territories to control and allocate economic resources. In this regard, there are a number of empirical themes that merit further investigation. In the presence of emerging discrepancies between economy, institutions, and territories, the new forms of the international division of labour, the inter-penetrations between informal and formal practices (i.e., recurrent or systemic corruption), and the formation of elitist social groups located in the gaps between politics and economy should be seriously considered.

1.6 The Challenge of Peripherality

Periphery is not an absolute element, neither epistemologically nor ontologically. The notion of periphery is not in fact a static one (Pezzi and Urso, 2016, 2017). It of course incorporates an interpretation in terms of geographical distance from a centre and a location on the fringes of a country or a region, i.e., a spatial peripherality (see Herrschel, 2011). In this sense, it is primarily associated with distance decay from a presumed core; that is with remoteness or "edgeness" (de Souza, 2018; Herrschel, 2012; Pezzi and Urso, 2016, 2017). A 'relational' character (core and periphery) is intrinsic in the concept of periphery, and it is implicitly characterized by connotations of power and/or inequality (Crone, 2012). Peripheral areas are not only linked to fixed geographical features (Armondi, 2020), but are "the outcome of complex processes of change in the economy, demography, political decision-making, and sociocultural norms and values" (Naumann and Fischer-Tahir, 2013: 9). Thus, the concept also encompasses an evolutive dimension, which discloses some power-emanating processual dynamics (de Souza, 2018) with the periphery functionally, economically, and politically subordinated to the core.

Apart from locational disadvantages connected to low accessibility, which very often couple with low population density, high levels of young, high-skilled out-migration and ageing, low absorptive capacity of new knowledge, and deterioration in quantity and quality of essential services, peripheries are interpreted also as a social configuration (Kühn, 2015) and are understood to be actively produced. As effectively articulated by Danson and de Souza (2012), "periphery is created, experienced and continuously present" (p. 4). The words used by the authors – "create", "experience" and "being present" – account for crucial phenomena pertaining to both a condition of peripherality and the processes of peripheralization. Peripherality and peripheralization are not synonyms – the former is a condition, the latter is a process – but they often go together (Pezzi and Urso, 2017). When talking of periphery, one key element for its characterization is the role of proximity versus distance at least along three dimensions: one merely geographical in nature (spatial/Euclidian distance); one related to the issue of the circulation of knowledge and, hence, innovation (cognitive distance); one that relates to access to policy-making networks and the voice that places can express within them (communicative, participative distance) (see Herrschel, 2012).

As for the first point, from a trade perspective, it is well acknowledged that proximity favours access to resources and markets, lowers transport costs, enables value chain flexibility and reliability, that is a set of advantages offset by cost and specialization benefits accruing from distant sourcing (Wang et al., 2019). This spatial distance, which cuts off isolated and/or low-accessibility areas from the main global flows of people, goods, and services, implies distance from the sources of knowledge. In what may be relatively small, remote environments, the lack of circulation of new knowledge is likely to lead to cognitive or institutional lock-in and limited productivity and growth (Pezzi and Urso, 2017). Peripheral regions in fact are assumed to lack most of the local conditions widely recognized as favourable for an innovative *milieu* to emerge due to the presence of several barriers to innovation (Doloreux and Dionne, 2008) – starting from accessibility to networks where knowledge is produced to a lack of the cognitive proximity enabling them to absorb and fruitfully exploit that knowledge. To use a very powerful metaphor in Economic Geography, they often lack both the "local buzz" and the "global pipelines" (Bathelt et al., 2004) to be innovative. Remote areas, in fact, "neither have the internal critical mass nor the capacity to generate external contacts and networks to compete with core areas" (Rodríguez-Pose and Fitjar, 2013: 355–356).

Peripherality is then a condition, but is also a process (i.e., peripheralization). This means that it is also actively produced (Pezzi and Urso, 2016) as a result of (shifts in) unbalanced (power) relationships and more or less unintended marginalization processes at various scales due to market forces and side effects of political interventions (see Herrschel, 2011, 2012). Peripheralization is thus directly linked to exclusion from markets, networks, and decision-making, and hence to the lack of agentic capabilities of economic agents, social actors, and institutions. As explained by Herrschel (2011: 98), marginalised actors, in their varied forms, may find it difficult to join, so as not to upset the existing relationships and balances of power negotiated between those who are part of the system and thus 'included' in the process of shaping and implementing decisions and control, and those who are not.

These territories being outside the primary networks of economic and political power (which are mainly the expression of urban elites) are then in a "weaker bargaining position, potentially being ignored, 'shut out'" (Ibid.: 98). Peripheral actors' priorities are deemed of little interest to – and therefore rarely enter – the agendas of the key decision-makers belonging to the dominant policy-making networks (Urso, 2021). As may be rather easily inferred from the reflections developed in Section 1.4, the functioning itself of markets and policy-making, by entailing a selection, creates cores and non-core areas, hence structuring a hierarchical space resulting in "geographies of centrality and marginality" (Paasi, 2006: 194). Economic agents compete over resources and access to markets; groups compete over value prioritization of issues and agenda setting policy-wise. Markets and both formal and informal institutions, being selective forces are, by implication, producers of peripheries at any geographical scale, creating new, or reinforcing old, hierarchies of connectivity and access to, and relevance in, the economy (and the trajectories of development) and in decision-making processes, corresponding to unevenness in likely scope and opportunity of participation between those who are "inside" and those who are "outside" the relevant flows, between winners and losers (see Section 1.4). Compared to proper geographical ones, these broadly understood peripheries are more difficult to clearly detect and, hence, strategically addressed (and changed) on a policy level. If the issue of peripherality has been mostly addressed so far through investments in physical infrastructure (i.e., transport systems), increasing local accessibility and thereby altering perceptions of remoteness, "social-political connectivities are much less obvious. They are thus more difficult to gauge and predict in their likely impact. They are also much less easy to alter or, indeed, utilise" (Herrschel, 2009: 248). Soft infrastructures, such as skills, expertise, attitudes, business culture, place image along with its structural economic composition, are much less likely to be modified, especially in the long run. We still need an academic and policy reflection on peripheries because of the huge consequences for specific places and for particular categories of actors. This implies going beyond a competitiveness-driven approach in regional development policies – which has further reiterated the longstanding core-periphery divide – mainly associated with a conventional view of peripheries understood as subordinate of the core (Anderson, 2000), as "laggards" which need to be encouraged to "catch up" with the core areas or "carriages" which are to be pulled along in their wake by cities, that is the "locomotives" of economic growth (Harrison and Heley, 2015). These lagging territories, if empowered, may well become active and design and pursue their own policy responses, based on their disconnected condition and associated values (Herrschel, 2009), through policies seeking to counterbalance the polarizing and atomizing effects of pursuing competitiveness and at the same time fostering cohesion across territories. The challenge is thus balancing the interests of urban and rural stakeholders and economic agents ensuring a large representational mix of both parties. In the absence of policies dedicated to the development of social, cultural, economic and (hard and soft) infrastructural conditions able to support entrepreneurship and growth in peripheral areas, processes of further marginalization and decay may be inescapable. With reference to the third term used by Danson and de Souza (2012) to conceptualize peripheries, namely "experience", periphery is also a perception of exclusion, that is of being left out and too far away or marginal in relation to sources of power (Syrett, 2012). As we have stated in the previous section, this sentiment was made evident, as never before, by the recent surge of populism, which has abruptly brought to the fore the issue of peripheries as "left behind" places by globalization and by city-first policies, which – perceiving, according to some interpretations, some distance from institutions, policy-makers and traditional parties – have finally struck back in the ballot boxes (Faggian et al., 2021; Rodríguez-Pose, 2018). What is more, the COVID-19 pandemic has shown us that peripheral areas and communities are uniquely vulnerable to the impact of this large-scale exogenous shock, needing tailor-made holistic policies to resist the present crisis and possible future crises (Cabana et al., 2021; Wang, 2021). As underlined by Faggian (2021) rural areas should more than ever be helped in addressing old, long-standing issues, especially the ones that underlie their marginalization and, among others, have led to population loss (namely improving essential services and digitalization).

1.7 Book Structure: Sections and Chapters

This edited book takes up the questions posed by the complex interplays between economy, institutions, and territories, tries to lay out several perspectives to look at these interconnections, and focuses empirically on the four main thematic areas we have highlighted in the previous sections. The thematic areas constitute, in fact, the four sections of the book.

The first chapters deal with issues related to the "Integration between the Economy and Society: Innovation, Tensions, and Dilemmas".

In the chapter "Change and Innovation within Florida's Food System in Response to COVID 19", David Outerbridge et al. focus on the huge impact of COVID-19 on local society, production systems and institutional frameworks. By observing the food system in Florida, U.S., they highlight significant changes in food consumption patterns as well as shifts, innovations and adaptations in product movements. These changes in the morphology of such an essential production system have generated unexpected advantages for some economic actors and, conversely, have increased the difficulties for others.

The chapter by Andrea Marpillero-Colomina, "Street Shock: How a Bike Lane Redefined a Neighbourhood", analyses a micro-change occurring within the city of New York and shows how the attempt to install a bike lane in Williamsburg (Brooklyn) fomented a new cultural fissure in the rapidly gentrifying neighbourhood. The author provides a situated analysis of political, economic, and social conditions that led the Bedford Avenue bike lane to become a space of conflict, tension, and experiment of innovation in the use of the space. By looking at micro-dynamics, the chapter shows broad interplay between institutions, economy and territories relating to urban environments.

Haoying Wang in "Building Drought Resilience in the US Southwest: The Institutional and Economic Challenges in Rural Communities" argues that climate changes tend to affect regions differently around the world. This can have huge impacts on rural communities, which are vulnerable to exogenous shocks related to climate variability that create dilemmatic situations in terms of socio-economic well-being. The author proposes a conceptual model to explore innovative strategies adopted by rural and local communities to face drought issues, thus supporting both their social organization and economic activities.

The second group of chapters relates to the "Coordination between State and Market: Emerging Problems".

In "A History of Modern European Monetary Unions as Territories, Regions, and Institutions" Rebecca Jean Emigh, Michelle Marinello, and Zachary DeGroot examine how monetary unions can be conceived as an "analytical gateway" to look at the trade-off between the expansion of markets at the expense of the loss of monetary adjustment as a policy tool. To understand the establishment of monetary unions, one should consider territories as cultural units that share basic conceptions of how transactions work, the institutional arrangements such as political ones that enforce monetary and fiscal policies, and the networks of people actually using the currency. Through a comparative analysis, the authors show that the functioning of monetary unions depend on creating stable territorial entities, institutions that have rules for the use of money, and markets in which people use money.

The chapter by Francesco Duina and Hermione Xiaoqing Zhou, "Brussels Under Pressure: Compliance, the Single Market, and National Purpose in the EU", analyses the puzzles concerning the EU law as the main institutional vehicle for the establishment of the single market and, with that, Europeanization more broadly, and member states' compliance with EU law, which is far from guaranteed. Inspired by recent calls for spatially informed EU analyses, the chapter shows how the member states can pressure a disinterested, and even resistant, EU for compliance support to reduce the centripetal tendencies of the European common market. Hence, the chapter highlights how the centre-periphery relations might affect the market regulation in a paradoxical and non-linear way.

Nuri Yavan, Şükrü Yılmaz, and Aykut Anic's chapter, "Institutional Context and Territorial Policy: Analysing the New Regional Policy and Regional Development Agencies in Turkey", focuses on regional development policy in Turkey. The chapter analyses the attempt by institutions to support flourishing markets at the local level. The authors show that Turkey has been partly changing its traditional regional development policy from top-down decision-making to a more participatory and bottom-up approach. In short, the chapter reconstructs the emergence of local economies at the intersection between regulatory practices exercised by a central government, on the one hand, and local dynamics characterized by greater participation and horizontality of decision-making, on the other hand – a mix that has often failed.

The third part of the book is about the multiple nexuses between economy-institutions-territories that can be identified by looking at "Social Inequalities, Displacement, and Conflicts between Social Groups".

Ling Zhou and Xueguang Zhou's chapter, "Bureaucrats, Local Elites, and Economic Development: Evidence from Chinese Counties", presents the multiple and complex patterns of "movers" and "stayers" in bureaucratic mobility in over 100 counties (districts) in Jiangsu Province and identifies the location and distribution of those local officials as local elites in administrative jurisdictions. On this basis, the authors examine the effect of local elites on economic development and argue that local officials in China's local governments play the role of local elites, with the double identity as the state agent and as representatives of local interests.

In "Working at the Nexus of Global Markets and Gig Work: US Gig Workers, Credential Capitalization, and Wealthy International Clienteles", Alexandrea J. Ravenelle and Ken Cai Kowalski challenge one of the common understandings of the gig economy as a land of opportunities. The authors deal with the experience of international competitive pressures within different groups of workers in the gig economy. The study examines how American workers perceive the global market and leverage the cachet of their U.S. credentials to command higher wages. This perception justifies migration to platforms that cater to higher-end clients, including foreign governments and multinational firms.

Isabelle Nilsson and Elizabeth C. Delmelle's chapter, "Understanding Residential Sorting through Property Listings: A Case Study of Neighborhood Change in Charlotte, NC 1993–2018", presents a classic topic about how neighbourhood dynamics are largely driven by the in- and out-migration of residents in conjunction with changes to the built environment. More precisely, the chapter shows the role of various amenities in shaping neighbourhood change (e.g., walkability, distance to natural amenities, breweries, coffee shops, etc.). Displacement patterns within cities are, thus, the outcome of a web of factors relating to economic resources, push/pull factors, social network features. These resources are asymmetrically distributed between social groups.

Drawing on a rich empirical analysis, Diana Dakhlallah's chapter, "Making the Right Move: How Effective Matching on the Frontlines Maintains the Market for Bribes", investigates the dynamics between groups and individuals who interact informally and illegally exchange specific resources or access to certain services. In detail, the author reconstructs how people choose to engage in bribery, what their choices imply for individuals, for the markets, and the local institutional frameworks. The author identifies the initiating factors of the exchange of bribes, illustrating how the initial moves and countermoves are defined by signals concerning the clients' socioeconomic status.

Finally, yet importantly, the fourth part of the book includes chapters on "The Challenge of Peripherality".

In the chapter entitled "Measuring the Interaction between Interregional Accessibility and the Geography of Economy and Institutions: The Case of Greece", Dimitrios Tsiotas and Vassilis Tselios touch upon the actual economic-institutional and spatial areas arising from the interregional multimodal transport accessibility interconnecting territories. Considering different aspects of institutional configuration in Greece, the authors examine the degree to which the spatial configuration of these institutional characteristics are related to the transportation network structure. The analysis promotes interdisciplinary research and provides insights into the effect of transport accessibility on the configuration of peripheral vs. central areas.

Laura Ryser, Neil Argent, Greg Halseth, Fiona Haslam-McKenzie, and Sean Markey's chapter "Marginal Returns? Institutional Dynamics, Peripherality, and Place-based Development in Canada's and Australia's Natural Resource-Dependent Regions" aims to shed a better light on the Coal Seam Gas (CSG) and the Liquefied Natural Gas (LNG) boom of the 2010s and on the institutional relationships within the rural communities and economies of the Peace River region of British Columbia, Canada, and the Surat Basin, Queensland, Australia. The authors chart the socio-spatially uneven distribution of benefits, harms, and responsibilities associated with CSG–LNG expansion in both regions, focusing on the extent to which formal and informal government and governance institutions were able to overcome conditions of lock-in and inertia and capitalise on the benefits. In doing so, the authors highlight the different territorial trajectories in these places. Some areas adapt and capture long-term economic benefits. Other regions predominantly suffer the negative consequences of CSG–LNG development, and feel largely abandoned by regional and central governments, thereby becoming peripheral.

Notes

1 There are three main approaches in Economic Sociology: New Economic Sociology (NES), which takes into account the different kind of relational embeddedness of economic action; Political Economy (PE), which explores the political regulation of the economic sphere; and Neo-Institutionalism (NIS), which pays attention to cultural embeddedness of economic action, i.e., the role played by informal institutions in shaping economic processes (Swedberg, 2007).
2 We are deeply grateful to Nicola Negri, who has suggested an analogy between the Bohr atom model and the model of actor we may find in the social sciences. For social scientists, actors are not conceived as persons but rather as agents that can be observed while interacting.

References

Acemoglu, D., Johnson, S., and Robinson, J. (2001). The colonial origins of comparative development: An empirical investigation. *American Economic Review*, 91 (5): 1369–1401.

Acemoglu, D., Johnson, S., and Robinson, J. A. (2005). Institutions as a fundamental cause of long-run growth. In P. Aghion and S. N. Durlauf (Eds.), *Handbook of Economic Growth*, Volume 1, Part A (pp. 385–472) Holland: Elsevier.

Acemoglu, D. and Robinson, J. (2012). *Why Nations Fail: The Origins of Power, Prosperity and Poverty*. London: Profile Books.

Anderson, A. (2000). Paradox in the periphery: An entrepreneurial reconstruction? *Entrepreneurship and Regional Development*, 12 (2): 91–109.

Armondi, S. (2020). Towards geopolitical reading of 'periphery' in state spatial strategies: Concepts and controversies. *Geopolitics*, 27 (2): 526–545. Published online: 26 July 2020.

Bathelt, H. and Glückler, J. (2011). *The Relational Economy: Geographies of Knowing and Learning*. Oxford: Oxford University Press.

Bathelt, H. and Glückler, J. (2014). Institutional change in economic geography. *Progress in Human Geography*, 38 (3): 340–363.

Bathelt, H., Malmberg, A., and Maskell, P. (2004). Clusters and knowledge: Local buzz, global pipelines and the process of knowledge creation. *Progress in Human Geography*, 28 (1): 31–56.

Beckert, J. (1996). What is sociological about economic sociology? Uncertainty and the embeddedness of economic action. *Theory and Society*, 25 (6): 803–840.

Boudon, R. (1986). *Theories of Social Change. A Critical Appraisal*. Berkeley and Los Angeles: University of California Press.

Brinton, M. C. and Nee, V. (Eds.) (1998). *The New Institutionalism in Sociology*. New York: Russel Sage Foundations.

Brody, J. (2020). *Quantum Entanglement*. Cambridge, MA: MIT Press.

Burt, R. (1995). *Structural Holes. The Social Structure of Competition*. Cambridge, MA: Harvard University Press.

Burt, R. (2000). The network entrepreneur. In R. Swedberg (Ed.), *Entrepreneurship: The Social Science View* (pp. 281–307). Oxford: Oxford University Press.

Busso, S. and Storti, L. (2014). Economic development and social cohesion: Some considerations on the Italian Case. *Modern Italy*, 18 (2): 197–217.

Butler, T. and Lees, L. (2006). Super-gentrification in Barnsbury, London: Globalization and gentrifying global elites at the neighbourhood level. *Transactions of the Institute of British Geographers, New Series*, 31 (4): 467–487.

Cabana, Y. D. C., Malone, A., Zeballos, E. Z., Huaranca, N. O. H., Tinta, M. T., Beltran, S. A. G., Arosquipa, A. A., and Pindero, D. (2021). Pandemic response in rural Peru: Multi-scale institutional analysis of the COVID-19 crisis. *Applied Geography*, 134. https://doi.org/10.1016/j.apgeog.2021.102519

Cairncross, F. (1997). *The Death of Distance: How the Communications Revolution is Changing Our Lives*. London: Texere Publishing Limited.

Christopherson, S. (2002). Why do national labor market practices continue to diverge in the global economy? The 'missing link' of investment rules. *Economic Geography*, 78 (1): 1–20.

Cooper, M. (2008). The inequality of security: Winners and losers in the risk society. *Human Relations*, 61 (9): 1229–1258.

Crone, M. (2012). Re-thinking 'peripherality' in the context of a knowledge-intensive, service-dominated economy. In M. Danson and P. de Souza (Eds.), *Regional Development in Northern Europe. Peripherality, Marginality and Border Issues* (pp. 49–64). London and New York: Routledge.

Crouch, C. (2019). *Will the Gig Economy Prevail?* Cambridge: Polity Press.

Crouch, C., Le Galès, P., Trigilia, C., and Voeltzkov, H. (2001). *Local Production Systems*. Oxford: Oxford University Press.

Danson, M. and de Souza, P. (2012). *Regional Development in Northern Europe. Peripherality, Marginality and Border Issues*. London and New York: Routledge.

de Souza, P. (2018). *The Rural and Peripheral in Regional Development. An Alternative Perspective*. London and New York: Routledge.

Dequech, D. (2001). Bounded rationality, institutions, and uncertainty. *Journal of Economic Issues*, 35 (4): 911–929.

Doloreux, D. and Dionne, S. (2008). Is regional innovation system development possible in peripheral regions? Some evidence from the case of La Pocatière, Canada. *Entrepreneurship & Regional Development: An International Journal*, 20 (3): 259–283.

Douglas, M. (1986). *How Institutions Think*. Syracuse, NY: Syracuse University Press.

Duarte, F. (2017). *Space, Place and Territory. A Critical Review on Spatialities*. London and New York: Routledge.

Duina, F. (2011). *Institutions and the Economy*. Cambridge: Polity Press.

Faggian, A. (2021). Resilient rural futures. *EU Rural Review*, 32: 22–27.

Faggian, A., Modica, M., Modrego, F., and Urso, G. (2021). One country, two populist parties. Voting patterns of the 2018 Italian elections and their determinants. *Regional Science Policy and Practice*, 3 (2): 397–413.

Farole, T., Rodríguez-Pose, A., and Storper, M. (2010). Human geography and the institutions that underlie economic growth. *Progress in Human Geography*, 35 (1): 58–80.

Feldman, G. (2013). *The Irony of the Solid South: Democrats, Republicans, and Race, 1864–1944*. Tuscaloosa, AL: University of Alabama Press.

Feldman, G. (2015). *The Great Melding: War, the Dixiecrat Rebellion, and the Southern Model for America's New Conservatism*. Tuscaloosa, AL: University of Alabama Press.

Ferrary, M. and Granovetter, M. (2009). The role of venture capital firms in Silicon Valley's Complex innovation network. *Economy and Society*, 18: 326–359.

Friedman, T. L. (2005). *The World Is Flat: A Brief History of the Twenty-First Century*. New York: Farrar, Straus and Giroux.

Gertler, M. S. (2010). Rules of the game: The place of institutions in regional economic change. *Regional Studies*, 44 (1): 1–15.

Gidron, N. and Hall, P. A. (2020). Populism as a problem of social integration. *Comparative Political Studies*, 53 (7): 1027–1059.

Granovetter, M. (2017). *Society and Economy. Framework and Principles*. Cambridge, MA: The Belknap Press of Harvard University Press.

Hanson, A. (2021). Taxes and economic development: An update on the state of the economics literature. *Economic Development Quarterly*, 35 (3): 232–253.

Harkins, A. (2016). The Midwest and the evolution of "Flyover Country". *Middle West Review*, 3 (1): 97–121.

Harrison, J. and Heley, J. (2015). Governing beyond the metropolis: Placing the rural in city-region development. *Urban Studies*, 52: 1113–1133.

Hayek, F. A. (1969). *Studies in Philosophy, Politics, and Economics*. London and New York: Routledge.

Hedström, P. (2005). *Dissecting the Social: On the Principles of Analytical Sociology.* Cambridge: Cambridge University Press.

Heidegger, M. (1962) [1927]. *Being and Time.* Oxford: Blackwell Publishing.

Herrschel, T. (2009). City regions, polycentricity and the construction of peripheralities through governance. *Urban Research & Practice*, 2: 240–250.

Herrschel, T. (2011). Regional development, peripheralisation and marginalisation – and the role of governance. In T. Herrschel and P. Tallberg (Eds.), *The Role of Regions? Networks, Scale, Territory* (pp. 85–102). Kristianstad: Kristianstad Boktryckeri.

Herrschel, T. (2012). Regionalisation and marginalisation. Bridging old and new divisions in regional governance. In M. Danson and P. De Souza (Eds.), *Regional Development in Northern Europe. Peripherality, Marginality and Border Issues* (pp. 30–48). London and New York: Routledge.

Keating, M. (Ed.) (2020). *The Oxford Handbook of Scottish Politics.* Oxford: Oxford University Press.

Kendzior, S. (2015). *The View from Flyover Country: Dispatches from the Forgotten America.* New York: Flatiron Books.

Ketterer, T. D. and Rodríguez-Pose, A. (2018). Institutions vs. 'first-nature' geography: What drives economic growth in Europe's regions? *Papers in Regional Science*, 97, S1: S25–S62.

Kühn, M. (2015). Peripheralization: Theoretical concepts explaining socio-spatial inequalities. *European Planning Studies*, 23 (2): 1–12.

Lecours, A. (Ed.) (2005). *New Institutionalism: Theory and Analysis.* Toronto: Toronto University Press.

Lees, L. (2003). Super-gentrification: The Case of Brooklyn Heightsm New York City. *Urban Studies*, 40 (12): 2487–2509.

Mahoney, J. and Thelen, K. (Eds.) (2009). *Explaining Institutional Change. Ambiguity, Agency, and Power.* Cambridge: Cambridge University Press.

Martin, R. (2000). Institutional approaches in economic geography. In S. Sheppard and T. Barnes (Eds.), *A Companion to Economic Geography* (pp. 77–94). Oxford: Basil Blackwell.

Morelli, N. (2019). Creating urban sociality in middle-class neighborhoods in Milan and Bologna: A study on the social streets phenomenon. *City and Community*, 18 (3): 834–852.

Motoyama, Y and Hui, I. (2015). How do business owners perceive the state business climate? Using hierarchical models to examine the business climate perceptions, state rankings, and tax rates. *Economic Development Quarterly*, 29 (3): 262–274.

Naumann, M. and Fischer-Tahir, A. (2013). Introduction: Peripheralization as the social production of spatial dependencies and injustice. In A. Fischer-Tahir and M. Naumann (Eds.), *Peripheralization: The Making of Spatial Dependencies and Social Injustice* (pp. 9–26). Wiesbaden, Germany: Springer VS.

North, D. (1990). *Institutions, Institutional Change and Economic Performance.* New York: Cambridge University Press.

North, D. (1991). Institutions. *Journal of Economic Perspectives*, 5 1: 97–112.

North, D. (2005). *Understanding the Process of Economic Change.* Princeton, NJ: Princeton University Press.

Paasi, A. (2006). *Cities in a World Economy.* London: Pine Forge Press.

Pezzi, M. G. and Urso, G. (2016). Peripheral areas: Conceptualizations and policies: Introduction and editorial note. *Italian Journal of Planning Practice*, 6 (1): 1–19.

Pezzi, M. G. and Urso, G. (2017). Coping with peripherality: Local resilience between policies and practices. Editorial note. *Italian Journal of Planning Practice*, 7 (1): 1–23.

Pike, A., Rodríguez-Pose, A., and Tomaney, J. (2007). What kind of local and regional development and for whom? *Regional Studies*, 41 (9): 1253–1269.

Piketty, T. (2014). *Capital in the Twenty First Century*. Cambridge, MA: Belknap Press.

Polanyi, K. (2002) [1944]. *The Great Transformation: The Political and Economic Origins of Our Time*. Boston: Beacon Press.

Powell, W. W. and DiMaggio, P. J. (1983). The Iron Cage revisited: Institutional iso-morphism and collective rationality in organizational fields. *American Sociological Review*, 48 (2): 147–160.

Powell, W. W. and DiMaggio, P. J. (Eds.) (1991). *The New-Institutionalism in Organizational Analysis*. Chicago: Chicago University Press.

Ramella, F. (2019). *Sociology of Economic Innovation*. London and New York: Routledge.

Rickardsson, J., Mellander, C., and Bjerke, L. (2021). The Stockholm Syndrome: The view of the capital by the "Places Left Behind". *Cambridge Journal of Regions, Economy and Society*, 14 (3): 601–617. https://doi.org/10.1093/cjres/rsab013

Rodríguez-Pose, A. (2018). The revenge of the places that don't matter (and what to do about it). *Cambridge Journal of Regions, Economy and Society*, 11 (1): 189–209.

Rodríguez-Pose, A. and Fitjar, R. D. (2013). Buzz, Archipelago economies and the future of intermediate and peripheral areas in a Spiky World. *European Planning Studies*, 21 (3): 355–372.

Rodríguez-Pose, A. and Storper, M. (2006). Better rules or stronger communities? On the social foundations of institutional change and its economic effects. *Economic Geography*, 82 (1): 1–25.

Rodrik, D. (1999). *Making Openness Work: The New Global Economy and the Developing Countries*. Washington, DC: Overseas Development Council.

Rodrik, D. (2000). Institutions for high-quality growth: What they are and how to acquire them. *Studies in Comparative International Development*, 35: 3–31.

Rodrik, D., Subramanian, A., and Trebbi, F. (2004). Institutions rule: The primacy of institutions over geography and integration in economic development. *Journal of Economic Growth*, 9: 131–165.

Rovelli, C. (2019). *The Order of Time*. London: Penguin Books.

Sciarrone, R. (2021). Tra storia e scienze sociali: ponti, porte e finestre. *Meridiana*, 100: 9–34.

Shetty, S. and Reid, N. (2014). Dealing with decline in old industrial cities in Europe and the United States: Problems and policies. *Built Environment*, 40 (4): 458–474.

Stark, D. C. (2009). *The Sense of Dissonance. Accounts of Worth in Economic Life*. Princeton: Princeton University Press.

Storti, L. (2018). Deepening the liaison: Mixed embeddedness and economic sociol-ogy. *Sociologica. International Journal for Sociological Debate*, 2: 23–37.

Storti, L. and Dagnes, J. (2021). Super-rich: Origin, reproduction and social accept-ance. *Sociologica. International Journal for Sociological Debate*, 15 (2): 5–23. https://doi.org/10.6092/issn.1971-8853/13546

Streek, W. and Schäfer, A. (2013). *The Politics of Austerity*. Cambridge: Polity Press.

Swedberg, R. (2007). *Principles of Economic Sociology*. Princeton: Princeton University Press.

Syrett, S. (2012). Conceptualising marginalisation in cities and regions. In M. Danson and P. de Souza (Eds.), *Regional Development in Northern Europe. Peripherality, Marginality and Border Issues* (pp. 65–77). London and New York: Routledge.

Tomaney, J. (2014). Region and place I: Institutions. *Progress in Human Geography*, 38 (1): 131–140.

Tomaney, J., Rodríguez-Pose, A., and Pike, A. (2010). Local and regional development in times of crisis. *Environment and Planning A*, 42 (4): 771–779.

Trabalzi, F. (2007). Crossing conventions in localized food networks: Insights from Southern Italy. *Environment and Planning A: Economy and Space*, 39: 283–300.

Trigilia, C. (2002). *Economic Sociology: State, Market, and Society in Modern Capitalism*. Oxford: Blackwell Pub.

Trigilia, C. (2020). *Capitalismi e Democrazie. Si possono cancellare crescita e uguaglianza?* Bologna: Il Mulino.

Tuan, Y-F. (1977). *Space and Place: The Perspective of Experience*. Minneapolis: University of Minnesota Press.

Urso, G. (2021). Metropolisation and the challenge of rural-urban dichotomies. *Urban Geography*, 42 (1): 37–57.

Urso, G., Storti, L., and Reid, N. (2021). Shocking events. Institutional reactions to abrupt changes. *Applied Geography*, 137 (10586): 1–9.

Uzzi, B. (1996). The sources and consequences of embeddedness for the economic performance of organizations: The network effect. *American Sociological Review*, 61 (4): 674–698.

Uzzi, B. (1997). Social structure and competition in interfirm networks: The paradox of embeddedness. *Administrative Science Quarterly*, 42 (1): 35–67.

Viesti, G. (2021). *Centri e periferie. Europa, Italia, Mezzogiorno dal XX al XXI secolo*. Roma-Bari: Laterza.

Wang, H. (2021). Why the Navajo Nation was hit so hard by coronavirus: Understanding the disproportionate impact of the COVID-19 pandemic. *Applied Geography*, 134. https://doi.org/10.1016/j.apgeog.2021.102526

Wang, M., Derudder, B., and Liu, X. (2019). Polycentric urban development and economic productivity in China: A multiscalar analysis. *Environment and Planning A: Economy and Space*, 51 (8): 1622–1643.

Warf, B. and Arias, S. (Eds.) (2009). *The Spatial Turn. Interdisciplinary Perspectives*. London and New York: Routledge.

Wood, A. (2011). The politics of local and regional development. In A. Pike, A. Rodríguez-Pose, and J. Tomaney (Eds.), *Handbook of Local and Regional Development* (pp. 306–317). Abingdon: Routledge.

Part I

Integration between the Economy and Society

Innovation, Tensions and Dilemmas

2 Change and Innovation within Florida's Food System in Response to COVID-19[1]

David Outerbridge, Christa D. Court, Laura Birou, Catherine Campbell, Liz Felter, Sebastian Galindo, Jennifer Hagen, Brittany Hall-Scharf, Lisa Hickey, Molly Jameson, Elias T. Kirche, Alexandre Magnier, and Fritz Roka

2.1 Introduction

The resilience of an economic or social system to a hazard event depends heavily on the size and scope of the event as well as the vulnerability conditions of the affected system and conveys the extent to which the system can quickly recover. The experience of a hazard event evolves into a disaster event when the impacts are so sizable and/or so extensive that a vulnerable system cannot quickly recover from the experience. The confluence of natural forces with the structural and philosophical frameworks that govern human society guarantees various temporal, geographical, and context-specific implications resulting from a hazard event (Wisner, 2001). When evaluating societal changes induced by hazard or disaster events, it is critical to determine the distinct inception point of change, i.e. the initial physical change caused by a natural force resulting in anthropogenic shifts or human reactions to the physical change (Fuchs et al., 2017). South Florida has experienced direct physical damage from natural events, such as major hurricanes, that have accelerated ethnic population shifts, affected land-use change, and spurred significant changes to the rules and regulations that govern new construction in the state (Solecki and Walker, 2001). These events and their impacts can redefine how economic, regulatory, and social systems function, and the study of these events allows for realization, adaptation, and greater understanding of this process of change.

In the case of a biological hazard such as the COVID-19 pandemic, the physical point of inception is more abstract and perhaps more ambiguous than the darkened skies of a foreboding tropical cyclone or the liquefaction of the ground during an earthquake (Morens et al., 2009). Possibilities include:

- Initial discovery of the pathogen.
- Confirmed community spread.
- Initiation of public health measures that aim to mitigate community spread.

DOI: 10.4324/9781003191049-3

Regardless, the resulting behavioral shifts in response to a pandemic are influenced by cultural, social, and political factors. Ultimately, these shifts will largely depend on individuals' risk perception and risk preferences within a local area, and how risk is institutionally managed locally (Fuchs et al., 2017; Vasavari, 2015). Risk perception is based on information provided to the general public and how individuals modify their behaviors as a result (Renn, 1998). Risk preference refers to an individual's level of risk tolerance. Risk management is the regulatory framework through which decision makers manage societal risk, sometimes affecting changes in risk perception and preference to affect behavioral change (Romang et al., 2009). Indeed, the individual risk perception and preferences of decision makers can determine how overall risk management decisions are made and enacted (Vasavari, 2015).

Measurements of how risk perceptions, preferences, and management strategies intersect are made more difficult by the fact that perceptions and preferences exist along a spectrum. This is especially true when the effects of the pandemic are delayed or are not witnessed directly (Abraham, 2011; Slovic, 2020). Nevertheless, a global driver of change, such as the COVID-19 pandemic, will likely create heterogeneous local stressors, the local community's response to which can have global implications (Rozell, 2015; Hertel et al., 2019).

The threat of a pandemic has been a reality repeated many times during the development of human society. The size and scope of the COVID-19 pandemic, however, has not occurred since the 1918–1920 influenza pandemic, more than 100 years ago. In the decades following, economic systems have evolved to include highly efficient global supply chains (Abraham, 2011; Huremovic, 2019; Piret & Boivin, 2021). By early 2020, in response to the developing COVID-19 crisis, supply chains began to shift and adapt to an invisible threat, primarily human borne. A significant portion of human interactions at work, at school, and in the marketplace rapidly converted to digital environments to limit potential exposure to the SARS-CoV-2 virus that caused COVID-19 (Spicer, 2020; Kniffin et al., 2021). Public health measures were enacted and, in some cases, regulations were altered to define the parameters of operations for these modified systems to ensure continued support of the healthcare system, the economy, and the education system (Aiyar & Pingali, 2020). For many, working from home, remote-learning, and the formation of family and community support systems for vulnerable (health-related and economic) populations became the "new normal" and a seemingly ever-growing list of ramifications started to develop. For segments of the economy that were deemed essential, owners and/or supervisors had to adapt workplace environments and practices to mitigate risk of viral transmission amongst workers and between workers and customers. These local behavioral changes eventually had global economic consequences. The COVID-19 pandemic offered a unique opportunity to analyze the impact of public health measures on global supply chains (Court et al., 2021; Ferreira et al., 2021).

The COVID-19 pandemic heavily impacted food systems – systems within which food products are grown, harvested, processed, distributed, and consumed over broad geographic areas (Tamburino et al., 2020; Huff et al., 2015;

Tendall et al., 2015). Food is fundamental for human survival and within industrialized economies, food consumption is often detached from its production. Food items cross numerous borders where various contexts of culturally refined regulations determine the passage, safety requirements, and possibilities of timely distribution (Foran et al., 2014; Pingali et al., 2005).

In the United States (U.S.), less than 2 percent of the working-age adult population are directly involved with crop and/or livestock production (Lusk, 2016). However, nearly 23 million adults, or 15 percent of the total U.S. workforce, are employed within the food-supply chain network (American Farm Bureau Federation, 2019). The food system represents a unique confluence of necessity, adaptation, and restriction, making it an ideal case study of the impacts of COVID-19 (Bron et al., 2021; Kummu et al., 2020). Significant changes in food consumption patterns (e.g., food at home versus food away from home) caused large, abrupt shifts in product movements. Everyday interactions within commerce, food production, manufacturing, and distribution were reformulated almost overnight. Faculty members from the University of Florida (UF) and Florida Gulf Coast University (FGCU) used a two-phase mixed methodology approach to evaluate the modifications, changes, adaptations, and innovations occurring within Florida's food system. Phase 1 consisted of a survey gathering quantitative information on the impacts of COVID-19 in Florida during the early months of the pandemic (March–Mid-May). Results indicated Florida's agricultural operations experienced nearly $900 million in production losses due to COVID-19 during this 2.5-month period alone. While the losses are staggering, it is worth noting that operations reported widely varying experiences across commodities produced, geography, and markets served (Court et al., 2020).

Recognizing the complexity within which these losses were taking place, the research team set out on the second phase of the mixed-methodology approach to qualitatively evaluate the shifts, changes, adaptations, and innovations occurring within Florida's food system. The qualitative approach has been used to evaluate both economic impacts of hazards (e.g., Dwyer & Horney, 2014; Phillips, 2014; Ostadtaghizadeh, et al., 2016) and food systems (e.g., Webber & Dollahite, 2008; Toth, Rendall, & Reitsma, 2016). Regarding qualitative works in food systems, they have helped provide context to understanding the complex relationships and have outlined how relationships and behavior define the interactions between players within the food system sectors.

The remainder of this chapter provides an overview of the multi-institutional effort to qualitatively analyze the impacts of COVID-19 on Florida's food system (Phase 2), with a particular focus on their ability and willingness to adapt economically and socially and the influence of the regulatory framework within which each sector operates. In the next section, additional contextual information is provided on Florida's food system, the project team, and the evolution of the COVID-19 pandemic within the State of Florida. The following sections provide details on the second phase of the mixed-methodology approach employed, findings for three distinct components of Florida's food system, and a discussion of the broader implications of findings.

2.2 Context

The global food system is a wonder of recent technological innovations, allowing many individuals to enjoy a wide variety of food options while the bounty is often harvested many miles away, perhaps even in another hemisphere. On average, food products in the U.S. travel 1,500 miles between the point of harvest and the consumer's plate (Lang, 2006; Hill, 2008). The term food system (Figure 2.1) refers to the activities involved in food production, processing, manufacturing, distribution, and consumption, as well as the infrastructure and institutions involved in or having an external influence on feeding a population (Ericksen, 2008; Born & Purcell, 2006; Feagan, 2007).

The concept of local food is broad, but there are three main perspectives through which local food systems can be characterized (Feagan, 2007; Foran et al., 2014). First, local can be defined geographically in miles, distance from the growth of food to a consumer, or production within a particular administrative or political boundary (Feagan, 2007; Low et al., 2015). Next, local can be categorized in direct sales in the form of farm-to-consumer direct links or of the place where it is sold or prepared (Low et al., 2015). Finally,

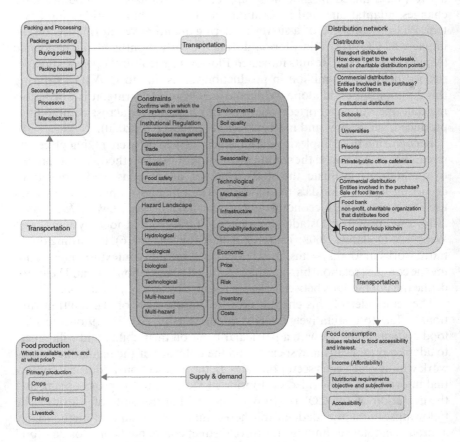

Figure 2.1 Conceptual diagram of the food system.

local food can be defined by the ecology of a region and the types of food that can grow in that climate (Low et al., 2015). In short, a local food system depends on availability, market reach, and a supply network, and the term can be applied quite broadly depending on the context of examination (Feenstra, 1997; Foran et al., 2014). For this chapter, the local food system refers to the geographic and institutional setting within which food products are produced in the State of Florida.

2.2.1 Florida's Food System

The State of Florida's unique geography and history provides for diverse climatic, ecological, and cultural features. Spanning several climatic zones from North to South, many types of crops, livestock, and aquaculture products can be grown successfully. Production agriculture is present in all 67 Florida counties along with commercial fishing landings in the state's coastal counties. Much of Florida's agricultural and fishing-related food products are exported out of state. Well over 80 percent of the citrus juice and a major portion of Florida's vegetable production supplies markets along the east coast of the U.S., especially throughout the winter months (UF/IFAS, 2021; USDA, 2019). Commercial fishers as well as aquaculturists provide fish and shellfish to consumers around the world (National Marine Fisheries Service, 2022). Florida's growers and fishers produce between 200 and 300 different commodities, including field crops, fruits and tree nuts, vegetables, livestock, aquaculture products, fish and shellfish, greenhouse and nursery products, and timber products. Collectively, these products contributed more than $10 billion in direct sales in 2018 (UF/IFAS, 2021). Additionally, direct economic activity associated with agricultural and wood product processing and manufacturing contributed another $40 billion, while wholesale and retail food operations accounted for another $101 billion dollars of sales. When multiplier effects are considered, Florida's agricultural and food system accounts for nearly $254 billion in total industry output and more than two million full-time and part-time jobs within the state's economy (UF/IFAS, 2021). With an abundance of diverse activities associated with the food system, Florida offers excellent potential to compare and contrast the impacts of the COVID-19 pandemic on many different agricultural products with varying cycles of production and duration, as well as the same agricultural products at various stages and contexts in the production and distribution process (Clancy, 2013).

2.2.2 Multi-institutional Collaboration

Mutual concern and interest in evaluating and learning from the unprecedented situation affecting the Florida food system spurred an equally unprecedented multi-institutional collaboration between Extension educators and state specialists[2] at UF's Institute of Food and Agricultural Sciences (UF/IFAS) and faculty members from the Lutgert College of Business at FGCU.

2.2.2.1 UF/IFAS Extension

UF/IFAS Extension is present within all 67 counties of the state of Florida, including 362 Extension educators and 279 statewide specialists on the main campus in Gainesville, FL, as of Fall 2020 (UF/IFAS, 2020). In the food system context, the Cooperative Extension System (CES) exists within the confluence between research conveyed by the University, local county governments, and food system stakeholders (business owners, producers, labor, and distribution/processing operations). As such, UF/IFAS Extension provides a touchpoint between research, implementation of regulations, and experience of producers and food system players within the local and regional food systems, providing science-based information to stakeholders, and relaying concerns and questions to UF/IFAS researchers when appropriate (UF/IFAS, 2020). More than 30 UF/IFAS Extension agents and state specialists collaborated to design and implement the project discussed in this chapter.

2.2.3 *Florida Gulf Coast University*

Opened in 1997 to serve Florida's Southwest region, FGCU is located between Ft. Myers and Naples. Seven faculty members from the Lutgert College of Business representing the Agribusiness, Economics and Finance, and Information Systems and Operations Management Departments helped to develop and implement the project framework, providing additional expertise in citrus production, agriculture, supply chain management, and marketing.

2.2.4 *COVID-19 in Florida*

Like nearly every other state in the U.S., Florida confronted the realities of the COVID-19 pandemic in early March 2020. Within two weeks, businesses across a broad swath of the state's economy were shuttered as "stay-at-home" orders were implemented in the hopes of mitigating the virus' spread and preventing unnecessary strain on the state's healthcare system. Unemployment numbers soared; business revenues and sales tax collections plummeted. The agricultural sector, however, was classified an "essential industry." Consequently, farms, food processing plants, distributors, trucking services, and grocery stores continued to operate. Despite being labeled as an essential industry, agricultural operations in Florida experienced significant financial and logistical impacts as restaurants, hotels, cruise lines, and theme parks were closed and sales of food products to the food service sector precipitously declined, almost overnight. Since the product size, packaging, and logistics involved in supply chains for food products intended for consumption at home and those intended for consumption away from home are quite different, many agricultural operations, large and small, specializing in production for food service had to pivot to find new outlets for and means of distributing their products. With respect to internal operations, personnel managers of agricultural companies had to institute new protocols to physically separate employees or relocate employees from company offices to their individual homes.

Upon hearing many stories of the devastating impacts of COVID-19 on agricultural businesses across the state, the UF/IFAS Economic Impact Analysis Program along with colleagues in the Food and Resource Economics Department distributed a survey to measure the impacts of COVID-19 on agriculture and fisheries related entities. This quantitative effort came to be known as Phase 1 of the UF/IFAS effort to document COVID-19's impact on Florida's food system. Although many researchers and extension educators shifted to working from home, many remained in contact with stakeholders and gathered first-hand information of emerging challenges within the food system. The formal collection of stories and experiences from food system stakeholders became known as Phase 2. These individual stories complemented the quantitative data, allowing for a more complete examination of the relationships that were severed, created, or changed by the pandemic and the impact that these changes had on regulatory decision-making and economic resilience. The idea was to expand the understanding and implications of the measured sales revenue impacts to determine lessons learned and evaluate improvements for regulatory bodies, producers, local and state governments, school systems, non-profit organizations, business owners, and the agribusiness supply chain. In short, the qualitative gathering of information provides a deeper understanding of the varying financial, regulatory, cultural, and behavioral implications of COVID-19 to better inform related decision making.

2.3 Methodology

To capture and analyze the shifts, adaptations, and innovations occurring within Florida's food system, faculty members at UF/IFAS and FGCU devised a research framework (Figure 2.2) that aimed to collect qualitative data via interviews following an explanatory sequential design (Creswell & Plano Clark, 2017). The project team leaders engaged Florida's CES network to collect data and stories from food system stakeholders to determine how the nexus of multivariate decisions in the regulatory, economic, and social spheres created various contexts whereby the changes within Florida's food system could be evaluated.

During Phase 2 of the project, the research team developed an interview guide (i.e. question bank) for each food system sector (see Figure 2.2) to provide a singular narrative to guide the conversations between stakeholders and the research team. The selected food system sectors were based on consultations between researchers in the fields of economics, agronomics, business, supply chain, sustainable food systems, and Cooperative Extension educators that specialize in family and consumer sciences, agriculture, livestock, dairy, small farms, and horticulture. There were 15 different interview guides

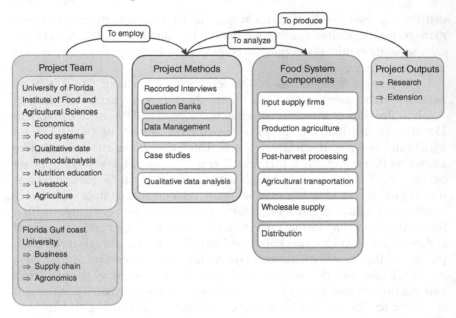

Figure 2.2 Qualitative research framework for assessing the impacts of COVID-19 on Florida's food system.

covering the following food system sectors: Input supply, Production agriculture, Processing/Packing, Transportation/ Logistics, Wholesaling, Retail distribution (local, state, and external), School and nutrition services, Community gardens, Food-related non-governmental operations (NGOs), Fisheries and aquaculture, Restaurants, Financial/Insurance/Real Estate, and Agricultural associations. The questions were categorized to provide information on specific areas of interest to the research team. The interviews covered general, overarching subjects such as risk, impacts, labor, new partnerships, regulation, and observed market changes that applied to all participants along with second-tier questions that were more specific to a particular sector. Interview guides for Production agriculture (same questions used for Small Farms) and Food-related NGOs are provided in Appendix 2.1 for reference.

Interview data was collected from July through November 2020. Participants were recruited via "word of mouth" through mentions of the effort in news media (e.g. Romaguera, 2020), webinars presenting quantitative results from Phase 1 of the project (e.g. Court et al., 2020), and the established stakeholder networks of UF/IFAS Extension and FGCU. Participants were owners or high-level managers who could speak knowingly of their respective company's experiences during the COVID-19 pandemic and were not monetarily compensated for their participation. These participants represented diverse business operations along the food supply chain in Florida including input supply companies, farming operations, food processing units, trucking and logistic firms, food-related NGOs, and food banks. This chapter summarizes the insights gained from interviews with 40 individuals representing production agriculture (9 participants), small farms (14 participants), and food-related NGOs (17 participants). On average, the interviews were conducted by individual project team members[3] via Zoom web conferencing technology and lasted an average of 35 minutes. Both institutions followed all guidelines and rules imposed on human subjects research by their respective Institutional Review Board.

The technological framework of digital meetings, transcription, analysis, and storage enabled this broad data gathering effort across the state. The digital meeting platform, Zoom, enabled a recorded interview data set to be formed without direct contact, while significantly reducing travel time and increasing reach and participation. Transcription software that automatically created editable transcripts reduced the effort required to document the interviews and offered the potential of a more extensive data set to be uniformly gathered by many individuals.

Following the collection of data, the individual interviewers edited the Zoom transcription files and submitted the files to the core research group. This core group of five to six developed a coding structure (Saldaña & Omasta, 2022) based on the desired research outcomes and codified the interviews in batches according to the food system sector. The categorized, redacted, edited transcripts were also retained for other research purposes based on individual research interests.

2.4 Findings

Findings for select sectors within Florida's food system are presented here as a multiple case study research initiative on the impacts of COVID-19 (Yin, 2016). The case study approach was selected because it facilitates exploration of a phenomenon within its context using a variety of data sources. The use of a mixed methods approach and the disciplinarity diversity of the research team ensure that the issue is not explored through one lens but rather a variety of lenses, which allows for multiple facets of the phenomenon to be revealed and understood (Baxter & Jack 2008). The COVID-19 pandemic and its impacts on food systems are both multidimensional; therefore, the case study framework allows for individual, sector-level, and overall food system impacts to be assessed. The creation of a qualitative data set uniformly collected to document a significant change event within the food system can therefore help to identify specific changes that both benefitted and inhibited the production, distribution, regulation, and consumption aspects of the food system. This can also help formulate a set of data, stakeholders, and circumstances that can be evaluated in later works to investigate the ongoing implications of the regulatory decisions related to COVID-19, the economy, and the food system.

Case study findings for production agriculture, small farms, and food-related NGOs were well-established within the data set and provide unique examinations of the relationships amongst the food system, the broader economy, governmental regulation, and the pandemic. Findings highlight not only the ways that the COVID-19 pandemic directly impacted these types of operations but also how these sectors were impacted indirectly as a result of COVID-19 impacts at all points along the food supply chain.

2.4.1 Production Agriculture

Here, production agriculture is an all-encompassing term for large commercial farm operations, of which nine stakeholders were interviewed. Production agriculture was deemed essential, allowing for continued operation throughout the pandemic, and it didn't take long for food safety experts around the world to determine that there were no direct food safety concerns associated with COVID-19 (WHO, 2020). Findings indicate, however, that across all commodities represented in the interview phase, the most significant and immediate impact of the COVID-19 pandemic stemmed from the ordered closure (or imposed capacity limitations) of non-essential businesses and stay-at-home orders enacted by the Governor in Florida during the Spring of 2020.

In Florida, pandemic-related shutdowns within the food service sector occurred at the peak of spring harvest season for many fruit and vegetable commodities. Hundreds of acres of highly perishable crops such as tomatoes, green beans, and sweet corn were destroyed as crop maintenance and harvest costs loomed with little to no market remaining for sales. Several farms that catered exclusively to restaurants and the food service sector reported losing more than 90 percent of their sales. Milk and other perishable dairy products had to be dumped as their immediate distribution points closed.

Production agriculture interviewees all detailed adjustments made to ensure the safety of the employees and farmworkers. Agribusiness operations invested time and money into designing new protocols to govern how employees and on-farm personnel interact. Most, if not all, office workers were asked to work from home, vendors and other non-farm/company personnel were prohibited from entering the farms to limit outside exposure to the virus, and business meetings once held face-to-face were recast as video conferences using Zoom, Microsoft Teams, and other software applications. Also, many operations instituted "hazard pay," usually an additional dollar per hour as an incentive for their employees to continue to work. Since production agriculture was classified as an essential industry in the U.S., making farm employees essential workers, farmworkers were not laid off and forced to seek unemployment benefits at nearly the same rates as workers in non-essential industries that were effectively shut down.

Seasonal farmworkers were, and will likely continue to be, the most challenging aspect of personnel management during the COVID-19 pandemic. Citrus and most row-crops depend on a sizable number of seasonal farmworkers, primarily to do the harvesting of fresh vegetables and citrus crops. Most Florida farmworkers are being hired through the foreign agricultural guest worker program known as H-2A. These workers are housed in employer-owned units and transported in 15-passenger vans or converted 55-passenger school buses. Unfortunately, the pandemic began after H-2A arrangements had been in place for the 2019–2020 season. H-2A housing and transportation facilities were not easily adaptable for social distancing. Where they could, housing managers rearranged living quarters to create more space among the workers. "Quarantine" units were established to isolate any worker suspected of having a COVID-19 infection. The number of workers riding a bus was reduced by 50 percent, thereby effectively doubling the transport time of workers between their housing facility and the work sites.

Despite the fact that personal protective equipment (PPE) and COVID-19 testing capabilities were limited during the early stages of the pandemic, there was no evidence of widespread infection rates of COVID-19 infections among Florida farmworkers. Yet, perceptions and concerns about COVID-19 were prevalent among H-2A workers. In a few cases, H-2A workers took advantage of an early return-home option to prevent becoming infected or due to concern for how COVID-19 would affect their families in Mexico.

Beyond the demand-side shock of the regulations aimed at mitigating the spread of the virus that causes COVID-19, there were several ways in which pandemic-related impacts to the broader food supply chain affected production agriculture in Florida. Rising case numbers of COVID-19 among workers in several large meat processing facilities caused slaughterhouses in western states to stop or slow production and receipt of finished cattle, which in turn delayed the shipment of calves from Florida ranches to feed lots in these western states and disrupted calving cycles. These impacts highlight the fact that the Florida food system depends on post-production processing and distribution networks stretching across several states.

Situational adaptations and outcomes varied across commodities, seemingly dependent on the broader supply chain effects and regulatory frameworks within which commodity production occurs. What was remarkable, however, was to observe how quickly many of the large farming operations were able to adapt and mitigate early pandemic losses. A notable portion of the excess produce supplies on hand at the very beginning of the pandemic was redirected to food-related NGOs, food banks, and food pantries in time for them to service a growing number of households needing food assistance. While food service firms closed down, grocery store prices surged for several food products. but in many cases producers were still experiencing declining farmgate prices (Mead et al., 2020; Johansson, 2021). Assistance from federal programs such as the United States Department of Agriculture's Coronavirus Food Assistance Program (CFAP) and more recent surges in foreign exports for select commodities might also offset some of the losses from the first quarter of 2020 (Johansson, 2021).

In some cases, COVID-19 provided an unexpected financial benefit. Consumers across the country associate orange juice and fresh citrus fruit with healthier immune systems and, as a result, increased their consumption of orange juice (Heng et al., 2020). Findings suggest that the losses citrus processors suffered from food service sales were more than offset by juice sold in grocery stores as higher-valued "not-from-concentrate" (NFC). Also, unlike livestock slaughterhouses and meat processing facilities in the western

U.S., workers in citrus processing plants and packing houses do not work in close physical proximity to one another, so the industry did not experience COVID-19 outbreaks among their workers and was able to continue steady processing and fresh fruit packing operations.

On the surface, the relationships between commercial farming operations and their input suppliers did not suffer extensive financial strains. COVID-19, however, did affect interactions between input suppliers and their farmer-customers. In Florida, like anywhere else, interactions between farmers and input suppliers are generally complex. Input suppliers provide not only physical inputs, such as seeds, fertilizers, and plant protection materials, they also share information and provide services of various types. Examples of services and information are agronomic and technical support, market and regulatory information, advice on risk-management strategies, plan of approach for disease-specific problems, and financing solutions. The services can be extremely valuable and critical to the success of the agricultural operation. Prior to the pandemic, input suppliers relied on direct contact with their farmer-customers. Farm visits allow the sales personnel to build a trusting relationship with the farmer and to accurately assess their customers' needs. To add to this complexity, this relationship and exchange of information between input suppliers and farmers are typically intermediated by an input retailer, a distributor, or a cooperative.

As expected, input suppliers had to limit face-to-face interactions with their distributors and farmers to limit the spread of the virus. Likewise, management and employees were required to follow strict sanitary protocols to avoid internal transmission of the virus. As a result, most if not all of the interactions described above had to be moved online using videoconferencing tools. From what our interviewees recounted, the transition to online interactions was fairly smooth, even if a minority of farmers were somewhat reluctant to use video conferencing tools. There were some concerns for their newly hired sales representatives and trainees because they could not get a chance to meet any potential customers face-to-face and establish a meaningful rapport with them. Long-established relationships, however, were not affected in any way and very few setbacks were experienced. However, it is important to note that farmers experienced the brunt of the COVID-19 crisis at the end of the growing seasons for most crops grown in Florida. The situation could have been very different if the containment measures had to be implemented at a different time of the year, for instance, when farmers have to prepare their fields or in the middle of the production season when farmers have to address production issues of various kinds.

In summary, heterogeneous adaptations demonstrated that the impacts of the COVID-19 pandemic were contingent on the type of agricultural product. Negative experiences were focused on the initial lull in demand, overhead costs associated with the implementation of physical distancing and other safety measures, and uncertainty regarding the stringency and duration of regulations or timing and scope of federal disaster relief. There does not appear to be a significant impact on labor or processing within production agriculture of row crops and citrus. Some products with longer cycles, such as livestock, did suffer processing, transport, and calving cycle constraints according to those interviewed, but it was not clear whether the durational impacts would be beneficial or detrimental to sales revenues or profits. Positive experiences that were shared included the speed of approval and distribution for economic stimuli from the federal government including stimulus checks for consumers and CFAP and the realization that regulatory decisions regarding COVID-19 restrictions did not impact their operations directly but through a shift in demand. Several interviewees indicated that federal relief funds were used to support staff that would otherwise be on furlough and to adopt new technologies, typically aimed at reaching new customers.

2.4.2 Small Farms

Fourteen small farms were interviewed as part of Phase 2. A small farm can be defined by the amount of area under production, scale, diversity of product (or lack thereof), and local market area that the farm or production system is focused on. In this respect, a small farm can be seen as serving a specialty niche within an urban area where demand will be high enough and the market large enough to provide customers for the farm to produce products focused on quality rather than quantity, largely with a consumer-direct business model with secondary wholesale to restaurants in a local setting. There are many different terms that are used with small farms, cooperative, local food, market garden, urban farm and peri-urban farm being just a few. In our interviews, we had small farm participants throughout the state that were directly serving urban areas, at varying scales of production from in-home microgreens to several acres of land. Although these small farms are producing a diverse array of agricultural products under an equally diverse set of business models, findings will focus on their economic experiences and adaptation to regulation regarding the markets served.

While local ordinances changed the parameters within which some small farms operated, regulations did not often result in the complete closure of small farms as food production was deemed essential. Differences in how farmers engaged with this relative freedom to operate throughout the pandemic was largely based on their perception of risk. Farms that had the capacity to host outdoor distanced gatherings, food preparation, and/or farm shops took advantage and remained open for their local and seasonal residents that largely remained in place due to the inability or reluctance to return to their summer abodes. U-pick operations noticed a significant increase in attendance as concerned parents and families confined to homes searched for outdoor activities to engage in.

Many small farms were also impacted by regulations and decisions affecting farmers' markets. Farmers' market managers and the owners of the property on which farmers' markets operate had varying perspectives on and reactions to the COVID-19 pandemic, often founded on the interpretation of local regulatory decisions, which differed by county and municipalities across the state. In some areas, these markets were not allowed to continue and either stopped operating during the restricted period or shifted to an online platform, Community Supported Agriculture (CSA) model, or expanded to provide delivery services. Other counties were able to keep their markets open but shifted the models to allow for physical distancing, required facial coverings (i.e. masks), provided hand washing stations, or set up drive-through markets to reduce contacts in unison with the adoption of Centers for Disease Control (CDC) guidelines.

Expedited dispersal of federal-level stimulus payments also provided a significant economic windfall for many small producers, enabling continued employment of support staff, technological improvements, and the identification of alternative revenue streams during times of uncertainty. Many of the producers interviewed kept employees or were able to support furloughs of employees during the summer months as a direct result of stimulus payments. The combination of stimulus payments and a brief lull in normal operations also allowed many producers to take a step back to make changes, adaptations, and innovations to their businesses, a step back that they might not have taken under normal circumstances. Participants indicated that they made changes within their operations that they had been considering for some time but had not found the time or funds to implement or adopted new practices that they became aware of during COVID-19 to keep their businesses running. Many small farms invested in technological improvements such as websites and the use of social media to engage new customers. The initial successes of these changes were significant as many producers identified increased brand awareness within their local communities and substantial increases in interaction and awareness of the digital presence and offerings of their small farm businesses.

The participants within this sector had very different experiences depending on the type of markets they served, and their ability to adapt by forming new partnerships or adopting new technologies. Several participants reported an overall positive experience, with some businesses reporting the best summer season ever in 2020. According to participants, this is due to several factors including heightened awareness of local production and small farms amid early pandemic shutdowns, the closure of restaurants and other food service operations, and media coverage related to food supply chain issues. With many people cooking at home, the demand for fresh local produce to prepare meals increased. For many small farms, the shift to CSA boxes and partnerships with other local small producers helped sustain direct sales. Some producers that largely provided products to restaurants via wholesale distribution and had limited experience with direct sales prior to the pandemic, continued to sell directly to consumers upon loosening of public health restrictions as the revenue stream is reportedly more lucrative and direct sales provides a great deal of goodwill (and free marketing) within their communities. The local farm as a destination to continue safe outdoor gatherings also significantly helped those producers who created markets, U-pick operations, or distribution areas for their products.

Negative experiences during the COVID-19 pandemic were shared by small producers that were extremely reliant on farmers' markets and restaurant customers that were shut down due to public health measures enacted to slow the spread of the virus. Many of these producers were selling highly perishable products and did not have, or were unable to shift to consumer direct models, resulting in significant losses in the first several weeks of COVID-19 shutdowns. Increases in input supply pricing (seeds, packaging, raw materials, fertilizers, pesticides, etc.) also had ramifications for small farmers during COVID-19. In particular, the rise in seed prices was interesting as it highlighted one of the many indirect effects of COVID-19 on small farms. Increased demand for seeds from the general public for gardening led to increasing seed prices, and this increase in demand was compounded by supply chain issues related to manufacturing, labor, and distribution.

2.4.3 Food-related Non-governmental Organizations

The food-related NGOs that were the focus of this research were entities that serve populations that have limited access to affordable and nutritious food. The 17 participants interviewed in this category represented food banks, food pantries, school foodservice programs, and community gardens. These groups are spread throughout the state and provide support for limited resource populations with less risk tolerance and higher food insecurity.

The interviews within this area of the food system were particularly insightful as the duration and impacts helped outline the significant frailties in access to food and resources. Economically, this sector saw as much as a 500 percent increase in demand. Food pantries and banks were aided by the economic support of federal disaster funds and donations to increase the frequency and volume of distribution. The increases in demand for services did not subside as restrictions shifted to allow for more movement and interaction, and businesses were allowed to re-open. According to recent follow-up discussions with an interview participant, in the fall of 2021, with all pandemic restrictions lifted in the state, demand has only reduced slightly, which points to broader economic implications from COVID-19 for vulnerable populations.

There were both challenges and benefits to regulatory changes as economic support in the form of federal-level stimulus payments and emergency distributions helped meet the required financial support. However, the social distancing requirements delegated significant changes in how these vital services operated with the need for more labor, space, innovative packaging, and consolidation of food to allow for efficient distribution. There was also the added challenge of reduced volunteer participation as many of those who have the available time and resources to contribute were retirees who were classified as high-risk for adverse health implications from COVID-19.

2.5 Discussion

The COVID-19 pandemic has provided and continues to provide lessons for Florida's food system. Food supply chains evolved quickly in 2020, shifting and adapting in response to fluid pandemic conditions as well as associated regulatory and economic adjustments. This evolution is ongoing as the food supply chain continues to adjust to supply and demand shifts related to the ongoing pandemic situation. Interviews with food supply chain stakeholders identified that one particularly significant outcome of this transition has been the increased awareness and adoption of technologies and technology-supported practices brought about by situational necessity and economic stimulus. Other positive outcomes include a greater focus on building local awareness of local production through on-site sales and food box offerings and on philanthropy including the distribution of some crops to the food bank system. While these efforts were perhaps successful in promoting awareness and community engagement within a food supply chain where the producer and end-user are typically substantially removed from one another, the volume of movement through these efforts was often insignificant in terms of shifting large volumes of produce from national and international markets.

Interviews also uncovered some negative outcomes that likely require attention and further study. The increase in demand experienced by food banks, food pantries, and community gardens demonstrates the severe implications on vulnerable populations of shutdowns and other institutional measures intended to mitigate the spread of COVID-19, highlighting the economic precarity under which many people live. The continuation of high levels of demand for food assistance highlights inequalities that can be exacerbated by characteristics of the local food system and the need for food systems that are both efficient and resilient to ensure food access as well as economic and social stability in the face of future hazard or disaster events. The inability to quickly pivot the production and/or processing of agricultural crops under the initial COVID-19 shutdowns suggests that capacity for emergency food processing and distribution infrastructure might be beneficial in emergent situations to improve resilience of the local or regional food system and to reduce food waste.

Long-term implications of the pandemic on both producer and consumer behaviors are yet to be determined. Some adaptations that were adopted as a necessary response to COVID-19 by consumers (e.g. online grocery shopping, increased demand for local food products, curb-side pickup of restaurant meals, etc.) and producers (online ordering platforms, curbside pickup and delivery options, smalls farm adoption of direct sales methods, use of online platforms and social media, etc.) have continued in the short to medium term, at least beyond relaxation of the initial shutdown measures (Hobbs, 2020; Shirvell, 2021; Reid, 2021). The Florida Department of Agriculture and Consumer Services, in collaboration with several academic institutions and agriculture associations, has committed to the improvement and continuation of the Florida Farm to You website, which was created during the COVID-19 pandemic to connect producers with buyers and consumers. While larger operations within the food supply chain are more cost-efficient and offer a wider variety of food products, continued concern about food security as well as the popularity of the "local food movement" might be enough to sustain awareness of and interest in local food and small farm providers in the short to medium term post-COVID-19 (Hobbs, 2020). Several ongoing compounding factors that continue to affect Florida's food supply chain were also mentioned by interview participants and will likely affect the long-term impacts and adaptations in food production that result from the COVID-19 pandemic. Compounding factors mentioned include issues related to international trade, labor supply, food safety regulations, and land use. Finally, the use of, termination, and potential future availability of federal-level economic stimulus payments and relief packages associated with COVID-19 for producers and consumers will also affect the long-term impacts on and changes within Florida's food system, including the potential continuation of higher levels of demand for services from food-related NGOs as a result of continued impacts for vulnerable populations.

Ultimately, a complete understanding of the impacts of and adaptations to the COVID-19 pandemic experienced by Florida's food system will require continued research efforts. One topic that will require significantly more study is the sentiment of the participants since many of our human systems are dependent on perception-based behavioral responses. While sentiment is a "soft measure," the perspective and mental models of stakeholders have certainly impacted the evolution of the food system during the pandemic – driving efficiencies in communication and adoption of technology that might contribute to longer-term economic benefits. Much of the information gleaned from the interviews also points to the need for a reevaluation of the food value chain, both locally and on larger scales. Operations solely focused on efficiency, economies of scale, and cost-based decision-making have created frailties within the Florida food system. Having experienced the COVID-19 pandemic, we have learned that this focus on efficiency created a lack of resiliency. However, there also appears to be an increased awareness of the current risk landscape and recognition of future potential risk that might bode well for more effective disaster risk management in the long-run.

Notes

1 Valuable assistance in the preparation of the interview guides, recruitment of participants, and implementation of interviews was provided by Lauren Butler, County Extension Director and Livestock Agent II, Okeechobee County Cooperative Extension Service, UF/IFAS; Angela Corona, Public Health Specialist, Family Nutrition Program, Osceola and Orange Counties, UF/IFAS; Dr. Vincent Encomio, Florida Sea Grant Agent, Martin and St. Lucie Counties, UF/IFAS; Carlita Fiestas-Nunez, Food Systems Specialist, Family Nutrition Program, UF/IFAS; Yvette Goodiel, Sustainability and Commercial Horticulture Agent, Martin County, UF/IFAS; Ron Hamel, Agribusiness Consultant, Center

for Agribusiness, Lutgert College of Business, FGCU; Colleen Larson, Regional Dairy Agent, Okeechobee, Highlands, DeSoto, and Hardee Counties, UF/IFAS; Gene McAvoy, Associate Director for Stakeholder Relations, Southwest Florida Research and Education Center, UF/IFAS; Jeannie Necessary, Food Systems Specialist, Family Nutrition Program, UF/IFAS; Chris Prevatt, State Specialized Extension Agent II—Beef Cattle and Forage Enterprise Budgeting and Marketing, Range Cattle Research and Education Center, UF/IFAS; Jessica Ryals, Agriculture and Sustainable Food Systems Agent, Collier County, UF/IFAS; Stuart Van Auken, Eminent Scholar, Marketing, Lutgert College of Business, FGCU; Andrea Moron Vasquez, Program Manager, Family Nutrition Program, Lee County, UF/IFAS; and Kelly Wilson, Food Systems Specialist, Family Nutrition Program, Lee County, UF/IFAS.

2 Extension educators, often colloquially known as Extension agents within the UF/IFAS Extension system, and research-focused state specialists are both part of the Cooperative Extension System (CES) in the United States, which is a collaborative effort between the Land-Grant University System and the federal, state, and local governments. State specialists and Extension educators collaborate to translate science-based research results into educational materials appropriate for targeted audiences, often working with community members to solve problems and to gather input to prioritize future research.

3 In a few rare instances, interviews included more than one project team member as an interviewer or more than one interviewee.

Appendix 2.1 Impact of Covid-19 on Florida and Agribusiness Food Supply Chain

Production Agriculture

1 Briefly describe your farming operation in Florida *pre-COVID-19*.
   Prompts, if necessary:
   What do you grow/produce?
   Who are your customers, or how do you sell your products?
   Where are your markets or customers located?
   How many people do you directly employ on your operation?

2 How has (have) your market(s) changed as a result of COVID-19?
   Prompts, if necessary:
   Did your operation expand or contract?
   Are you employing more or fewer people than before the pandemic?
   Describe how your relationships with your customers AND suppliers have changed because of the pandemic.

3 What adjustments did you make in your farming and/or marketing in reaction to COVID-19?
   Prompts, if necessary:
   Did you adopt any new innovations or form new partnerships?

4 Have there been any regulations or other constraints that have enabled or prevented you from making adjustments you wanted to make within your operations in response to COVID-19?

5 How do you think COVID-19 will change the future of your business?

6 Describe how you managed risk before COVID-19 (i.e. weather, market volatility).

7   Where you able to prepare for risks associated with the COVID-19 pandemic?
8   Has the COVID-19 pandemic changed your approach to risk management? If so, how?

Food-Related NGOs

1   Briefly describe your non-profit *pre COVID-19*.
    Prompts, if necessary:
    What communities do you serve?
    What is the structure of organization?
    How many employees?
    How many participants?
2   What area do you serve geographically?
3   How and how much has demand for your services changed?
4   Where have you seen increases in demand geographically? Has this changed the way you distribute to these areas?
5   How has your volunteer base changed?
6   Have you made changes to how and where you procure food? If so, what are they?
7   What funding impacts do you anticipate as a result of COVID-19?
8   Do you anticipate lasting operational changes as a result of COVID-19? If so, what are they?
9   Have there been any regulations or other constraints that have enabled or prevented you from making adjustments you wanted to make within your operations in response to COVID-19?
10 What role have partnerships played in responding to COVID-19? Have you formed new partnerships?
11 What risks has COVID-19 highlighted for your operation?

References

Abraham, T. (2011). Lessons from the pandemic: The need for new tools for risk and outbreak communication. *Emerging Health Threats Journal*, 4(1), 7160. https://doi.org/10.3402/ehtj.v4i0.7160

Aiyar, A., & Pingali, P. (2020). Pandemics and food systems—towards a proactive food safety approach to disease prevention & management. *Food Security*, 12(4), 749–756. https://doi.org/10.1007/s12571-020-01074-3

American Farm Bureau Federation. (2019). Feeding the economy: Agricultural jobs by state. March 21. Available at: http://fb.org/market-intel/feeding-the-economy-agricultural-jobs-by-state. Accessed August 25, 2021.

Baxter, P., & Jack, S. (2008). Qualitative case study methodology: Study design and implementation for novice researchers. *The Qualitative Report*, 13(4), 544–559. https://doi.org/10.46743/2160-3715/2008.1573

Born, B., & Purcell, M. (2006). Avoiding the local trap: Scale and food systems in planning research. *Journal of Planning Education and Research*, 26(2), 195–207. https://doi.org/10.1177/0739456X06291389

Bron, G. M., Siebenga, J. J., & Fresco, L. O. (2021). *In the Age of Pandemics, Connecting Food Systems and Health: A Global One Health Approach.* UN Food Systems Summit Brief February 15, 2021.

Clancy, K. (2013). Digging deeper bringing a systems approach to food systems: Feedback loops. *Journal of Agriculture, Food Systems, and Community Development,* 3(3), 5–7. Available at: www.agdevjournal.com/attachments/article/342/JAFSCD_Clancy_Column_June-2013.pdf

Court, C. D., Lai, J., & Ropicki, A. (2020). Assessing the impacts of COVID-19 on Florida's Agriculture and Marine Industries. June 8. Webinar. Available at: https://fred.ifas.ufl.edu/extension/economic-impact-analysis-program/webinars/

Court, C. D., Ferreira, J. P., Hewings, G. J. D., & Lahr, M. L. (2021). Accounting for global value chains: Rising global inequality in the wake of COVID-19? *International Review of Applied Economics,* 35(6): 813–831. https://doi.org/10.1080/02692171.2021.1912716

Creswell, J. W. & Plano Clark, V. L. (2017). *Designing and Conducting Mixed Methods Research* (3rd ed.). Thousand Oaks, CA: Sage.

Huremovic, D. (2019). Psychiatry of pandemics. In *Psychiatry of Pandemics* (pp. 7–35). Cham: Springer. https://doi.org/10.1007/978-3-030-15346-5

Dwyer, C. & Horney, J. (2014). Validating indicators of disaster recovery with qualitative research. *PLoS Currents,* 6. https://doi.org/10.1371/currents.dis.ec60859ff436919e096d51ef7d50736f

Ericksen, P.J. (2008). Conceptualizing food systems for global environmental change research. *Global Environmental Change,* 18(1): 234–245. https://doi.org/10.1016/j.gloenvcha.2007.09.002

Feagan, R. (2007). The place of food: Mapping out the "local" in local food systems. *Progress in Human Geography,* 31(1), 23–42. https://doi.org/10.1177/0309132507073527

Feenstra, G. W. (1997). Local food systems and sustainable communities. *American Journal of Alternative Agriculture,* 12(1), 28–36. https://doi.org/10.1017/s0889189300007165

Ferreira, J. P., Ramos, P., Barata, E. Court, C.D., & Cruz, L. (2021). The impact of COVID-19 on global value chains: Disruption in nonessential goods consumption. *Regional Science Policy & Practice,* 13(S1): 32–54. https://doi.org/10.1111/rsp3.12416

Foran, T., Butler, J. R. A., Williams, L. J., Wanjura, W. J., Hall, A., Carter, L., & Carberry, P. S. (2014). Taking complexity in food systems seriously: An interdisciplinary analysis. *World Development,* 61, 85–101. https://doi.org/10.1016/j.worlddev.2014.03.023

Fuchs, S., Röthlisberger, V., Thaler, T., Zischg, A., Fuchs, S., Veronika, R., & Thaler, T. (2017). Natural hazard management from a coevolutionary perspective: Exposure and policy response in the European Alps. *Annals of the American Association of Geographers,* 107(2), 382–392. https://doi.org/10.1080/24694452.2016.1235494

Heng, Y., Zansler, M., & House, L. (2020). Orange juice consumers' response to the COVID-19 pandemic. EDIS Document FE1082. University of Florida-IFAS, Food & Resource Economics, Gainesville, FL. Available at: https://edis.ifas.ufl.edu/publication/FE1082

Hertel, Thomas W., West, T. A. P., Börner, J., & Villoria, N. B. (2019). A review of global-local-global linkages in economic land-use/cover change models. *Environmental Research Letters* 14 053003. https://iopscience.iop.org/article/10.1088/1748-9326/ab0d33/pdf

Hill, H. (2008). Food miles: Background and marketing. IP312, Slot 311, Version 010708. Available at: https://attra.ncat.org/product/food-miles-background-and-marketing/. Accessed September 12, 2021.

Hobbs, J. E. (2020). Food supply chains during the COVID-19 pandemic. *Canadian Journal of Agricultural Economics*, 68, 171–176. https://doi.org/10.1111/cjag.12237

Huff, A. G., Beyeler, W. E., Kelley, N. S., & McNitt, J. A. (2015). How resilient is the United States' food system to pandemics? *Journal of Environmental Studies and Sciences*, 5(3), 337–347. https://doi.org/10.1007/s13412-015-0275-3

Johansson, R. (2021). Another look at availability and prices of food amid the COVID-19 pandemic. United States Department of Agriculture. www.usda.gov/media/blog/2020/05/28/another-look-availability-and-prices-food-amid-covid-19-pandemic

Kniffin, K. M., Narayanan, J., Anseel, F., Antonakis, J., Ashford, S. P., Bakker, A. B., Bamberger, P., Bapuji, H., Bhave, D. P., Choi, V. K., Creary, S. J., Demerouti, E., Flynn, F. J., Gelfand, M. J., Greer, L. L., Johns, G., Kesebir, S., Klein, P. G., Lee, S. Y., ... Vugt, M. van. (2021). COVID-19 and the workplace: Implications, issues, and insights for future research and action. *American Psychologist*, 76(1), 63–77. https://doi.org/10.1037/amp0000716

Kummu, M., Kinnunen, P., Lehikoinen, E., Porkka, M., Queiroz, C., Röös, E., Troell, M., & Weil, C. (2020). Interplay of trade and food system resilience: Gains on supply diversity over time at the cost of trade independency. *Global Food Security*, 24(November 2019), 100360. https://doi.org/10.1016/j.gfs.2020.100360

Lang, T. (2006). Locale/globale (food miles). *Slow Food*, May, 94–97.

Low, S. A., Adalja, A., Beaulieu, E., Key, N., Martinez, S., Melton, A., Perez, A., Ralston, K., Stewart, H., Suttles, S., & Vogel, S. (2015). Trends in U.S. local and regional food systems. *Local and Regional Food Systems: Trends, Resources and Federal Initiatives*, 068, 87–195.

Lusk, Jayson. (2016). Evolution of American agriculture. Blog June 27, 2016. Available at: http://jaysonlusk.com/blog/2016/6/20/the=evolutions-of-american-agriculture. Accessed August 25, 2021.

Mead, D., Ransom, K., Reed, S. B., & Sager, S. (2020). The impact of the COVID-19 pandemic on food price indexes and data collection. *Monthly Labor Review*, U.S. Bureau of Labor Statistics, August, https://doi.org/10.21916/mlr.2020.18

Morens, D. M., Folkers, G. K., & Fauci, A. S. (2009). What is a pandemic? *Journal of Infectious Diseases*, 200(7), 1018–1021. https://doi.org/10.1086/644537

National Marine Fisheries Service (2022). Fisheries Economics of the United States, 2019. US Dept. of Commerce, NOAA Tech. Memo. NMFS-F/SPO-229, 236 p. www.fisheries.noaa.gov/national/sustainable-fisheries/fisheries-economics-united-states

Ostadtaghizadeh, A., Ardalan, A., Paton, D. et al. (2016). Community disaster resilience: A qualitative study on Iranian concepts and indicators. *Natural Hazards* 83, 1843–1861. https://doi.org/10.1007/s11069-016-2377-y

Phillips, B. D. (2014). *Qualitative Disaster Research: Understanding Qualitative Research*. New York: Oxford University Press.

Pingali, P., Alinovi, L., & Sutton, J. (2005). Food security in complex emergencies: Enhancing food system resilience. *Disasters*, 29(suppl.), 5–24. https://doi.org/10.1111/j.0361-3666.2005.00282.x

Piret, J., & Boivin, G. (2021). Pandemics throughout history. *Frontiers in Microbiology*, 11(January). https://doi.org/10.3389/fmicb.2020.631736

Reid, N. (2021). A toast to the survival skills of craft breweries during American craft beer week. *The Brewer Magazine*, May 13. Available at: https://thebrewermagazine. com/a-toast-to-the-survival-skills-of-craft-breweries-during-american-craft-beer-week/. Accessed September 26, 2021.

Renn, O. (1998). The role of risk perception for risk management. *Reliability Engineering and System Safety*, 59, 49–62.

Romaguera, K. (2020). Response requested from ag and marine industry professionals for new round of surveys on pandemic effects. UF/IFAS Blogs. Posted August 21, 2020. Available at: http://blogs.ifas.ufl.edu/news/2020/08/21/response-requested-from-ag-and-marine-industry-professionals-for-new-round-of-surveys-on-pandemic-effects/. Accessed September 25, 2021.

Romang, H. E., Bischof, N., & Rheinberger, C. M. (2009). The risk concept and its application in natural hazard risk management in Switzerland. *Natural Hazards and Earth System Science*, 9, 801–813.

Rozell, D. J. (2015). Assessing and managing the risks of potential pandemic pathogen research. *MBio*, 6(4), 18–21. https://doi.org/10.1128/mBio.01075-15

Saldaña, J. & Omasta, M. (2022). *Qualitative Research: Analyzing Life* (2nd ed.). Thousand Oaks, CA; Sage Publications.

Shirvell, B. (2021). Will the CSA boom survive beyon the pandemic? *Civil Eats*. Available at: https://civileats.com/2021/03/10/will-the-csa-boom-survive-beyond-the-pandemic/. Accessed September 21, 2021.

Slovic, P. (2020). Risk perception and risk analysis in a hyperpartisan and virtuously violent world. *Risk Analysis*, 40, 2231–2239. https://doi.org/10.1111/risa.13606

Solecki, W. & Walker, R. T. (2001). Transformation of the South Florida Landscape. In *Growing Populations, Changing Landscapes: Studies from India, China, and the United States* (pp. 237–274). Washington, DC: The National Academies Press. https://doi.org/10.17226/10144

Spicer, A. (2020). Organizational culture and COVID-19. *Journal of Management Studies*, 57(8), 1737–1740. https://doi.org/10.1111/joms.12625

Tamburino, L., Bravo, G., Clough, Y., & Nicholas, K. A. (2020). From population to production: 50 years of scientific literature on how to feed the world. *Global Food Security*, 24(December 2019). https://doi.org/10.1016/j.gfs.2019.100346

Tendall, D. M., Joerin, J., Kopainsky, B., Edwards, P., Shreck, A., Le, Q. B., Kruetli, P., Grant, M., & Six, J. (2015). Food system resilience: Defining the concept. *Global Food Security*, 6, 17–23. https://doi.org/10.1016/j.gfs.2015.08.001

Toth, A., Rendall, S., & Reitsma, F. (2016). Resilient food systems: A qualitative tool for measuring food resilience. *Urban Ecosystems*, 19:19–43. https://doi.org/10.1007/s11252-015-0489-x

UF/IFAS. (2020). The science of better living: Extension overview booklet. UF/IFAS. Available at: https://branding.ifas.ufl.edu/media/brandingifasufledu/brochures/ExtensionOverviewBooklet2020.pdf. Accessed July 23, 2021.

UF/IFAS. (2021). Agriculture and food system fast facts 2021. *Our World*.

USDA. (2019). Florida agricultural overview 2020. *Web* 1(1).

Vasavari, T. (2015, April). Risk, risk perception, risk management. *Public Finance Quarterly*, April 2015, LX, 29–48.

Webber, C. B., & Dollahite, J. S. (2008). Attitudes and behaviors of low-income food heads of households toward sustainable food systems concepts. *Journal of Hunger & Environmental Nutrition*, 3(2–3), 186–205. doi: 10.1080/19320240802243266.

Wisner, B. (2001). Capitalism and the shifting spatial and social distribution of hazard and vulnerability. *Australian Journal of Emergency Management*, 16(2), 44–50. Available at: http://search.informit.com.au/documentSummary;dn=378174191869592;res=IELHSS

World Health Organization (2020). COVID-19 and food safety: guidance for food businesses. World Health Organization. Interim Guidance 7 April 2020. www.who.int/publications/i/item/covid-19-and-food-safety-guidance-for-food-businesses

Yin, R. K. (2016). *Qualitative Research from Start to Finish* (2nd ed.). New York: The Guilford Press.

3 Street Shock

How a Bike Lane Redefined a Neighborhood

Andrea Marpillero-Colomina

3.1 Introduction

Between October 2007 and December 2009, the City of New York abruptly installed then uninstalled what became a very contentious bike lane, unproductively spent taxpayer funds, and mediated a topless protest. The bike lane fomented a new cultural fissure in the rapidly gentrifying neighborhood of South Williamsburg, Brooklyn. This chapter is a microanalysis of political, economic, and social conditions that led the Bedford Avenue bike lane to become a space of conflict. History matters here. For generations, Williamsburg was the tensely shared territory of two groups – Latinxs[1] and Hasidic Jews[2] – who settled there in the mid-twentieth century. In the late 1990s, Williamsburg became the forefront of New York's rapid gentrification.

This chapter examines the hyperlocal lived reality of cultural groups living in a rapidly transforming urban neighborhood shaped by the policymaking decisions of outsiders and how spatial change to place disrupts existing dynamics of power. It explores how local change at the economic, territorial, and temporal level link to the City's role as an institutional mechanism. The study asks: How can policymaking that transforms public space provoke sociocultural conflict in a gentrifying neighborhood? Findings consider how municipal policymakers can create effective, equitable change, while disassembling mechanisms that obstruct positive territorial transformation.

The study contributes to debates that explore dynamics of power in contested space, focusing on shifts in local cultural identity (or identities) as neighborhoods are redefined and redesigned during processes of gentrification. It adds to a growing body cross-disciplinary scholarship exploring symbolism assigned to new infrastructure and offers a new perspective about how changes to public infrastructure affect trajectories of neighborhood change.

The research explores lasting effects of policymaking that forces communities to absorb shocks of rapid change in space. Increasingly, as cities densify (and gentrify), cultural groups with disparate ethical, religious, and moral values are compelled to live side by side, sharing spaces and resources. Policy change does not exist in a values vacuum. To thrive together, these communities need the support of effective public mechanisms that enable their capacity to equitably participate in local policymaking.

DOI: 10.4324/9781003191049-4

The chapter starts with a brief review of extant literatures in gentrification studies and urban policymaking that explore impacts of state-implemented spatial change on neighborhood ecologies. Next, I present the mixed methods qualitative methodology and the culturally complex neighborhood context in South Williamsburg. Following is an outline to the policy context for the implementation of the Bedford Avenue bike lane, including the "street shock" it induced. The chapter closes with a discussion about how more effective participatory mechanisms could have mitigated the impact of the bike lane on neighborhood dynamics.

3.2 "Power Laden Fields" of Urban Space and Infrastructure

Cities are constantly being upended and reimagined. Gentrification is a fundamental restructuring of urban space defined by "productive capital returns" of new neighborhood inhabitants (Smith, 1979; Smith, 1996). Gentrification is the "embourgeoisement of the inner city" (Ley, 2010: 108). The actor here is the "bourgeois bohemian," or *bobo*: "a *bobo* acts cool but has money, pretends to make art but buys shoes, pretends to be radical but hangs on to privilege" (Greif et al., 2010: 36). Globally, the *bobo* – known to some as the hipster – has become a villainous caricature of people who flock to places like Williamsburg and other like neighborhoods around the globe.

These newcomers are white-collar professionals who produce "new forms or designs that are readily transferable and broadly useful" (Florida, 2003: 8). The creative class drives the economy of culture, of trends, and consumption. Their economic power and social organizational structures differ greatly from other urban cultural groups, although they follow similar settlement patterns by forming enclaves and growing identity-based social capital. The "new economic geography of creativity" takes root in places that possess the three Ts: technology, talent, and tolerance (Florida, 2003: 10). These places are often also characterized by a fourth T: access to transportation (Jones and Ley, 2016).

A spatial amenity like transportation infrastructure, enabling easy travel to central business districts and jobs, is crucial for the gentrifier (Beauregard, 2010: 20). The creative class establishes new linkages between home, work, and leisure, "imagineering" an alternative urbanism (Ley, 1997). Every change in the system of goods and services induces a change in tastes, which can induce transformation of the field of production (Bordieu, 1984: 231).

The ecology of urban neighborhoods – their density, diversity, and development – is often fraught. In *Black Corona*, sociologist Steven Gregory (1998) examines the fragility of the very notion of community in the Corona neighborhood, in Queens, New York: "From my perspective, community describes not a static, place-based social collective but a power-laden field of social relations whose meanings, structures, and frontiers are continually produced and reworked in relation to a complex range of sociopolitical attachments and antagonisms" (Gregory, 1998: 11). Networks are constantly evolving. In *The Social Construction of Communities* (1972), an empirical study of Chicago's South Side, sociologist Gerald Suttles describes the "defended neighborhood," where "cognitive maps" that residents have do not necessarily correspond with physical structures, and identity is acquired through ongoing commentary between themselves and outsiders, media coverage, and claims by developers and city officials (Suttles, 1972). As neighborhoods gentrify, place becomes increasingly contested as new spatial and cultural priorities are introduced, and

> each unit of a community strives, at the expense of the others, to enhance the land-use potential of the parcels with which it is associated... fighting over highway routes, airport locations, campus developments, defense contracts, traffic lights, one-way street designations, and park developments.
> (Molotch, 1976: 311)

In *The Cultures of Cities*, urban sociologist Sharon Zukin examines "cultural strategies of development," describing how in nineteenth-century New York, "cultural products were inspired, then cross-fertilized, by social elites, business leaders, and constant streams of immigrants." By the late twentieth century, cultural spaces "anchored markets in upscale real estate development" (Zukin, 1995: 121). In later work, Zukin argues that "investment in urban infrastructure of any sort only represents a temporary solution to [the] search for higher profits" (Zukin, 2006: 112).

When the state creates incentives for investment, this can bring "local politics to a critical point." In gentrifying places, "public space is sometimes a battleground" (Shaw, 2002). "Capitalism is always creating new places, new environments designed for profit and accumulation, in the process devalorizing previous investments and landscapes" (Lees et al., 2008: 51). Urban geographer John Stehlin describes the "moment when the changing political economy of many cities makes more localized patterns of mobility viable for highly educated, flexibly employed segments of the labor market, and livability serves as a capitalizable municipal asset." Specifically, "bicycling plays a key role in the production of ecologically responsible and politically progressive gentrifying subjects" (Stehlin, 2015: 122). In *Bike Lanes Are White Lanes*, communications scholar Melody Hoffman argues that the meaning of bicycles and bike infrastructure changes in different spaces, with different people, and in different cultures. The socioeconomic and political schematics of transportation funding prioritizes needs of the middle to upper classes (Hoffman, 2016: 87).

Municipal governments leverage authority by funding assets like bike lanes (Smith, 2002; Atkinson and Bridge, 2005). Such transformations are socially constructed processes that create fields of power. The spatiality of policymaking is not a neutral "transaction space," instead its "a three-dimensional mosaic" constantly being remade by relational connections of power between sites of policymaking and policy actors (Peck and Theodore: 2010: 170).

3.3 Methodology

I used a mixed methods qualitative approach to investigate lived experiences, histories, and conflict among cultural communities living in South Williamsburg during the late 1990s and 2000s. This purpose of this approach was to build knowledge about how policymaking affected neighborhood dynamics, shaped gentrification, and ultimately set the stage for the bike lane conflict. The methods were each deployed at different junctures to deepen the analysis. Participant observation was used as an approach to inquiry and for initial data-gathering. Interviews were used to address informational gaps. During my research, it became increasingly clear to me that lived experience plays a critical role in how communities respond to policymaking. For each interview, I used a semi-structured approach and developed an individualized protocol. A policy and legislative review identified key policies that shaped South Williamsburg's development in the early 2000s, including judicial cases that shaped land use. I also extensively reviewed news media coverage. Traditional and, increasingly, social media (including social networks, blogs, and comment boards) play a pivotal role in shaping the public's understanding and can change the discourse by creating new analytical frames and introducing values that influence how people interpret and respond to policymaking (Kingdon, 1994; Howlett and Ramesh, 2009; Soroka et al., 2013; Stone, 2012). Throughout, I used a journal to create "a research trail' of gradually altering methodologies and analysis (Ortlipp, 2008: 696).

3.4 Rational Plan, Human Reaction

Cultural communities often do not respond rationally to policy change that impacts their normative rights (Stone, 2012: 332–333). This is not to say that policymakers who enact change are acting rationality. Rather than integrating with existing systems of order, state actors create new mechanisms that are disconnected from existing ways of life and power structures (Gregory, 1998; Cooke, 2002; Simandan, 2012). Yet, public infrastructure can be intrinsic to how cultural groups define themselves and how they are defined by others in the places where they live.

When the City of New York installed a bike lane on Bedford Avenue, it intended to create better connectivity to the Williamsburg Bridge and accommodate a growing number of bicycle commuters. Instead, the new bike lane became a flash point in a simmering conflict between the neighborhood's cultural groups. In South Williamsburg (also known as Southside), attitudes towards public policymaking are deeply rooted in cultural values. For more than half a century, the neighborhood was shared by two leading cultural groups, Latinxs (primarily Puerto Rican and Dominican immigrants and their descendants) and Hasidic Jews.

For Orthodox Jewish communities, neighborhood boundaries are precise and literal. Wire or cord is used to demarcate an *eruv*, the physical boundary within which they are allowed to carry objects on the Sabbath. In the Hasidic section of Williamsburg, the *eruv* marks a precise neighborhood boundary. For Southside's Latinx community, territorial demarcation is less formal. No wire surrounds the enclave's boundaries, although Los Sures (named for the neighborhood's 'South' numbered streets; South First, South Second, etc.) is as close knit and fiercely protective as their Hasidic neighbors. Ties in Los Sures are bound not by religious practices but by cultural identity. Windows draped with Puerto Rican and Dominican flags exalt the occupants' heritage. Paralleling the Hasidim's investment in the neighborhood by building synagogues, Talmudic academies, and other anchors to support their faith, the community in Los Sures has built and sustained history-making institutions like community-led nursery school Nuestros Niños. In 1993, local leaders founded El Puente Academy for Peace and Justice, the first-ever public school created by a community-based organization.

Between the early 1990s and late 2000s, policies enacted at the federal, state, and municipal level transformed the urban landscape. When creative class newcomers began moving to the neighborhood during the 1990s, the migration coincided with the dismantling of federal and state safety nets providing social services and economic support to Southside's most vulnerable residents. New sets of competing cultural values were overlaid onto existing tensions, further complicating sociopolitical relationships and spatial realities. As gentrification became more visible in all aspects of life – from housing prices to the kinds of groceries sold at the local bodega – long-term residents began to organize against the tide of change. By the late 1990s, the Latinx and Hasidic communities brokered an unsteady peace through efforts like creating a neighborhood master plan to gain new affordable housing and preventing a new waste incinerator from opening.

But gentrification continued to encroach. According to *The New York Times* (Haughney, 2008), more than 2,000 new condo units were built in Williamsburg between 2003 and 2008. These *Artisten* were described as "Manhattan transplants who brought high rents and loose morals" (Greif et al., 2010) (Figure 3.1).

When the Bedford Avenue bike lane was installed in late 2007, Southside was swiftly changing from a low-income, ethnically diverse neighborhood to being increasingly populated by *bobos* or hipsters, who disrupted and distorted hard-fought power structures of existing communities. What ground Latinxs and Hasidim shared – deep roots in the neighborhood, relationships with public social services, effective activism, and an urgent need for affordable housing – the *bobos* did not. Instead, they were well positioned to disrupt South Williamsburg's delicate social, economic, and political balance by introducing new cultural norms and territorial lines.

The bike lane became a physical embodiment of this potential threat. After the bike lane was installed, the Hasidic community mounted a two-year opposition campaign that protested both its physical existence and dispossession that it represented to them. At first their efforts seemed futile – or so

Figure 3.1 Map of Williamsburg.

were portrayed by the press – especially given that the new bike lane was part of a citywide policy agenda. Neighborhoods across the city were adapting to the new bike lane infrastructure. Southside's predicament was not unique, but its cultural history and hyperlocal politics set it apart. Then, in late 2009, the City suddenly reversed course and removed a 14-block portion of the bike lane, disrupting direct access for bicyclists to the Williamsburg Bridge. Logically and logistically, the bike lane's removal appeared antithetical to the city's own policy goals. Not lost on anyone was the fact that the removed portion ran through the nucleus of Hasidic Williamsburg. In New York, Hasidim are revered for their power as a voting bloc. The bike lane's removal took place just a few weeks after then-Mayor Michael Bloomberg's reelection to a third term. A theory quickly emerged that the bike lane's removal was politically engineered in exchange for the Hasidic vote. A wave of protests followed, fomenting an atmosphere of hostility between the Hasidim and hipsters.

3.5 Policies and Politics of Street Shock

The Bedford Avenue bike lane was installed during the administration of Mayor Michael Bloomberg and leadership of New York City Department of Transportation (NYC DOT) Commissioner Janette Sadik-Khan. In 2009, New York became the first U.S. city to win the Sustainable Transport Award.[3] The award recognized how City leaders "took 49 acres of road space, traffic lanes and parking spots away from cars and gave it back to the public for bike lanes, pedestrian areas and public plazas. Bike ridership increased by 35 percent from [2008 to 2009]." Sadik-Khan led the creation of over 400 miles of bike lanes in five years at NYC DOT.

A bike lane reconfigures the functionality of a street and transforms dynamics between bicyclists and users of other modes. Constructing a bike lane is part of a hegemonic process that identifies some bodies as belonging in that space, while encouraging others to stay out (Hoffman, 2016). While usually cars still occupy most of the roadway and pedestrians continue using the sidewalk, creating a bike lane sends a clear message that bicycles also have a right to the street. For drivers and pedestrians, this recalibration can feel like an infringement of their spatial rights.

Cities court bicyclists as part of the "creative class" demographic (Hoffman, 2016). It's not a coincidence that during moments of revitalization, "planners start paying attention and catering to the urban, white, educated cyclist" (Hoffman, 2016: 89). White cyclists have the power to gentrify neighborhoods and are recognized as predictors of displacement. In neighborhoods already coping with changes to established territorial boundaries, a bike lane – a publicly sanctioned invitation for outsiders to pass through – can be deeply shocking. A community's sense of place is more complex than public infrastructure and services, it extends to "symbolic meanings" and discrete boundaries that demarcate claims to space (Suttles, 1972; Logan and Molotch, 1987).

By 2007, gentrification in Williamsburg threatened to upend existing cultural and economic ecosystems. The Bedford Avenue bike lane created a new thruway that disregarded territorial boundaries, showcasing newcomers streaming into the neighborhood. Teresa Toro, then chair of Brooklyn Community Board 1's transportation committee, recalled minimal community consultation before the bike lane was installed:

> At the point we had a new mayor with a new commissioner who had a mandate to lay down, I forget, like x miles of bike lanes to create a bike network through the city, particularly with an eye towards connectivity for commuters. And Bedford Avenue, to and from the Williamsburg Bridge, was a key street. [NYC] DOT is funny, the City's funny. They use community boards, you know... Sometimes they ask and sometimes they don't. When it's controversial, they like to hide behind the community board and say, "Well, we asked." And sometimes they just muscle things through. So I'm trying to remember if they asked us what we felt. I kinda think they didn't. It was just part of the network that they put through.[4]

Phil Andrews, a resident of Williamsburg in the 2000s and a bike activist with Time's Up! and the Rude Mechanical Orchestra, shared how people interacted with bicyclists in Southside:

> You really got the sense, either as a group or as an individual, that the folks in South Williamsburg would rather you not have been biking through their community. Just like the cars seem to not notice your existence in a way that was more extreme than most drivers who also don't care about cyclists, but would just happily cut you off. If you'd be riding on the side of the road and a pack of cars would pass you, as soon as the cars would pass, pedestrians would look right through you and then just flood the street, even if you were five feet away from them going at top speed.[5]

Not everyone shared this perspective. City Councilperson Diana Reyna (in office 2002–2013) disagreed with the idea that residents preferred that the bike lane not exist.

> I had pushed for there to be equitable transportation in the South Williamsburg area. And I continued to have that conversation from the beginning until the end. But people see [the Hasidim] as very powerful. And make them more powerful by giving in to… What was bothering them was that their homogenous culture was now broken into because [of the bike lane].[6]

At a neighborhood meeting, NYC DOT Bicycle Program Coordinator Joshua Benson described the new Bedford Avenue bike lane as part of a boroughwide "network." Benson suggested that, in time, Southside's residents might come to embrace this new infrastructure: "Change is hard, and when we change the way the streets work, there is always an adjustment period."[7]

3.6 Street Shock: Erasing the Bike Lane

After two years of pushback from the Hasidic community and others opposed to the bike lane, the City appeared to suddenly reverse course. One morning in late 2009 a portion of the Bedford Avenue bike lane was literally erased. Bike riders and advocates were incensed. A *Streetsblog* post (December 1, 2009) announced the news; there are 102 comments: "This is BS," wrote the very first commenter. Almost a quarter of commentors suggest forming a protest. Some comments give instructions for how to call NYC DOT, other commenters report back on their experience calling NYC DOT to complain, others propose more radical interventions; "Whens [sic] the Bedford 'scantily clad' ride going to happen? Lets make it happen," wrote user "Pete," on December 2 (Figure 3.2).

By the week's end, activists had organized. On Friday, December 4, 2009, a group created stencils and repainted a portion of the erased bike lane. Users of the (defunct) community forum *Williamsboard* took note: "A few hipsters took on the initiative and repainted the Bedford av bike lane with paint and

Figure 3.2 NYC DOT removes the Bedford Avenue bike lane.

Source: Streetsblog/Elizabeth Press, December 1, 2009.

brushes late Friday night and it looks pretty charming," wrote a user. Encouraged by a lack of fallout and messages of support, the activists decided to finish the job on Sunday night. This time was more eventful: At around 4 a.m., a group of Hasidic men surrounded the activists and called the *Shomrim* (Hasidic community patrol) to detain them until police arrived.[8]

Shortly afterward, a short video was posted to *YouTube* by user occupye-verything, showing activists repainting the bike lane. The video is edited and has an electric guitar riff added as a pump-up soundtrack. The opening shot shows the intersection of Bedford Avenue and Williamsburg Street East, exactly where Hasidim had intervened and detained activists. The video shows the action in progress; one frame shows four people in black sweatshirts and hats, their faces obscured, repainting the bike lane next to a row of parked yellow school buses with Hebrew lettering. Activists use a homemade cardboard stencil replicating the official NYC DOT bike lane symbol. At the end of the video, a message appears in four parts, written in white type on a black screen:

1   "We are New York city bicyclists,
    and our message is clear:
    Don't take away our bike lanes."
2   "We use this stretch of Bedford Avenue
    because it is a direct route to the
    Williamsburg bridge."
3   "We will continue to use it whether or not
    there is a bike lane there, but not having one
    puts us at greater risk from cars."
4   "That's why bike lanes exist – for safety.
    Do not try to remove them, or we will
    put them back for our own safety."

There are no numbers in the original message – the numbers and spacing indicate the break-up of the text onto four screen frames in the video (Author Note).

Media forums were abuzz with news of the bike lane's removal and the repainting. Quickly, theories began to emerge about the City's rationale, unsubstantiated by anything but hearsay. On local sites like *Streetsblog* and *Gothamist*, commenters suggested that Hasidim had promised to vote in high numbers in exchange for the Bloomberg administration removing the bike lane when he won reelection:

> The removal of the bike lane is perhaps a small "reward" for voting how they were told to in the city council and mayoral races? Best guess.[9]
>
> my guess is that the Hasidim got Bloomberg to remove the lanes because they dont like people in tight lyrca [sic] riding through their neighborhood.[10]

This narrative was legitimized when the *New York Post* (Oishan, December 8, 2009) published a story about the repainting: "A source close to Mayor Bloomberg said removing the [bike] lanes was an effort to appease the Hasidic community just before last month's election."

> On the same day, activist collective Time's Up! announced a planned action in a press release:
>
> On Sunday, December 13th, the Time's Up! Bicycle Clown Brigade, defenders of NYC bike lanes, will join other cyclists in a New Orleans-style "funeral procession" followed by a vigil, for the 14 blocks removed from the Bedford bike lane. Harnessing the power of the cycling community, they will perform a mock ceremony to resurrect the section of the missing bike lane between Flushing & Division avenues in South Williamsburg.[11]

The press release header conveys their motivation: "Mayor Bloomberg panders to motorists who prefer free parking over safe bikes lanes for Williamsburg's large cycling community."

City authorities were evidently aware of growing hostility in both physical and virtual public forums. More cops than protestors reportedly attended the funeral procession protest; NYPD detectives took photos from unmarked vehicles and traffic cops dressed in full riot gear (*Brooklyn Paper*, December 14, 2009). But activists succeeded in making the case to the press that the issue at hand was their right to public space: "We're talking about a public street. It's for everybody," a participant told a *New York Daily News* reporter (December 13, 2009). Members of the Hasidic community shared their own perspective; a father of five children was quoted in *Gothamist* (December 16, 2009) saying:

> The main concern is the safety of our kids. There are lot of institutions and families on that Bedford Avenue stretch, and we are always really

concerned about the kids being picked up and dropped off. There are sometimes small accidents where the cyclists are violating the law because they don't stop for flashing school buses.

A few days after the funeral procession protest, activist Heather Loop announced another action: She and at least 50 other bicyclists would ride through the area where the lane had been removed, clad only in underwear. In conversation with me a decade later, she said the protest was a statement: "Screw everyone for taking this away from us. Because it was like the most direct and most safe route for cyclists."[12]

Orchestrated via word of mouth and Facebook, and planned intentionally for the Sabbath, the Jewish holy day of rest, the underwear ride was a pointed response to reports that some Hasidic leaders wanted the bike lane removed because of how some riders dressed. "They removed it because the Hasidic residents just don't want to see scantily clad women," Loop told *Brooklyn Paper* (December 20, 2009). This remark was printed by major local outlets including *New York Post* (December 16, 2009) and *NBC New York* (December 17, 2009). Loop's words were amplified in the global Orthodox publication, *The Yeshiva World* (December 19, 2009) and spread even more widely by the *Associated Press* (December 20, 2009) (Figure 3.3).

Figure 3.3 Flowers are laid on the street during the New Orleans-style funeral procession to mourn the loss of the Bedford Avenue bike lane.

Source: TIME'S UP! Environmental Organization, December 13, 2009.

After Loop's remarks were published, Transportation Alternatives (TA) made a statement, published in *Brooklyn Paper* (December 20, 2009) condemning the protest's agenda:

> A bike lane on Bedford Avenue is about transportation and road safety... Rhetoric or acts that pit neighbors against one another are not just irrelevant to this discussion, they are flat-out offensive. A bike ride of people in provocative undress doesn't make Bedford any safer, and undermines efforts to bring north Brooklynites together to solve this problem.

Factionalism emerged as advocacy organizations like TA were finding themselves at odds with guerrilla tactics deployed by independent activists. Hours after TA released their statement, Hasidic leader Isaac Abraham wrote an open letter to bike activists, printed in *Brooklyn Paper* (December 18, 2009) echoing similar concerns:

> We urge and plead with you to reconsider the way the people want to protest and respect the religious believes [sic] of our community as well as the moral and respect of all New Yorkers, proceeding with your plan is and will be an insult to the entire Jewish community... Don't take out your anger and frustration of the bike lane removal on the entire community, when it was only one individual who stated that 'the dress code' was the problem. The entire community [has] always stated that safety and parking was the issue.

Despite this opposition, Loop and her cadre continued to plan the topless ride. But, despite the buildup, fewer than 20 people participated, all fully clothed due to inclement weather, while the media turnout was almost twice that. Journalists were disappointed that it was "not the NC-17 provocation" they had hoped for (*Gothamist*, December 19, 2009). Instead, fully clothed protestors rode on Bedford Avenue, obeying traffic signals while under watch by NYPD officers. From interviews, I learned that it was common practice for police to issue traffic violation tickets or conduct mass arrests during group rides and protests, which forced activists to pedal a thin line – demonstrating their right to the street space, while also being vigilant to not break any minor laws that could trigger police intervention. To be effective, they had to strike a delicate balance between dissident and law-abiding citizens (Figure 3.4).

In the months that followed, distrust continued to grow between the hipsters and Hasidim as they uneasily cohabitated in Southside. Interviewed a decade later, Loop opined that how the bike lane's removal was handled amounted to a missed opportunity:

> I just wish that we would have come to a compromise... It was just the way that it was done; it was just so like this is the decision. And it was done so sneakily too.... I knew a lot of younger Hasidics and they all said, "We don't really agree with our elders, we're more progressive." But they still had to abide by the elder's rules."[13]

Figure 3.4 The "topless" protest became a small fully clothed protest due to inclement weather.

Source: Katie Sokoler / Gothamist.com, December 19, 2009.

Despite Loop's retrospective desire to reach a compromise, actions by bike activists shaped how they were portrayed by the media and interpreted by policymakers. Their demands were delegitimized because they were perceived as interlopers disrupting established Hasidic territory, despite their capacity to effectively deploy organizational tactics. Ultimately, the Hasidim won the 'hipster versus Hasid' battle that the media created and framed, and the bike lane remained removed.

3.7 Shaping Power in Place

Media has the power to shape the attention paid to policy issues and the solutions sought. Sensationalism highlights only aspects of a perceived reality by promoting a particular problem definition and causal interpretations, which in turn impacts political and policy agendas (Forester, 1984; Soroka et al., 2013).

Confrontations in public space are explicit assertions of power. Policymaking that changes shared street space sends a clear message that one interest group is being prioritized over another. When policy implementation excludes – or seemingly excludes – certain groups of people from using a shared good, backlash is sure to follow. This can lead to the "non-prioritized" group organizing around their threatened or perceived loss and mounting a campaign to regain access to what they previously considered theirs. In turn, this conflict reduces the efficiency and effectiveness of policy implementation. Mobilization of interest groups demanding access to shared good forces policymakers to show their cards since at some point they must prioritize one group's demands over another. This becomes more complicated when the demand is to eliminate a spatial policy outcome, like a bike lane. If we look at the space of the street as a marketplace, the value assigned to different uses is subjective and largely symbolic. Finding a satisfactory compromise can be almost impossible, as activist Austin Horse explained to me:

> Here's really the problem you see oftentimes in New York and every other city when they implement bike infrastructure is that because it's such a political battle... There's 50 percent of their constituency that parks for free, and 5 percent of the constituency that doesn't want to die on their bike. And so the bikes lose.[14]

Clashing claims over street space are granted legitimacy by the government based on how harms are mediated. For public goods to effectively function in shared public space, community buy-in is necessary. But transformative goods like new transportation infrastructure often face formidable stakeholder resistance due to perceived psychic harms, which can make stakeholders feel that policies like installing a bike lane are coercive and harmful (Stone, 2012: 76). Complicating this is the fact that in many neighborhoods, there are multiple constituencies, with complex and shifting (and shifty) alliances largely "defined in contradistinction to other another" (Suttles, 1972: 50). Cultural groups, in part, gain their identity through their most apparent differences from other groups, and are often shaped by current or past struggles to create a sense of community. Groups "who share an arm's length of space are vulnerable to one another" (Suttles, 1972: 234). To mitigate this, policymakers must make spatial trade-offs clear.

3.8 Discussion

This case study contributes new findings and analytic frameworks to bodies of literature investigating how physical infrastructure can transform processes of gentrification. The research also adds to literature that examines how capital-driven policymaking disrupts cultural ecologies and spur conflict, exacerbating tensions, creating new dilemmas, and inhibiting sustainable innovation in place and space.

The Bedford Avenue case study highlights the need for policymaking to actively protect public space from the inconsistencies of political will and assure continued implementation of a planned connective infrastructure. Cities must find a way to strike an effective balance between the many contradictory wills of local stakeholders and the logical necessity of consistent, connective infrastructure. The Bedford Avenue bike lane was a product of the Bloomberg administration's agenda to increase the number of bike lanes citywide, which was largely carried out by implementing plans and policies that had been developed during the 1990s. Yet, the very same leaders later removed the bike lane when it became clear that their own policymaking was a political liability.

On Bedford Avenue, stakeholders did not share equal power. The entrenched power of the Hasidic community made them an effective opposition to the bike lane, as they deployed multilateral tactics to defend their space against unwelcome change. The sidelining of the Latinx community, due to shifting community power structures and media coverage that diminished their voice, resulted in their stake in this conflict being undervalued and unheard. Conversely, the hipsters' power grew as they became more visible and developed strategic alliances with advocacy groups like Time's Up!

When the city removed the Bedford Avenue bike lane, advocates felt it necessary to reclaim the space, even if it meant using guerilla tactics. In some ways, immediately illegally repainting the lane amounted to a missed opportunity for bike advocates and their allies to establish and sustain meaningful ownership of the street space. City officials fanned the flames of conflict by removing the guerilla-installed bike lane without any discussion or substantive acknowledgement of the action. This legitimized the position that bicycle riders should not have a right to the space of the street, sparking ire from bike lane advocates. Rather than escalating this point of conflict, the City could have facilitated a collaborative envisioning of the street space – badly needed in this rapidly gentrifying neighborhood. With a stronger participative structure, stakeholders in Southside might have developed a functional compromise to consider needs of drivers, bicycle riders, and pedestrians.

Given the opportunity to work directly with city officials, community leaders could have created a more equitable and accessible Bedford Avenue, rather than the street becoming a turf war. Trust and understanding between cultural communities are necessary for spatial infrastructure to work effectively; a bike lane that is blocked by cars or used as a sidewalk extension by pedestrians cannot function as intended. To develop and sustain successful spatial interventions, cities must support inclusion in processes that shape public space.

Urban planners and policymakers have an ever-growing understanding of the value that public infrastructure has in making cities better places to live if installed using mechanisms that invite meaningful stakeholder participation. But let's be clear: To be effective, bike lanes must sustain connectivity through neighborhoods and across territorial lines. Occasionally, controversial decisions are necessary to create more equitable street space. Indeed, NYC DOT Commissioner Janette Sadik-Khan wrote an op-ed in *New York* (2016) triumphantly titled, "The Bike Wars Are Over, and the Bikes Have Won." But when discussing Bedford Avenue, she conceded, "Politicians learned the lesson: Bikes were bad politics."

Place is constantly transformed. The capacity of cultural communities to leverage change is crucial for sustaining inclusion as physical features metamorphose. Regardless of whether users perceive change to be for better or for worse, contributing to the development of new spatial infrastructure is crucial for people to feel they have ownership of the space. Removing barriers to participation by enabling a wider range of people to design, program, and enjoy shared public space is key to creating more equitable communities.[15]

3.9 Conclusion

The production of this research was a seven-year journey (2012–2019). During this time, the urgency with which cities need to generate new mechanisms for transportation infrastructure policymaking has become increasingly clear. Today, over 30 U.S. cities including New York have adopted Vision Zero strategies to make streets more equitable and safer for all users. When I began this work in 2012, the Vision Zero Network that advocates for implementing such policies did not yet exist.[16]

In 2019, following a sharp spike in cyclist deaths, the New York City Council passed a new law improving safety for city streets, sidewalks, and pedestrian spaces. The law requires NYC DOT to issue and implement a master plan every five years that prioritizes street safety for all users, at a cost of $1.7 billion over ten years.[17] The law specifies that the first plan be published in December 2021 and include 250 miles of protected bike lanes over five years.[18] When the first plan was published in December 2021, fully one-third (33%) of online comments received by NYC DOT concerned "challenging bicycle conditions."[19] Ten miles of new protected lanes are planned for Brooklyn. There are no plans to replace the removed portion of the Bedford Avenue bike lane, despite the street's continued frequent use as a bicycling route.[20,21]

Notes

1 The terminology used to describe populations of Latin American origin varies, encompassing Hispanic, Latino, Latinx, Latin@, and more, all of which bear distinct connotations and for which there is no common agreement. In the chapter, I adopt the gender-neutral term Latinx.

2 The term *hasid* means pious in Hebrew. Today, Hasidic practitioners are comprised of several sects, each which adhere closely to Orthodox Jewish practice and use Yiddish as a primary language. To keep it simple, the term Hasidim is used here to describe all ultra-Orthodox Jewish people living in Williamsburg. Occasionally, I use the term Orthodox is used in place of Hasidic or Hasidim; they are equivalent unless otherwise denoted.

3 Each year, Institute for Transportation and Development Policy and the Sustainable Transport Award Committee select a city that has implemented innovative sustainable transportation projects in the preceding year, improving mobility for all residents, reducing transportation greenhouse and air pollution emissions, and improving safety and access for cyclists and pedestrians.

4 Toro, Teresa. 2019. Interview with author, February 7.

5 Andrews, Phil. 2019. Interview with author, February 26.

6 Reyna, Diana. 2018. Interview with author, November 13.

7 This conversation likely took place at the Brooklyn CB 1 transportation committee meeting on November 24, 2008, but the meeting's record does not make this apparent.

8 Muessig, Ben. 2009. "Bike Lane Repainters: "We're Self-Hating Jewish Hipsters." *Gothamist*. December 9. Online: https://gothamist.com/news/bike-lane-repainters-were-self-hating-jewish-hipsters

9 Commenter "Meredith," on the article: Muessig, Ben. 2009. "UPDATE: City To Remove 14 Blocks Of Bike Lanes On Bedford Ave." *Gothamist*, December 9. Online: http://gothamist.com/2009/12/01/city_to_remove_14_blocks_of_bedford.php

10 Commenter "myman," in the article: Fried, Ben. 2009. "Guerrilla Stripers Paint Back Bedford Avenue Bike Lane." *Streetsblog*, December 7. Online: https://nyc. streetsblog.org/2009/12/07/guerrilla-stripers-paint-back-bedford-avenue-bike-lane/

11 "2009-12-08 Removal of Bedford Bike Lanes." 2009. Press Release. Time's Up! December8.Online:www.times-up.org/2009-releases/2009-12-08-removal-bedford-bike-lanes

12 Loop, Heather. 2019. Interview with author, March 14.

13 Loop, Heather. 2019. Interview with author, March 14.

14 Horse, Austin. 2019. Interview with author, March 6.

15 Gardner, Jennifer; Marpillero-Colomina, Andrea; and Begault, Larissa. 2018. "A Guide to Inclusion and Health in Public Space: Learning Globally to Transform Locally." Gehl Institute + The Robert Wood Johnson Foundation.

16 "About." Vision Zero Network, webpage. Accessed: April 30, 2019. Online: https://visionzeronetwork.org/about/vision-zero-network/

17 Fitzsimmons, Emma. 2019. "After Spike in Deaths, New York to Get 250 Miles of Protected Bike Lanes." *The New York Times*. October 28.

18 City of New York. 2019. Local law 2019/195 Title: A Local Law to amend the administrative code of the city of New York, in relation to five-year plans for city streets, sidewalks, and pedestrian spaces. Enacted: November 19. https://legistar. council.nyc.gov/LegislationDetail.aspx?ID=3954291&GUID=D37BA0B0-9AB 6-434B-A82E-E49A7895A1A4&Options=&Search=

19 New York City Department of Transportation. *NYC Streets Plan*. December 1, 2021. Page 50. www1.nyc.gov/html/dot/downloads/pdf/nyc-streets-plan.pdf

20 City of New York, Office of the Mayor. 2020. "Vision Zero: De Blasio Administration Announces 2020 Major Projected Bicycle Lane Projects in Brooklyn." *Press Release*. January 29. Online: www1.nyc.gov/office-of-the-mayor/news/049-20/vision-zero-de-blasio-administration-2020-major-projected-bicycle-lane-projects-in#/0

21 New York City Department of Transportation. *NYC Streets Plan*. December 1, 2021. Page 50. www1.nyc.gov/html/dot/downloads/pdf/nyc-streets-plan.pdf

References

Atkinson, Rowland G. and Gary Bridge. 2005. *Gentrification in a Global Context: The New Urban Colonialism*. London: Routledge.

Beauregard, Robert A. 2010. "The Chaos and Complexity of Gentrification." In *The Gentrification Reader*, edited by Loretta Lees, Tom Slater, and Elvin K. Wyly, 11–23. London & New York: Routledge.

Bordieu, Pierre. 1984. *Distinction*. Cambridge, MA: Harvard University Press.

Cooke, Fadzilah Majid. 2002. "Vulnerability, Control and Oil Palm in Sarawak: Globalization and a New Era?" *Development and Change*, 33 (2): 189–211.

Florida, Richard. 2003. "Cities and the Creative Class." *City & Community*, 2 (1): 3–19.

Forester, John. 1984. "Bounded Rationality and the Politics of Muddling Through." *Public Administration Review*, 44 (1): 23–31.

Gregory, Steven. 1998. *Black Corona: Race and the Politics of Place in an Urban Community*. Princeton, NJ: Princeton University Press.

Greif, Mark, Kathleen Ross, and Dayna Tortorici, eds. 2010. *What Was the Hipster: A Sociological Investigation*. New York: N+1 Foundation.

Haughney, Christine. 2008. "Old Europe and New Brooklyn in Williamsburg." *The New York Times*, November 21.

Hoffman, Melody L. 2016. *Bike Lanes Are White Lanes: Bicycle Advocacy and Urban Planning*. Lincoln, NE: University of Nebraska Press.

Howlett, Michael and Michael Ramesh. 2009. *Studying Public Policy: Policy Cycles and Policy Subsystems* (3rd Edition). Oxford University Press.

Jones, Craig E. and David Ley. 2016. "Transit-oriented Development and Gentrification along Metro Vancouver's Low-Income SkyTrain Corridor." *The Canadian Geographer*, 60 (1): 9–22.

Kingdon, John. 1994. "Agendas, Ideas, and Policy Change." In *New Perspectives on American Politics*, edited by Lawrence C. Dodd and Calvin Jillson, 215–229. Washington, DC: CQ Press.

Lees, Loretta, Tom Slater, and Elvin K. Wyly, eds. 2008. *Gentrification*. New York: Routledge, Taylor and Francis Group.

Ley, David. 1997. *The New Middle Class and the Remaking of the Central City*. New York: Oxford University Press.

Ley, David. 2010. "Introduction: Restructuring and Dislocations." In *The Gentrification Debates*, edited by Japonica Brown-Saracino, 103–112. New York: Routledge.

Logan, John R. and Harvey L. Molotch. 1987. *Urban Fortunes: The Political Economy of Place*. Berkeley: University of California Press.

Molotch, Harvey L. 1976. "City as a Growth Machine." *American Journal of Sociology*, 82 (2): 309–332.

Oishan, Jeremy. 2009. "Hipsters Repaint Bike Lanes in Brush Off to Hasids." *The New York Post*, December 8. Online: https://nypost.com/2009/12/08/hipsters-repaint-bike-lanes-in-brush-off-to-hasids/

Ortlipp, Michelle. 2008. "Keeping and Using Reflective Journals in the Qualitative Research Process." *The Qualitative Report*, 14 (4): 695–705.

Peck, Jamie and Nik Theodore. 2010. "Mobilizing Policy: Models, Methods, and Mutations." *Geoforum*, 41: 169–174.

Shaw, Kate. 2002. "Culture, Economics and Evolution in Gentrification." *Just Policy: A Journal of Australian Social Policy*, 28: 42–50.

Simandan, Dragos. 2012. "The Logical Status of Applied Geographical Reasoning." *The Geographical Journal*, 178 (1): 9–12.

Smith, Neil. 1979. "Toward a Theory of Gentrification: A Back to the City Movement by Capital, not People." *Journal of the American Planning Association*, 45 (4): 538–548.

Smith, Neil. 1996. *The New Urban Frontier: Gentrification and the Revanchist City*. London: Routledge.

Smith, Neil. 2002. "New Globalism, New Urbanism: Gentrification as Global Urban Strategy." *Antipode*, 34 (3): 427–450.

Soroka, Stuart, Andrea Lawlor, Stephen Farnsworth, and Lori Young. 2013. "Mass Media and Policymaking." In *The Routledge Handbook of the Public Policy*, edited by Eduardo Araral, Jr., Scott Fritzen, Michael Howlett, M. Ramesh, and Xun Wu, 204–214. Abingdon: Routledge.

Stehlin, John. 2015. "Cycles of Investment: Bicycle Infrastructure, Gentrification, and the Restructuring of the San Francisco Bay Area." *Environment and Planning A*, 47: 121–137.

Stone, Deborah. 2012. *Policy Paradox: The Art of Political Decision Making* 3rd Edition. New York: W.W. Norton.

Suttles, Gerald. 1972. *The Social Construction of Communities*. Chicago: University of Chicago Press.

Zukin, Sharon. 1995. *The Cultures of Cities*. Oxford: Blackwell Publishers.
Zukin, Sharon. 2006. "David Harvey on Cities." In *David Harvey: A Critical Reader*, edited by Noel Castree and David Gregory, 102–120. Malden, MA: Blackwell Publishing.

4 Building Drought Resilience in the US Southwest

The Institutional and Economic Challenges in Rural Communities

Haoying Wang

4.1 Introduction[1]

The 2021 Pacific Northwest drought and heatwaves are a recent reminder of the importance of drought preparedness and resilience. Whether it is perceived as a signal of accelerating climate change or a stochastic event of normal climate variability, its implication is clear: a drought (flash or persistent) can cause significant damages to society and the ecosystem. In the past several decades, drought has become a growing threat to social livelihood and ecosystem sustainability worldwide, especially in the context of population growth and the global freshwater crisis (Rogers 2008). The literature concerning drought risks (agricultural drought risk in particular) has focused mostly on developing economies (e.g., the Indian subcontinent and African countries). Developed countries have rarely been the study region of drought research until recently, except for Australia (Heathcote 1969; Horridge et al. 2005), likely due to its unique landscape and climate. In the United States (US), the Great Plains and the West tend to be the popular drought study regions as they are more vulnerable to heatwaves and persistent droughts. California is an example of that strain of literature. The recent 2012–2016 drought in California was historical and caused severe damages in many rural, agricultural, and wildfire-prone regions (Swain 2015). There are two important observations from the California drought study literature, that could shed light on managing drought risk and building long-term drought resilience. First, drought risk presents considerable spatial heterogeneities (Greene 2018; Hanak et al. 2015; Swain 2015). Second, urban areas tend to have higher levels of preparedness and hence are more resilient to drought risks (Hanak et al. 2015). Understanding these general patterns helps design better drought management solutions and strategies.

It is difficult or even impossible to adjust the natural resource endowment (e.g., water resources) in many cases. It can also be costly to reverse the changing environmental conditions in the short term. Therefore, building drought resilience becomes a passive but often effective strategy. The definition of resilience varies by discipline. Adger (2000) summarizes resilience into two related definitions. First, resilience is the buffer capacity or the ability of a system to absorb exogenous perturbations. Second, resilience is the speed

DOI: 10.4324/9781003191049-5

of recovery or the time taken to recover from an exogenous disturbance. The first definition is more applicable for defining and measuring drought resilience. It is the capacity of an agricultural production system or other ecosystems to absorb drought impact and maintain its function. In the case of crop agriculture, for example, the resilience capacity may be defined as being able to maintain a certain level of yield throughout the growing season. In the case of grassland, the resilience capacity may be defined as the ability to survive until the next growth cycle. A related concept here is vulnerability. It happens when the threshold of irreversible changes of the system is experienced, partially or wholly (Adger 2000). It is similar to the concept of tipping points (Reyer et al. 2015).

As far as building drought resilience is concerned, the short-run and long-run strategies can be very different. And it often relates to the concept of resilience. In the short run, available resources can be deployed to mitigate the impact of a drought. Irrigation is an ancient example of this. Another example is the use of air conditioners during hot weather or heat waves. Both are short-run solutions in the sense that they offer temporary alleviation of the problem but do not address it permanently. Their effectiveness is subject to the capacity of the associated infrastructures (e.g., canals and power grids). Improving infrastructure is a long-run strategy for building drought resilience, and also a way to enhance economic resilience. It is essentially an investment to increase the long-term resilience of the economic system. Here economic resilience can be simply defined as the ability to recover from the negative impacts of external economic shocks (Briguglio et al. 2006). Another necessary component for long-run drought resilience is institutional resilience. According to Herrfahrdt-Pähle and Pahl-Wostl (2012), an institutional system (formal and informal) is resilient if it can (1) withstand disturbances and thus provide stability and reduce uncertainty, and (2) change in the medium to long term to react to the uncertainties of environmental changes and/or changes in the social system. Efficient institutions and institutional resilience help reduce transaction costs in the economic system (De Stefano et al. 2012). Therefore, institutional resilience can improve the efficiency of adjusting to or recovering from external shocks. It is worth noting that the change of institutions, formal or informal, is usually a long process. Short-run strategies have to work under the current institutions. The prior appropriation doctrine of water rights and irrigation water use in the US West are a good example of the short-run vs. long-run difference. In the long run, accumulating environmental changes can lead to institutional adaptation and innovation. For example, Marchildon et al. (2008) showed that major droughts in the southern Great Plains of Canada in the early twentieth century provoked key institutional adaptations that have since improved the drought resilience of the region.

The broader literature has explored the mechanism of drought resilience from several aspects. Most of them are region-specific studies. Synthesizing their findings help shed light on how to move forward with the continuing global change. First, demographics and socio-economic characteristics are crucial for building drought resilience. For example, Simelton et al. (2009) found that the size of a region's rural population in China was negatively correlated with drought vulnerability. Udmale et al. (2015) and Coppock (2020) found similar results in India and Utah (USA), respectively. Meanwhile, drought stress and risk can have demographic consequences. For example, Guiney (2012) found that drought stress has been a potential contributor to farmer suicides in rural Australia. Second, as far as institutional resilience is concerned, both formal and informal institutions are important. For example, by analyzing the recent multi-year drought (2012–2016) in California, Tortajada et al. (2017) highlighted the role of both formal and informal institutions in building drought resilience. Also, as Ojima (2021) argued, formal institutions and regulatory entities are usually less well-suited to working with underlying social factors that determine drought vulnerability, which implies the critical role of informal institutions. Additionally, it is worth noting that formal institutions (e.g., water use laws and regulations) tend to be effective in addressing short-run problems (Mullin and Rubado 2017). To develop long-run strategies for drought management, it is necessary to integrate formal and informal institutions. This is consistent with the vertically integrated drought management approach proposed by Rey et al. (2017). The third aspect is related. Building drought resilience should take a long-run perspective. For example, Hanak et al. (2015) argued that California should start a long-term effort to building drought resilience in its vulnerable areas. There are two important things to note related to the long-run perspective of building drought resilience. One is embedding short-run solutions like community-based drought early warning systems in long-term resilience-building (Ewbank et al. 2019). Most likely, such an early warning system relies on formal institutions to work. But it is possible for it to operate based on informal institutions, for example, through a cooperative with shared expectations on member responsibilities. Coppock's (2020) study in Utah (USA) also highlights the importance of such community-based solutions. Another one is the role of technological and management innovations, for example, adding flexibility to the current water resources management system (Singletary and Sterle 2017). Based on a study in Ethiopia, MacDonald et al. (2019) showed that, given the hydrogeological conditions, certain water-fetching technologies are more drought-resilient than others. Therefore, technology and management can play critical roles in the long-run sustainable use of water resources and hence building drought resilience.

In the remainder of this chapter, I focus on seeking a better understanding of challenges and opportunities for building drought resilience in the rural US Southwest. We define the Southwest as the four-state region of Arizona, Utah, Colorado, and New Mexico (see Figure 4.1). It has been an understudied region in terms of climate-related extreme weather hazards like droughts and floods. The region has a diversified set of communities, including large urban metropolitan areas, rural irrigated agricultural areas, rural tourism areas, and Native American communities. Because of the spatially heterogeneous nature of droughts, understanding the drought impacts on these

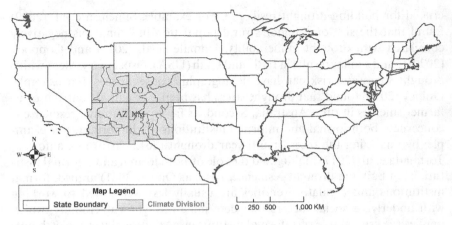

Figure 4.1 The US Southwest region.

Source: NASA Physical Sciences Laboratory, available at https://psl.noaa.gov/data/usclimdivs/, accessed September 1, 2021.

different communities and their coping mechanisms requires a mixed-methods approach and pooling multi-disciplinary literature. Next, I develop a conceptual model of community resilience and sustainability as the framework for the remaining analysis in this chapter. I then provide some background for the drought risk in the Southwest; and focus the discussion on the institutional and economic resilience in rural communities. Finally, I explore future opportunities for community resilience and sustainability.

4.2 A Conceptual Model of Community Resilience and Sustainability

There are no standard solutions for building resilience to environmental changes in a social system or an ecosystem. A particular environmental change can trigger different potential pathways of impact. Some of them can lead to disastrous outcomes like a persistent agricultural drought with soil moisture evaporation consistently exceeding recharge. Other pathways may lead to resilient yield outcomes because of adequate drought preparedness or efficient resource management. To explore these different potential pathways, first, conceptual model analysis is necessary to understand the mechanisms. In this section, I propose a conceptual model of community resilience and sustainability. The goal of the model is to explore different pathways of impact triggered by environmental changes and how they connect. Given these pathways, the model is used to explore strategies for building community resilience and sustainability. Figure 4.2 illustrates the structure of the model. A new concept introduced in the model is social resilience. Social resilience needs to be defined at the community level rather than being an individual-level capacity (Adger 2000). It is the resilience of a social system, similar to the concept of ecological resilience but not the same. In the proposed model, I define it as the integration of institutional resilience and

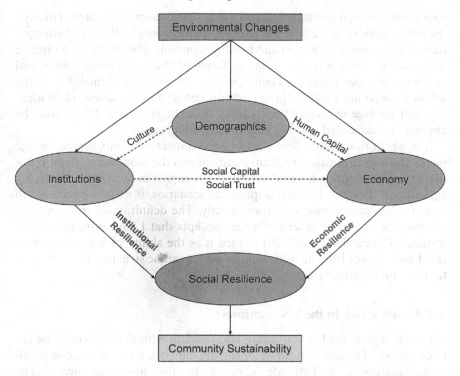

Figure 4.2 A conceptual model of community resilience and sustainability and pathways.

economic resilience in the given community. If both the institutional resilience and the economic resilience are measurable, the social resilience can be measured through a concave utility function (e.g., the Cobb–Douglas or constant elasticity of substitution [CES] function) to reflect the complementarity between the two resiliences.

In the model, a particular environmental change can trigger three potential pathways of impact: institutional, demographic, and economic. In the context of the US Southwest, one example of institutional impact is the New Mexico Produced Water Act and Water Data Act enacted in 2019. It was mainly a reaction to the growing environmental concerns associated with the produced water from the hydrocarbon production in the Permian Basin (Wang 2021). An example of demographic impact is international environmental migration due to drought (e.g., in Mexico, see Haeffner et al. 2018). An example of economic impact is the drought-induced yield reduction, which can further affect the commodity market and farm income. Figure 4.2 also illustrates a few connections between different pathways (in no way exhaustive). Institutional change can affect the functioning of social capital and social trust (e.g., in financial lending and borrowing, see Guiso et al. 2004) and hence the economic outcome. Demographic change can transform culture and social norms in the long run and eventually affect the (formal and informal) institutional environment. Demographic change can also affect community human capital stock and hence the economic outcome. Through the interactions of these different pathways, an efficient institutional environment (institutional resilience) and healthy economic development (economic resilience) contribute to the social resilience of the community. And social resilience provides a necessary condition for community sustainability. In the following sections, I use the proposed conceptual model framework to identify and analyze strategies for building drought resilience. It can also be applied and expanded to other resilience-building scenarios.

It is worth noting that here I assume the impact of demographic change works through the institutional environment and the economic system. It is a reasonable assumption for analyzing drought impact in rural areas, especially agricultural drought. In other application scenarios, it may be necessary to capture the demographic resilience directly. The definition of demographic resilience is similar to other resilience concepts that I have introduced. For instance, Capdevila et al. (2020) defined it as the ability of a population to resist and recover from alterations in its demographic structure (e.g., population size or age structure).

4.3 Drought risk in the US Southwest

My study region, the US Southwest, also overlaps with the US Southwest climate region.[2] The region features largely dry climates, with some moist continental/subtropical mid-latitude climates in the mountain areas. The Southwestern Monsoon (also called Mexican Monsoon or North American Monsoon) brings substantial precipitation to the region in the summer, especially for Arizona and New Mexico. According to the University of Arizona's CLIMAS project, Arizona and New Mexico receive up to half of their annual rainfall during the summer (July–August) monsoon season.[3] Recent studies have shown that global warming can significantly weaken the North American monsoon and lead to a drier Southwest (Pascale et al. 2017; Pascale et al. 2019). The change is expected to be accompanied by an increased frequency of extreme rainfall in the summer monsoon season (Pascale et al. 2019). The drying trend has already had significant impacts on vegetation and plant communities in the region (e.g., Brusca et al. 2013; Wang 2019a). Meanwhile, the region has already seen negative impacts on its agricultural production system from both the changing climate and the growing water scarcity (Steele et al. 2018). Figure 4.3 compares the drought monitor indices of the four Southwestern states and the average of the US continent in the past two decades. It is clear that the drought risk is significantly higher in the Southwest. What makes the situation worse is the lack of infrastructure in the rural Southwest, which increases its drought vulnerability. The region's irrigation and water resource management system is one example. Low efficiency in water transport, allocation, and use has worsened the water scarcity issue (Wang 2019b).

Although the potential drought risk in the region has been well documented, there is a lack of research on strategies for building drought resilience in the region. As mentioned in the introduction section, drought resilience in

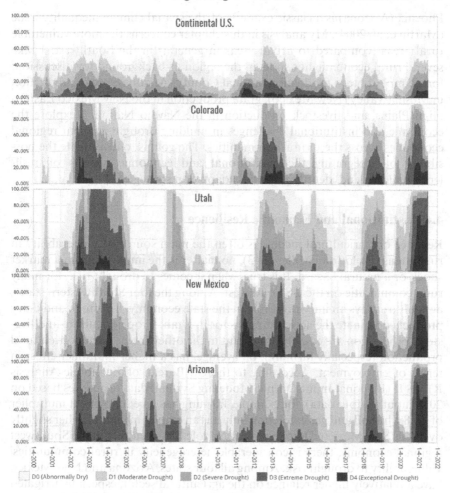

Figure 4.3 Percent area in US drought monitor categories: Southwestern states vs. US continental.

Data source: National Drought Mitigation Center, University of Nebraska-Lincoln. For definition of drought monitor categories: https://droughtmonitor.unl.edu/About/AbouttheData/DroughtClassification.aspx.

developed economies has been understudied until recent years. In rural areas, existing research focuses primarily on technical aspects like climate-yield relationships and crop insurances. The socio-economic aspects are often ignored. As Greene (2018) argued in her California's San Joaquin Valley study, the vulnerability of agricultural labor in the context of developed economies is largely overlooked in the climate impact and adaptation literature. In urban areas, uniformity is a typical assumption. In reality, even within the same city, drought preparedness and mitigation capacity vary significantly across communities. Communities with better infrastructure and more trees and open spaces tend to be more drought resilient and endure less heat in the summer.

Phoenix (Arizona) is a classic example of the spatial climate inequality issue (Martin et al. 2004). My analysis in this chapter concerns the more vulnerable rural areas (compared to urban areas in general) in the Southwest. I take several rural economic hot spots in the region as background for discussion; for example, cotton production in Pinal County (Arizona), pecan production in Mesilla Valley (New Mexico), grain production in the New Mexico Eastern High Plains, and livestock production in the Navajo Nation. I explore the economic and institutional challenges in building drought-resilient regional economies across these rural communities. The goal is to investigate the possibility of integrating the institutional and economic aspects of resilience-building and identify potential adaptation strategies.

4.4 Institutional and Economic Resilience

Reliance on agricultural income is often the main source of vulnerability in rural areas (Prelog and Miller 2013), no matter if the impact is from a market shock or a natural hazard. It is true, when it comes to drought risk, for many rural communities in the US Southwest. Among the four Southwestern states, agriculture plays an important role in the state economy. Livestock and dairy products dominate the agricultural sector in Utah. Crop agriculture and livestock production are equally important in the other three states. Agricultural production (especially farming), on average, has a higher multiplier effect in terms of employment. According to the US Bureau of Economic Analysis RIMS II (Regional Input-Output Modeling System) data (estimates based on 2018 input–output data), for example, farming has an employment multiplier of 13.8 while oil and gas extraction only has an employment multiplier slightly over 4 in New Mexico. Despite the critical role of agriculture in the Southwest regional economy, there is considerable heterogeneity in rural communities' vulnerability to drought risk. Among all four states, Arizona and New Mexico have a relatively higher reliance on the agricultural sector, especially irrigated crop production. Between the two states, the overall vulnerability of the state economy to drought is higher in New Mexico. The main reason is that New Mexico has a less diversified economy compared to Arizona. Its state economy relies heavily on agriculture and natural resource (oil and gas) extraction. Arizona has a significantly higher share of manufacturing in the state economy, which makes it less vulnerable to natural disasters. Table 4.1 lists the key industries in each of the Southwestern states. It is worth noting that the key industries are not necessarily the largest industries (in terms of employment or GDP). A key industry is usually a strategic industry identified by the state government for its long-run economic development potential.

Overall, the current institutional environment and regional economic systems in the rural Southwest are not resilient to environmental changes. It is evident from at least two aspects: the institutional lock-in of water rights and the lack of diversification in the regional economy. All four Southwestern states have adopted the pure prior appropriation water rights system.[4] It is usually referred to as a "first in time, first in right" system, with senior water

Table 4.1 The key industries in each of the Southwestern states

State	Key industries
Arizona (AZ)	Agriculture, Manufacturing, Service, Mining, Energy, Film
Colorado (CO)	Agriculture, Manufacturing, Energy, Services, Tourism
New Mexico (NM)	Agriculture, Manufacturing, Energy, Film
Utah (UT)	Manufacturing, Energy, Services, Tourism

Source: Arizona: www.azcommerce.com/industries/, www.azcommerce.com/programs/ rural-destinations/industries/; Colorado: https://choosecolorado.com/key-industries/; New Mexico: https://gonm.biz/why-new-mexico/key-industries; Utah: https://business.utah.gov/ uniquely-utah/targeted-industries/.

rights taking precedence when water flows are insufficient to serve all users. It is widely regarded as an inefficient water resource allocation system (e.g., Huffaker et al. 2000; Tarlock 2001). In other words, this particular water-related formal institution is unlikely to be drought-resilient. In the event of a persistent drought, the downstream or junior water users suffer the most, while a more efficient allocation system may minimize the overall drought damage. Although it has been emphasized in the literature that developing new institutions is an effective way to adapt and build drought resilience (e.g., Dinar and Jammalamadaka 2013), such an institutional lock-in of water rights is difficult to change even in the long run because of existing interests and lack of Pareto improvement opportunities.

The critical question is how to enhance drought resilience for irrigated agriculture without drastically changing the *status quo* formal institution – prior appropriation water rights system. The answer often boils down to technology and the market. As Huffaker et al. (2000) argued, continuing technological innovation in agriculture production has weakened the security of traditional appropriative water rights. One recent example is treating and reusing produced water from hydrocarbon production in agriculture. Substantial progress towards industrial-scale applications has already been made in Colorado and California (USA) (Dolan et al. 2018; Echchelh et al. 2018). New Mexico is actively exploring options to harness the economic value in produced water from the Permian Basin (Wang 2021). It is worth noting that reusing produced water for irrigation also involves changes of formal institutions, which is why it may improve institutional resilience. The jurisdictions of surface water and groundwater often belong to different state agencies. The prior appropriation doctrine of water rights applies only to surface water (including shallow groundwater due to direct hydraulic connections). In the case of produced water reuse, it concerns groundwater resources consisting mainly of formation water. Taking the state of New Mexico as an example, the Office of State Engineer (OSE) has jurisdiction over surface water (including shallow groundwater). Historically, the state Energy, Minerals, and Natural Resources Department (EMNRD) has jurisdiction over produced water because it is a byproduct of oil & gas production. In 2019, the State Legislature enacted a new bill – New Mexico Produced Water

Act (House Bill 546).[5] The new law granted the State Environment Department jurisdiction over the treatment and reuse of produced water for purposes outside the oil and gas sector. Given that one of the most likely potential beneficial reuses of produced water is agricultural irrigation (Echchelh et al. 2018), the legislative action has indirectly improved the formal institutions that govern irrigation water use.

Reusing produced water brings a new source of water into the agricultural sector. Water markets, on the other hand, work differently. It does not necessarily increase the total water supply for irrigation. Instead, it facilitates more efficient use of the existing water supply within the agricultural sector. If designed and managed efficiently, it can incentivize innovations in both informal institutions and formal institutions for irrigation water governance. The water market itself is usually regulated through the law or based on an enforceable contract. In theory, it can help capture potential economic benefits by reallocating water from low-value to high-value uses. That is the formal institutions part. In practice, however, a water market may not work as effectively as desired, for example, due to high transaction costs (Ghosh et al. 2014). This is where informal institutions can help. For instance, social trust is usually a critical component of informal institutions. In water markets, strong social trust between senior water rights holders and junior holders can reduce transaction' costs and facilitate timely water transfer to manage a drought situation. Another relevant component of informal institutions is traditional and local knowledge. As Fernald et al. (2015) showed, local knowledge about the hydro-social cycle of community water management is essential for managing exogenous disturbances like drought. In the case of water markets, traditional and local knowledge can facilitate negotiation and help create collaborative opportunities, which can also reduce the transaction costs of water markets. Overall, as Sullivan et al. (2019) argued, informal institutions can play a critical role in water governance and water resource management, especially when a diverse group of stakeholders is involved, including the Native American communities.

As far as diversification in the regional economy is concerned, it takes a portfolio management perspective on economic resilience. First, a given community belongs to the broader regional economy. A diversified regional economy provides multiple potential sources of income, which makes the income of community members less sensitive to exogenous shocks like drought. In reality, a rural community often relies on one or two industries (e.g., crop production, livestock production, or tourism). Fostering a new industry can be a regional development challenge. Still, diversification can happen within an existing industry. One straightforward example in rural communities is the integrated crop-livestock system (Bell et al. 2021). In the New Mexico Eastern High Plains, something similar happens. As its groundwater level continues to decline, farmers have increased the share of more drought-resilient sorghum in the crop mix. It follows the same portfolio diversification principle. Second, diversification helps detach rural communities from the so-called 'resource curse.' Resource curse usually refers to the observation that resource-abundant countries or regions tend to grow more slowly than resource-scarce ones, especially from a long-run sustainable development perspective. Institutions are commonly a factor to blame (Mehlum et al. 2006). In such cases, the dominant institutions are usually attached to an existing resource that is rich in the community or region. Economic diversification can bypass the 'locked-in' institutions and resources and develop new institutions. A necessary condition for such a bypass to happen is infrastructure. In the rural Southwest, the much-needed infrastructure for a 'bypass' development ranges from broadband access to financial credit systems. Third, diversification itself provides an enter-exit mechanism, which can also enhance community economic resilience. From a demographic perspective, it increases dynamics in the population, which further facilitates local knowledge creation and exchange. From an institutional perspective, it provides a constructive environment for risk-taking activities like entrepreneurship. And entrepreneurship is critical for rural community vitality (Akgün et al. 2011).

Lastly, it is worthwhile to stress the necessity of integrating institutional resilience and economic resilience. There is no standard formula for how to integrate. One way to think about it is the following. Institutional resilience prepares rural communities for expected and unexpected drought risks. Economic resilience helps reduce rural communities' vulnerability when impacted by drought. The former increases the community tolerance to drought, and the latter reduces damages from the drought. Therefore, such integration is critical for developing drought-resilient strategies for rural communities.

4.5 Future Opportunities for Community Resilience and Sustainability

In this chapter, I developed a conceptual model of community resilience and sustainability to explore strategies for building drought-resilient rural communities. I considered different pathways of impact and resilience in rural communities. My analysis and discussion focus on the rural US Southwest. In general, rural communities are more vulnerable to drought risks due to their reliance on agricultural income or a particular type of natural resource. The root reason for rural drought vulnerability is usually one or more of the following: institutional lock-in, lack of infrastructure, and lack of economic diversification. I have elaborated on examples for each of these challenges in the context of the rural Southwest. These challenges of peripherality do not necessarily imply a doomed future. Quite the opposite, challenges often provide foundations for tapping future opportunities. For instance, pessimistic views often spread around through the public media about the unsustainable use of water from the High Plains Aquifer. However, scientific studies have shown that with proper pumping and recharge management, society has an opportunity to sustain the livelihood that depends on the aquifer (e.g., Steward et al. 2013). Of course, 'proper' management regimes likely require addressing many institutional and technological barriers.

To address the challenge of peripherality and materialize future opportunities, first, institutional innovation is necessary to adapt to environmental changes, especially when formal institutions are concerned. It can take a top-down or bottom-up approach. A typical top-down approach is reforms of policy and the legal regime concerning water resources and conservation. The level of resistance and difficulty likely varies from community to community and from region to region. Another idea is to develop community-based institutional partnerships (Chhetri et al. 2012). Such partnerships emphasize collaboration among farmers, non-governmental organizations, and other stakeholders. It can help improve local knowledge networks and hence increase community resilience to environmental changes like drought. Second, investments in essential infrastructure and economic diversification are also critical. Many drought-stressed rural communities are either trapped by some form of institutional lock-in or on an undesired path dependency. Re-routing or boosting local economic development requires extraordinary measures, which usually translates to a capital investment problem. And it is also where many rural community development initiatives stagnate. Such community stagnations tend to be very vulnerable to exogenous shocks, regardless of whether it is due to an economic crisis or a natural disaster.

To break the stagnation and gain economic resilience, community-level out-of-the-box thinking and decision-making are indispensable. Rural broadband access is a good example. Broadband access is critical for many aspects of rural communities, from commodity production to public health. During the past two decades of network growth and bandwidth expansion, many rural communities across the country lagged. Large-scale federal subsidies on rural broadband access were not on the table until 2020. During the decade-long exploring process, many rural communities failed and stagnated in the dial-up age, but some succeeded (Levitz and Bauerlein 2017). A key factor of success has been stakeholder engagement through partnerships and cooperatives. When looking in retrospect and with the ongoing infrastructure investment in rural America during the Biden administration, the implications of those successes may seem trivial. However, their experiences provide valuable inputs for exploring the nexuses between economies, institutions, and territories.

Not to divert from the drought-centered discussion in this chapter, it is still worth pointing out that drought is not the only challenge of environmental changes faced by rural communities in the Southwest. Relatedly, and on the other extreme, the changing Southwest monsoon also brings intensive rainfall and dangerous flash floods to the region. When facing flooding risk, rural communities are also more vulnerable compared to their urban counterparts. The 2021 summer monsoon-induced flash floods in Arizona and New Mexico are hard evidence of this (e.g., Skabelund 2021; Smith 2021). The model framework and analysis presented above can shed light on these similar exogenous shocks threatening the rural communities in the Southwest.

Notes

1 This research is partially based upon work supported by the USDA National Institute of Food and Agriculture under Award 2022-67020-36265.
2 For definitions of the US climate regions, see the NOAA info page at www.ncdc. noaa.gov/monitoring-references/maps/us-climate-regions.php, accessed July 1, 2021.

3 See https://climas.arizona.edu/sw-climate/monsoon, accessed July 1, 2021.
4 See https://watermarkets.us/the-anatomy-of-a-water-right/, accessed June 10, 2021.
5 See https://nmlegis.gov/Sessions/19%20Regular/final/HB0546.pdf, accessed June 10, 2021.

References

Adger, W. Neil. 2000. "Social and ecological resilience: Are they related?" *Progress in Human Geography* 24, no. 3: 347–364.

Akgün, Ali ye A., Tüzi N. Baycan-Levent, Peter Nijkamp, and Jacques Poot. 2011. "Roles of local and newcomer entrepreneurs in rural development: A comparative meta-analytic study." *Regional Studies* 45, no. 9: 1207–1223.

Bell, L. W., A. D. Moore, and D. T. Thomas. 2021. "Diversified crop-livestock farms are risk-efficient in the face of price and production variability." *Agricultural Systems* 189: 103050.

Briguglio, Lino, Gordon Cordina, Nadia Farrugia, and Stephanie Vella. 2006. "Conceptualizing and measuring economic resilience." In L. Briguglio, G. Cordina, and E.J. Kisanga (Eds.), *Building the Economic Resilience of Small States*, 265–287. Blata I-Bajda: Formatek Ltd.

Brusca, Richard C., John F. Wiens, Wallace M. Meyer, Jeff Eble, Kim Franklin, Jonathan T. Overpeck, and Wendy Moore. 2013. "Dramatic response to climate change in the Southwest: Robert Whittaker's 1963 Arizona Mountain plant transect revisited." *Ecology and Evolution* 3, no. 10: 3307–3319.

Capdevila, Pol, Iain Stott, Maria Beger, and Roberto Salguero-Gómez. 2020. "Towards a comparative framework of demographic resilience." *Trends in Ecology & Evolution* 35, no. 9: 776–786.

Chhetri, Netra, Pashupati Chaudhary, Puspa Raj Tiwari, and Ram Baran Yadaw. 2012. "Institutional and technological innovation: Understanding agricultural adaptation to climate change in Nepal." *Applied Geography* 33: 142–150.

Coppock, D. Layne. 2020. "Improving drought preparedness among Utah cattle ranchers." *Rangeland Ecology & Management* 73, no. 6: 879–890.

Dinar, Ariel, and Uday Kumar Jammalamadaka. 2013. "Adaptation of irrigated agriculture to adversity and variability under conditions of drought and likely climate change: Interaction between water institutions and social norms." *International Journal of Water Governance* 1, no. 1–2: 41–64.

Dolan, Flannery C., Tzahi Y. Cath, and Terri S. Hogue. 2018. "Assessing the feasibility of using produced water for irrigation in Colorado." *Science of The Total Environment* 640: 619–628.

Echchelh, Alban, Tim Hess, and Ruben Sakrabani. 2018. "Reusing oil and gas produced water for irrigation of food crops in drylands." *Agricultural Water Management* 206: 124–134.

Ewbank, Richard, Carlos Perez, Hilary Cornish, Mulugeta Worku, and Solomon Woldetsadik. 2019. "Building resilience to El Niño-related drought: Experiences in early warning and early action from Nicaragua and Ethiopia." *Disasters* 43: S345–S367.

Fernald, A., S. Guldan, K. Boykin, A. Cibils, M. Gonzales, B. Hurd, S. Lopez et al. 2015. "Linked hydrologic and social systems that support resilience of traditional irrigation communities." *Hydrology and Earth System Sciences* 19, no. 1: 293–307.

Ghosh, Sanchari, Kelly M. Cobourn, and Levan Elbakidze. 2014. "Water banking, conjunctive administration, and drought: The interaction of water markets and prior appropriation in southeastern Idaho." *Water Resources Research* 50, no. 8: 6927–6949.

Greene, Christina. 2018. "Broadening understandings of drought–The climate vulnerability of farmworkers and rural communities in California (USA)." *Environmental Science & Policy* 89: 283–291.

Guiney, Robyn. 2012. "Farming suicides during the Victorian drought: 2001–2007." *Australian Journal of Rural Health* 20, no. 1: 11–15.

Guiso, Luigi, Paola Sapienza, and Luigi Zingales. 2004. "The role of social capital in financial development." *American Economic Review* 94, no. 3: 526–556.

Haeffner, Melissa, Jacopo A. Baggio, and Kathleen Galvin. 2018. "Investigating environmental migration and other rural drought adaptation strategies in Baja California Sur, Mexico." *Regional Environmental Change* 18, no. 5: 1495–1507.

Hanak, Ellen, Jeffrey Mount, Caitrin Chappelle, Jay Lund, Josué Medellín-Azuara, Peter Moyle, and Nathaniel Seavy. 2015. "What if California's drought continues." *Public Policy Institute of California Report:* 28–33.

Heathcote, Ronald Leslie. 1969. "Drought in Australia: A problem of perception." *Geographical Review* 59, no.2: 175–194.

Herrfahrdt-Pähle, Elke, and Claudia Pahl-Wostl. 2012. "Continuity and change in social-ecological systems: The role of institutional resilience." *Ecology and Society* 17, no. 2: 8.

Horridge, Mark, John Madden, and Glyn Wittwer. 2005. "The impact of the 2002–2003 drought on Australia." *Journal of Policy Modeling* 27, no. 3: 285–308.

Huffaker, Ray, Norman Whittlesey, and Joel R. Hamilton. 2000. "The role of prior appropriation in allocating water resources into the 21st century." *International Journal of Water Resources Development* 16, no. 2: 265–273.

Levitz, Jennifer, and Valerie Bauerlein. 2017. "Rural America is stranded in the Dial-Up Age." *The Wall Street Journal*, June 15, 2017. www.wsj.com/articles/rural-america-is-stranded-in-the-dial-up-age-1497535841, accessed July 2, 2021.

MacDonald, Alan M., Rachel A. Bell, Seifu Kebede, Tilahun Azagegn, Yehualaeshet Tadesse, Florence Pichon, Matthew Young et al. 2019. "Groundwater and resilience to drought in the Ethiopian Highlands." *Environmental Research Letters* 14, no. 9: 095003.

Marchildon, Gregory P., Suren Kulshreshtha, Elaine Wheaton, and Dave Sauchyn. 2008. "Drought and institutional adaptation in the Great Plains of Alberta and Saskatchewan, 1914–1939." *Natural Hazards* 45, no. 3: 391–411.

Martin, Chris A., Paige S. Warren, and Ann P. Kinzig. 2004. "Neighborhood socioeconomic status is a useful predictor of perennial landscape vegetation in residential neighborhoods and embedded small parks of Phoenix, AZ." *Landscape and Urban Planning* 69, no. 4: 355–368.

Mehlum, Halvor, Karl Moene, and Ragnar Torvik. 2006. "Institutions and the resource curse." *The Economic Journal* 116, no. 508: 1–20.

Mullin, Megan, and Meghan E. Rubado. 2017. "Local response to water crisis: Explaining variation in usage restrictions during a Texas drought." *Urban Affairs Review* 53, no. 4: 752–774.

Ojima, Dennis S. 2021. "Climate resilient management in response to flash droughts in the US Northern Great Plains." *Current Opinion in Environmental Sustainability* 48: 125–131.

Pascale, Salvatore, William R. Boos, Simona Bordoni, Thomas L. Delworth, Sarah B. Kapnick, Hiroyuki Murakami, Gabriel A. Vecchi, and Wei Zhang. 2017. "Weakening of the North American monsoon with global warming." *Nature Climate Change* 7, no. 11: 806–812.

Pascale, Salvatore, Leila M.V. Carvalho, David K. Adams, Christopher L. Castro, and Iracema F.A. Cavalcanti. 2019. "Current and future variations of the monsoons of the Americas in a warming climate." *Current Climate Change Reports* 5, no. 3: 125–144.

Prelog, Andrew J., and Lee M. Miller. 2013. "Perceptions of disaster risk and vulnerability in rural Texas." *Journal of Rural Social Sciences* 28, no. 3: 1–31.

Rey, Dolores, Ian P. Holman, and Jerry W. Knox. 2017. "Developing drought resilience in irrigated agriculture in the face of increasing water scarcity." *Regional Environmental Change* 17, no. 5: 1527–1540.

Reyer, Christopher P.O., Niels Brouwers, Anja Rammig, Barry W. Brook, Jackie Epila, Robert F. Grant, Milena Holmgren et al. 2015. "Forest resilience and tipping points at different spatio-temporal scales: Approaches and challenges." *Journal of Ecology* 103, no. 1: 5–15.

Rogers, Peter. 2008. "Facing the freshwater crisis." *Scientific American* 299, no. 2: 46–53.

Simelton, Elisabeth, Evan D.G. Fraser, Mette Termansen, Piers M. Forster, and Andrew J. Dougill. 2009. "Typologies of crop-drought vulnerability: An empirical analysis of the socio-economic factors that influence the sensitivity and resilience to drought of three major food crops in China (1961–2001)." *Environmental Science & Policy* 12, no. 4: 438–452.

Singletary, Loretta, and Kelley Sterle. 2017. "Collaborative modeling to assess drought resiliency of snow-fed river dependent communities in the Western United States: A case study in the Truckee-Carson River system." *Water* 9, no. 2: 99.

Skabelund, Adrian. 2021. "As Ducey issued emergency declaration, flooding pushed Flagstaff to forefront of climate debate." *Arizona Daily Sun*, July 16, 2021. https://azdailysun.com/news/local/as-ducey-issued-emergency-declaration-flooding-pushed-flagstaff-to-forefront-of-climate-debate/article_81a3f3e0-5f21-5d6f-b194-7d2cd351efce.html, accessed July 17, 2021.

Smith, Mike. 2021. "City of Carlsbad approves flood declaration after damaging rains." *Carlsbad Current-Argus*, July 14, 2021. www.currentargus.com/story/news/2021/07/14/carlsbad-city-council-approves-flood-declaration-1-m-damages/7949598002/, accessed July 17, 2021.

Steele, Caitriana, Julian Reyes, Emile Elias, Sierra Aney, and Albert Rango. 2018. "Cascading impacts of climate change on southwestern US cropland agriculture." *Climatic Change* 148, no. 3: 437–450.

De Stefano, Lucia, James Duncan, Shlomi Dinar, Kerstin Stahl, Kenneth M. Strzepek, and Aaron T. Wolf. 2012. "Climate change and the institutional resilience of international river basins." *Journal of Peace Research* 49, no. 1: 193–209.

Steward, David R., Paul J. Bruss, Xiaoying Yang, Scott A. Staggenborg, Stephen M. Welch, and Michael D. Apley. 2013. "Tapping unsustainable groundwater stores for agricultural production in the High Plains Aquifer of Kansas, projections to 2110." *Proceedings of the National Academy of Sciences* 110, no. 37: E3477–E3486.

Sullivan, Abigail, Dave D. White, and Michael Hanemann. 2019. "Designing collaborative governance: Insights from the drought contingency planning process for the lower Colorado River basin." *Environmental Science & Policy* 91: 39–49.

Swain, Daniel L. 2015. "A tale of two California droughts: Lessons amidst record warmth and dryness in a region of complex physical and human geography." *Geophysical Research Letters* 42, no. 22: 9999–99100.

Tarlock, A. Dan. 2001. "The future of prior appropriation in the new west." *Natural Resources Journal* 41, no. 4: 769–793.

Tortajada, Cecilia, Matthew J. Kastner, Joost Buurman, and Asit K. Biswas. 2017. "The California drought: Coping responses and resilience building." *Environmental Science & Policy* 78: 97–113.

Udmale, Parmeshwar D., Yutaka Ichikawa, Sujata Manandhar, Hiroshi Ishidaira, Anthony S. Kiem, Ning Shaowei, and Sudhindra N. Panda. 2015. "How did the 2012 drought affect rural livelihoods in vulnerable areas? Empirical evidence from India." *International Journal of Disaster Risk Reduction* 13: 454–469.

Wang, Haoying. 2019a. "Change of vegetation cover in the US–Mexico border region: Illegal activities or climatic variability?" *Environmental Research Letters* 14, no. 5: 054012.

Wang, Haoying. 2019b. "Irrigation efficiency and water withdrawal in US agriculture." *Water Policy* 21, no. 4: 768–786.

Wang, Haoying. 2021. "Shale oil production and groundwater: What can we learn from produced water data?" *PloS One* 16, no. 4: e0250791.

Part II

Coordination between State and Market

Emerging Problems

5 A History of Modern European Monetary Unions as Territories, Regions, and Institutions

Rebecca Jean Emigh, Michelle Marinello, and Zachary DeGroot

5.1 Introduction: Modern European Monetary Unions

Monetary unions are strategic research sites for considering the interconnections among regions, economies, and institutions. Today, especially in Europe and North America, money is used within well-established and legally defined spatial boundaries of regions, usually nation states. Yet, for most of history, money had a much looser relationship with political units, and, going forward, supranational currencies like the euro or spatially unbounded cryptocurrencies like Bitcoin may become more common (cf. Helleiner, 2003: 1; Maurer, 2017: 219–220). A monetary union attempts to create a bounded region of monetary use from an unbounded territory (on territories and regions, see Chapter 1 in this volume). However, these "regions-to-be-constructed" have different features: some are composed of other bounded regions (e.g., nation states), and some are composed of unbounded territories (e.g., parts of units such as empires, city states, or regional states).

We explore how these differences in these regions-to-be-constructed have intersected with economies and institutions by reviewing five European monetary unions: the Latin Monetary Union, the Scandinavian Monetary Union, the German Monetary Union, the Austro-Hungarian Monetary Union, and the European Monetary Union. These were the principal European monetary unions that were set up to work in territories, attempting to create a larger unit where one did not exist before. We argue that monetary unions coincided with stable, bounded regions when they were supported by relatively similar economic conditions and by shared institutions of culture, politics, and social institutions encouraging monetary use (cf. Davis, 2017: 1; Dodd, 1994: x, 88–89, 122; Ingham, 2004: 12; Karatzas, 2007: xiii; Krämer, 1970: 11; McNamara, 2016: 353). For definitional purposes, we note that a currency union is a geographical area throughout which a single currency circulates as the principal medium of exchange (Masson and Taylor, 1993: 3). Monetary unions or exchange-rate unions are geographical areas within which the exchange rates among currencies have a fixed relationship to each other (Masson and Taylor, 1993: 3). While these definitions are sharp in principle, in practice they may blur.

DOI: 10.4324/9781003191049-7

5.2 Money and Institutions

5.2.1 Economic Institutions

Formally, an optimum currency area (OCA) is an area within which the adoption of a single currency (or, as a close second alternative, a fixed exchange rate between guaranteed convertible currencies) would be economically efficient and beneficial (McKinnon, 1963: 717; see reviews in Masson and Taylor, 1993: 7–17; McNamara, 2015a: 29). First, an OCA requires a high degree of factor mobility (labor and capital), so that supply and demand distribute resources to prevent unemployment in the absence of a flexible exchange rate (De Grauwe, 2014: 5; Fleming, 1971: 472; Kenen, 1969: 43; Masson and Taylor, 1993: 7–11; McKinnon, 1963: 724; McNamara, 2015a: 29; Mundell, 1961: 661). Second, price and wage flexibility are necessary so that a shock in one subregion will not lead to sustained inflation or unemployment (De Grauwe, 2014: 5; Fleming, 1971: 471; Masson and Taylor, 1993: 16–17; McNamara, 2015a: 29; Mundell, 1961: 662). Third, the members of the OCA must have open markets to engage in interregional trade (De Grauwe, 2014: 68; Masson and Taylor, 1993: 11–13; McKinnon, 1963: 717; McNamara, 2015a: 29). Finally, diversified production reduces variations in employment, thereby promoting factor mobility (Kenen, 1969: 54; Masson and Taylor, 1993: 13–16).

Of course, there are no OCAs; most nation states with a single currency do not meet these conditions (e.g., Masson and Taylor, 1993: 10–11). Furthermore, these conditions can be created once a common currency is in place (review in Bergman, 1999: 364). Thus, economic theory focuses on comparing the costs and benefits of currency or monetary union (review in McCallum, 2003: 8–9). Its benefits include the reduction of transaction costs associated with buying and selling goods across borders, thereby facilitating trade (De Grauwe, 2014: 53; Frankel and Rose, 1997: 753; Minford, 2004: 76). It also increases price transparency and decreases the uncertainty and risk associated with exchange rates (De Grauwe, 2014: 55, 57, 61). Its costs include the elimination of adjusting exchange rates to cushion economic shocks (Masson and Taylor, 1993: 38; McCallum, 2003: 21; Minford, 2004: 82). Of course, the costs and benefits to union may be different for different actors (cf. De Grauwe, 2014: 90). The costs and benefits may also be tangible or intangible, which leads to the idea that actors' evaluations have social, political, and cultural dimensions.

5.2.2 Social Institutions

Money is a social relation of debt and credit (Ingham, 2004: 12, 37). It expresses the relationship between every individual in a society, not merely between the two involved in trade (Ingham, 2004: 74). Thus, money relies on social networks founded in trust, confidence, habit, or faith (Dodd, 1994: xi; Ingham, 2004: 12; Maurer, 2017: 217, 225–226). Information flows through these networks: money, therefore, is a socially constructed form of information based on and reflective of the social condition of the networks (Dodd, 1994: xxiv).

This information shapes the actual use of money, that is, its substantive validity, defined as the rate at which currency is accepted in exchange for commodities (Weber, 1978: 178). The state can influence substantive validity, but it does not necessarily control it (Weber, 1978: 178). Thus, monetary unions may fail unless the integrated areas share either the same, or strikingly similar, networks of social institutions. A new, unifying network may also be constructed alongside the union.

5.2.3 Political Institutions

In contrast to these informal networks of social institutions of monetary use, the state creates and sustains formal networks. States generally reserve the right to designate official money so that the burden of trust shifts from individuals onto the issuer of money (Ingham, 2004: 72, 75). Currency and the modern state are intrinsically linked, as the state controls the economy through monetary policy and becomes the largest maker and receiver of payments (Weber, 1978: 166–168; review in Otero-Iglesias, 2015: 352–354). The state, then, establishes the formal validity of money, that is, the value of money set by law (Weber, 1978: 178).

The four economic conditions necessary for an OCA correspond to political institutions (McNamara, 2015a: 29). First, to assure factor mobility, a lender of last resort or a central bank that will stabilize the currency and provide liquidity to markets, allowing them to function smoothly, is needed (McNamara, 2015a: 26–27). Second, unified fiscal institutions provide for fiscal redistribution and debt management to buffer shocks in the absence of an exchange rate (Masson and Taylor, 1993: 26; McNamara, 2015a: 27–28). Without fiscal redistribution (welfare or unemployment payments), governments may be tempted to resort to addressing such recessions with unsustainable deficit spending (Masson and Taylor, 1993: 24; McNamara, 2015a: 27). Third, a banking union creates a regulatory framework, security guarantees, and authoritative rules that reduce risk in markets and trade (McNamara, 2015a: 28). Finally, political union, with democratic institutions, is necessary. Democratic institutions create political solidarity and democratic legitimacy that assure that markets can operate smoothly across a diverse and wide range of economic activities (McNamara, 2015a: 29).

5.2.4 Cultural Institutions

More broadly, money depends on historically variable systems of meanings and structures of social relations (Zelizer, 1989: 371). Cultural forces restrict and control the use of money and distinguish among sources and modes of allocation (Zelizer, 1989: 351). States, and elites within them, developed unified, national currencies as a political tool to foster collective identity (Dyson and Maes, 2016: 2–3; Kaelberer, 2004: 163, 173–174; Marcussen, 2000: 3, 23; cf. Lachmann, 2000: 9). The imagery used on national coins and paper money, featuring political and cultural figures from a nation's past, foster pride in this identity (Kaelberer, 2004: 164). Below, we review these four conditions for the European monetary unions and summarize them in Table 5.1. We also note the spatial characteristics (region, territory) of the monetary union, roughly from when it was founded until its peak.

5.3 Case studies

5.3.1 German Monetary Union (GMU) ˎ

5.3.1.1 The History of the GMU

The German Monetary Union formed to solve the problems created by the plethora of currencies used by the many different German states. Historically, these states had full sovereignty to regulate and issue coinage, so many different coins, all with different metallic content, circulated, making exchange and trade difficult and expensive (Bordo and Jonung, 2003: 49–50; Holtfrerich, 1993: 520–521; Pentecost and van Poeck, 2001: 2; Price, 1949: 10–11). Starting in the fourteenth century, various monetary unions established common rules for the issuance of currency, the adoption of a common unit, and fixed exchange rates (Einaudi, 2000b: 91–92). Trade and commerce were also hindered by the numerous customs borders—perhaps some 1,800 of them, for example, at the end of the eighteenth century (Henderson, 1939: 21–23).

Most of these states were part of the Holy Roman Empire until its dissolution in 1806 (Nipperdey, [1983] 1996: 4; Whaley, 2012: 644). Napoleon organized many of these states, former members of the Empire located between Austria and Prussia, into the Confederation of the Rhineland (Schmitt, 1983: 11–12; Whaley, 2012: 637–638). After Napoleon's defeat in 1815, the Congress of Vienna created the German Confederation comprising many of these states (Nipperdey, [1983] 1996: 81).

Prussia used financial and monetary reform to integrate its new territorial acquisitions from the Congress of Vienna (Nipperdey, [1983] 1996: 74–75; Tilly, 2007: 44). Its customs union of 1818 eliminated internal trade barriers, and the coinage law of 1821 standardized the value and metallic content of the thaler (Price, 1949: 119; Tilly, 2007: 44). The Prussian government absorbed the costs of re-coining and assured the establishment of the thaler as a stable currency by the 1830s (Tilly, 2007: 44). Prussian success in monetary restructuring and trade integration inspired other, similar efforts. In the 1820s and 1830s, the numerous German states arranged (and sometimes rearranged) themselves into several different customs unions, sometimes allied with the Prussian one (Einaudi, 2000b: 94; Henderson, 1939: 57–94; Murphy, 1991: 288, 290, 302; Price, 1949: 143–158, 192–252). In 1834, the Prussian trade union merged with these German unions to establish the Zollverein, which encompassed many of the German states (Einaudi, 2000b: 94; Henderson, 1939: 94, 251; Keller and Shiue, 2014: 1175). More states joined the union throughout the nineteenth century (Keller and Shiue, 2014: 1175).

The multiple currencies—with different systems of hierarchies of coins, all with different weights and metallic content—used by the Zollverein's members still created problems (Tilly, 2007: 44–45). In 1837, the Munich Coinage Treaty specified standards and seigniorage rates for the gulden used by the southern German states, and they adopted the metallic content standards for the thaler established by Prussia in 1821, thereby promoting a thaler standard (Holtfrerich, 1993: 521; Tilly, 2007: 44–45). At the Dresden convention in 1838, the Zollverein members created a monetary union covering most of the German states by fixing the exchange rate at 1.75 gulden per thaler (Einaudi, 2000b: 94; Henderson, 1939: 139–141; Holtfrerich, 1993: 521; Pentecost and van Poeck, 2001: 2; Tilly, 2007: 45). Their agreement also obligated government mints to exchange small denomination coins for gold or silver, thereby preventing seigniorage profits on them and strengthening the monetary union (Holtfrerich, 1993: 521–522; Tilly, 2007: 45). The Dresden convention also created a common coin, valued at 2 thaler (3.5 gulden); although it was too heavy for practical use, it promoted the thaler standard (Einaudi, 2000b: 94; Henderson, 1939: 140; Holtfrerich, 1993: 522; Tilly, 2007: 45). Instead of this heavy coin, the Prussian one-thaler piece became commonly circulated and accepted as a common coin, even in the southern states where the gulden was officially adopted (Holtfrerich, 1993: 522).

In 1857, the Vienna Coin Treaty affirmed Austria's intention to join the Zollverein and secured the thaler as the central currency base, equating 1 thaler to 1.5 Austrian florins and 1.75 German gulden (Einaudi, 2000b: 95; Holtfrerich, 1993: 522; Krämer, 1970: 10; Tilly, 2007: 45–46). This treaty reinforced the silver standard and helped standardize the small coins (Holtfrerich, 1993: 522–523; Tilly, 2007: 45–46). Additionally, the printing of non-convertible paper money, which Austria had relied on, was forbidden (Pentecost and van Poeck, 2001: 3; Tilly, 2007: 46). Austria, however, suspended convertibility of its paper money because of a financially straining war with Italy shortly thereafter, so it never became a member of the Zollverein (Henderson, 1939: 251; Holtfrerich, 1993: 522; Tilly, 2007: 46).

Prussian victory in the Austro-Prussian War in 1866 allowed Prussia to create the North German Confederation in 1867, which assumed formal control of monetary policy (Bordo and Jonung, 2003: 50; Pentecost and van Poeck, 2001: 3; Sheehan, 1989: 908–909; Tilly, 2007: 46). In 1870, their control was extended to paper money (Bordo and Jonung, 2003: 50; Tilly, 2007: 46). Following the French defeat in the Franco-Prussian War in 1871, the German states were united into the German Empire under Prussian rule (Bordo and Jonung, 2003: 50; Einaudi, 2001: 184–185; Tilly, 2007: 46). A new national currency, the mark, replaced the thaler and gulden (1 mark equaled 1/3 of a thaler and 7/12 of a gulden) (Pentecost and van Poeck, 2001: 3; Tilly, 2007: 46). The laws of 1871 and 1873 moved the German Empire away from a silver standard and toward a gold standard (Bordo and Jonung, 2003: 50; Einaudi, 2001: 185–186; Tilly, 2007: 46–47). The transition to the new currency was facilitated by the five billion francs war indemnity placed on France, as it protected the new German Empire from political and financial tensions (Einaudi, 2001: 185; Pentecost and van Poeck, 2001: 3; Tilly, 2007: 47). The Coinage Act of 1873 and the Bank Act of 1875 regulated paper money and created a central bank (Bordo and Jonung, 2003: 50; Einaudi, 2001: 187; Tilly, 2007: 47).

5.3.1.2 Institutions, Territories, Regions, and the GMU

The various German states were hardly an optimum currency area, as they varied widely in terms of economic conditions (Nipperdey, [1983] 1996: 130–132, 56–159) (Table 5.1, Column 1, Row 1). Nevertheless, the various trade, monetary, and currency unions were highly advantageous to the participants, facilitating trade and commerce (review in Ploeckl, 2021: 306; Tilly, 2007: 57). The mark was not introduced as a common currency until the 1870s, but during the 1830s, the thaler was widely used in practice as a common unit,

Table 5.1 European monetary unions and their characteristics

	Name of the monetary union				
	GMU	*LMU*	*SMU*	*AHMU*	*EMU*
Institutional characteristics					
Forms an optimal currency area	No	No	No	No	No
Shared social institutions for use	Some, developed over time	Few	Some	Many	Many
Shared political institutions	Some, developed over time	Few	Some	Some	Many, though incomplete
Shared cultural institutions	High	Low	High	Low	Low, but explicitly created
Spatial characteristic	Territory→region	Territory (of regions) → territory (of regions)	Territory → territory	Territory → regions	Territory (of regions) → region
Date of dissolution	Never dissolved; eventually incorporated into broader union	1926	1924	1918	Still exists

creating a historical experience with a shared currency (Table 5.1, Column 1, Row 2). The gulden also circulated widely, creating more experience with a shared unit.

There is a considerable debate about whether political or economic union came first (Bordo and Jonung, 2003: 49; Einaudi, 2001: 177; Holtfrerich, 1993: 518; Pentecost and van Poeck, 2001: 3). Prussian political unification facilitated a currency union with the introduction of the mark and the establishment of a central bank (Tilly, 2007: 46). Nevertheless, the German states had a shared political history from membership in the Holy Roman Empire, the Confederation of the Rhine, the German Confederation, the North German Confederation, and the German Empire (although the states had very different political structures, e.g., Schmitt, 1983: 30) (Table 5.1, Column 1, Row 3). The success of the union depended to some extent on the political authority and economic prowess of Prussia (Tilly, 2007: 57). (The extent to which Prussia used monetary union explicitly for political control is also debated; reviews in Murphy, 1991: 285–288; Ploeckl, 2021: 306–307.) It held 60% of both the population and the national income of the German states, so it could afford to offset hefty unification start-up costs (Tilly, 2007: 57). Although they had different economies and internal political systems, the German states had long been culturally unified through language, history, and geography (though divided by religion, mostly Protestant in the north, and Catholic in the south), which facilitated the monetary union (Table 5.1, Column 1, Row 4).

In sum, the GMU worked primarily to reduce transaction costs. However, the GMU was facilitated by at least some elements of a shared culture, political institutions, and common currencies. The GMU was established in a territory that slowly became a region (Table 5.1, Column 1, Row 5).

5.3.2 Latin Monetary Union (LMU)

5.3.2.1 The History of the LMU

France, Italy, Belgium, and Switzerland signed an agreement on December 23, 1865 creating the Latin Monetary Union, which standardized the metallic content of their currencies (Einaudi, 2001: 38–39). French, Belgian, and Swiss francs and Italian lire were formally declared to be equivalent, and public administrators were legally obligated to accept the coins (Einaudi, 2001: 38).

The agreement formalized several decades of informal convergence among these countries' currencies. Belgium, France, Switzerland, and Piedmont (in Northern Italy) informally shared a common currency between 1850 and 1860 (Bordo and Jonung, 2003: 51; Einaudi, 2001: 37). After achieving independence from the Netherlands in 1831, the Belgians adopted the French monetary system, with the same metallic content in French and Belgian coins (Willis, 1901: 15). In 1848, the Swiss cantons adopted a new national constitution that granted sole control of monetary policy to the national government (Willis, 1901: 26–27). The Swiss first adopted a silver currency system based on the French one, and then adopted the gold currency as well, to create a bimetallic currency following the French model (Willis, 1901: 27, 31–32). Prior to unification in 1861, the Italian peninsula was composed of various states with their own currencies (Willis, 1901: 36–37). The Kingdom of Sardinia-Piedmont, home state of the ruler of unified Italy, Victor Emmanuel II, used the French system (de Cecco, 2007: 64; Willis, 1901: 37). The newly minted Italian lira followed the French model (de Cecco, 2007: 65; Willis, 1901: 37).

The informal convergence in monetary policy between the four nations continued until 1865, when the Belgian government called for a meeting of these nations because of a scarcity of small silver coins used for everyday transactions (Bordo and Jonung, 2003: 51; Redish, 1993: 76). The bimetallic system was ratified despite Belgium, Switzerland, and Italy all favoring a strict gold standard (Redish, 1993: 78). Greece joined the LMU in 1869 (Einaudi, 2001: 107; see translation of the treaties in Willis, 1901: appendices). In the following years, the Papal States (in central Italy), Austria, Finland, Spain, Romania, Serbia, Bulgaria, Chile, Venezuela, Peru, and Colombia also accepted some elements of the LMU monetary system without formally being parties in the treaties (de Cecco, 2007: 68–69; Einaudi, 2001: 98–105; Willis, 1901: 83–84, appendices).

The LMU faced two difficult problems. First, a bimetallic system functioned well when the relative price of silver and gold was stable, but changes in their relative values created incentives to melt the coins for their market value. During the second half of the 1800s, the relative values of silver and gold fluctuated frequently, making it difficult to maintain a stable currency (Bordo and Jonung, 2003: 52; de Cecco, 2007: 70; Ryan and Loughlin, 2018: 711; Willis, 1901: 86).

Second, member states printed paper money, inconvertible in coins, when they faced difficult economic conditions (often exacerbated by the changing relative values of silver and gold). Italy was perhaps the most problematic, though Greece, the Papal States, and France also issued paper money (de Cecco, 2007: 72; Einaudi, 2001: 98–110, 133–134; Willis, 1901: 81, 109, 175). The young Italian state had a weak economic base and high government expenditures (de Cecco, 2007: 68–69). In 1866, facing insolvency and a necessity to fund a war, the Italian National Bank issued inconvertible paper money (which was not forbidden by the terms of the LMU) (Bordo and Jonung, 2003: 52; de Cecco, 2007: 68–69; Einaudi, 2001: 90–91, 94; Krämer, 1970: 6). The increased money supply depreciated the lira, pushing Italian coins into other countries until only paper divisionary currency circulated in Italy (Bordo and Jonung, 2003: 52; de Cecco, 2007: 69; Krämer, 1970: 6; Willis, 1901: 69). Inconvertibility was maintained until the early 1880s (Bordo and Jonung, 2003: 52; Einaudi, 2001: 91; Willis, 1901: 205).

In 1878, a conference was convened to deal with these issues. The members agreed to suspend free coinage of silver 5-franc coins, mint only new gold coins, and keep the silver divisional coins in circulation (Bordo and Jonung, 2003: 52; de Cecco, 2007: 72; Redish, 1993: 79; Ryan and Loughlin, 2018: 711). This created, in effect, a gold standard (the "limping gold standard";

Bordo and Jonung, 2003: 52; Redish, 1993: 79; Ryan and Loughlin, 2018: 711). Upon renewing the treaty in 1885, France insisted on a liquidation clause that would oblige members to redeem each nation's respective currency to the host nation at face value to terminate the LMU (Bordo and Jonung, 2003: 52; de Cecco, 2007: 73–74; Willis, 1901: 231). This effectively secured the survival of the LMU because members could not afford to liquidate or repatriate foreign coins at the official, original gold-to-silver valuation (de Cecco, 2007: 74; Krämer, 1970: 7). The LMU faced additional crises with the outbreak of World War I in 1914, when the circulation of gold was suspended (Kramer, 1970: 3). It was not possible to revive the LMU after the war, and it was formally suspended in 1926 (Krämer, 1970: 3).

5.3.2.2 Institutions, Territories, Regions, and the LMU

Given that France was more developed economically than the other countries, it is doubtful that the LMU was an optimum currency area (Table 5.1, Column 2, Row 1). Nevertheless, Belgium, Switzerland, Greece, and Italy undoubtedly benefited economically from being able to draw on French strength, the opening of trade connections, and the stabilization of their relatively weak currencies. While some of the motivation for the LMU was geopolitical, stemming from France's attempts to secure political dominance, the adoption of a monetary agreement among these countries was also a reflection of the degree of integration of their economies and their previously existing trade patterns (de Cecco, 2007: 65, 67, 69; Einaudi, 2001: 40; Flandreau, 2000: 42; Krämer, 1970: 7; Willis, 1901: 55–60). Thus, the monetary union was beneficial in what was already, at least to some extent, a shared trading territory.

The LMU, as primarily an exchange rate agreement, of course, had few social networks of use, as there was no single currency (Table 5.1, Column 2, Row 2). Each member state struck its own coins (Einaudi, 2000a: 287). Individual citizens were not legally obligated to accept foreign currencies (only government institutions were obliged to do so). Furthermore, the national banks (e.g., the Bank of France and the National Bank of the Kingdom of Italy) were private institutions, so they also were not required to accept foreign currencies (Einaudi, 2000a: 287). The Bank of France, for example, sometimes opposed government policy and refused to accept Swiss, Italian, and Belgian coins (Einaudi, 2000a: 287).

Although the French provided leadership, and there were some enforcement mechanisms such as limits on issuance of currency and refusal of membership to applicants (Einaudi, 2001: 112), there were relatively few common social, political, or administrative institutions to support monetary union, and no new ones were created (Table 5.1, Column 2, Row 3). There was never a central bank, or even a network of national banks (Einaudi, 2000a: 287). No institution managed interest rates, and there was no system of fiscal transfers in case of asymmetrical shocks (Ryan and Loughlin, 2018: 713). No penalties were created for member states who over-issued currency or who suspended the convertibility of their paper currency into gold or silver

(Einaudi, 2000a: 287). The French Ministry of Foreign Affairs became the de facto coordinating institution (Einaudi, 2001: 51). It hosted the conferences, sent and received communications to and from the members, summarized their opinions, dealt with candidates and new members, and negotiated terms (Einaudi, 2001: 51). In essence, the French government allowed the French national bank to operate as a lender of last resort; it absorbed the other members' currencies as needed and amassed a large supply of silver that helped to stabilize the exchange rate (de Cecco, 2007: 75).

Little cultural unity among members pre-existed the union, and no efforts were made during the union to create it (Table 5.1, Column 2, Row 4). On the contrary, nationalist sentiments were rising in Europe at the time (Einaudi, 2000a: 304). The French were explicitly aware of the symbolic power of a national currency in allowing nations to keep their own coins (Einaudi, 2000a: 289).

In sum, the LMU, like the GMU, functioned primarily to reduce the transaction costs in a common trading territory, but other aspects of a region were not created. The LMU was established in a territory of distinct regions, and this changed little over time (Table 5.1, Column 2, Row 5).

5.3.3 Scandinavian Monetary Union (SMU)

5.3.3.1 The History of the SMU

Following the defeat of Napoleon and his Danish allies in 1813, Denmark was forced to cede control of Norway to Sweden. Following a brief military engagement, Sweden and Norway entered a union (Libæk and Stenersen, 1991: 65). Norway maintained its status as an independent legal state, maintaining all ministries and institutions separate from Sweden except shared rule by the same Swedish king and Foreign Service ministry. Sweden thus emerged as the politically dominant force in Scandinavia.

Their respective currencies had been circulating already in all three countries (Bordo and Jonung, 2003: 52; Henriksen and Kærgård, 1995: 93). This informal acceptance of the currencies generally disadvantaged the Swedes, whose coins had a higher silver content, so they were particularly interested in currency reform (Bordo and Jonung, 2003: 53; Henriksen and Kærgård, 1995: 93). From the beginning, the countries agreed on a decimal system and a gold standard (Bordo and Jonung, 2003: 53; Henriksen and Kærgård, 1995: 92; Jonung, 2007: 77–78).

In 1872, the Scandinavian Monetary Commission adopted a gold standard; in December 1873, Sweden and Denmark signed the treaty that formally created the SMU (Bergman, 1999: 365). The Norwegian parliament rejected the treaty, but opposition quelled after Norway adopted the gold standard in 1873, and Norway joined in October 1875 (Bergman, 1999: 365; Bergman et al., 1993: 508; Henriksen and Kærgård, 1995: 91, 94; Ryan and Loughlin, 2018: 712). They adopted a new common currency, the Scandinavian krona (in Sweden and Norway, krone in Denmark), which was equivalent to the old

Swedish riksdaler (Bergman, 1999: 365; Jonung, 2007: 79). The treaty specified the metallic content of the gold and of the subsidiary copper and silver coins (Bordo and Jonung, 2003: 53; Jonung, 2007: 79; Krämer, 1970: 8). The issuance of these small coins was not limited, so each country could produce the necessary amount (Bergman, 1999: 365). The rates between the small and gold coins were also fixed to prevent the over-issuing of national currencies (Bergman, 1999: 365). Stability of the currency was assured because the countries had to redeem their subsidiary coins in gold (Henriksen and Kærgård, 1995: 94).

The original agreement did not include notes, but the Bank of Sweden informally accepted Danish and Norwegian bank notes at par from the beginning of the union. Sweden and Norway signed formal agreements in 1894, and Denmark joined the agreement in 1901 (Bergman, 1999: 365; Henriksen and Kærgård, 1995: 91, 95; Jonung, 2007: 82; Ryan and Loughlin, 2018: 712). In 1885, the three central banks agreed that transactions among them would occur without interest or other charges (Jonung, 2007: 82). This agreement also specified that no country should seek seigniorage profits at the expense of the others (Jonung, 2007: 82).

In 1905, the political union between Sweden and Norway dissolved, ending the 1885 monetary policy of mutual drawing rights between banks, and a new agreement allowed the central banks to charge fees (Bergman, 1999: 367; Jonung, 2007: 82). Subsequently, World War I brought a rapid rise in inflation that was higher in Norway and Denmark than in Sweden (Jonung, 2007: 89). All three members declared their currencies unconvertible in gold, thereby abandoning the gold standard in 1914 (Bergman et al., 1993: 514; Jonung, 2007: 89). In 1915, the Swedish krona valuation rose above the Norwegian and Danish ones (Jonung, 2007: 92). Because SMU currencies remained legal tender, gold was exported from Denmark and Norway to Sweden despite the opposition of the Swedish Central Bank (Jonung, 2007: 92). Norway and Denmark wanted to remain in the union, however, so they conceded to Swedish demands to end gold exportation in 1917 (Bergman et al., 1993: 515; Jonung, 2007: 92). At the end of the war, because the Swedish coins had a higher value than the Danish and Norwegian ones, the lower value currencies poured into Sweden (Jonung, 2007: 92). The agreement of 1924 subsequently ended the legality of Scandinavian coins, and thus the SMU union (Jonung, 2007: 92–93; Talia, 2004: 178–179).

5.3.3.2 Institutions, Territories, Regions, and the SMU

It was unlikely that the SMU was an optimum currency area (Bergman, 1999: 374; Ryan and Loughlin, 2018: 712) (Table 5.1, Column 3, Row 1). There were major differences in economic structure, trade patterns, and population growth (Bergman, 1999: 366–367; Ryan and Loughlin, 2018: 712). The Danish economy was based primarily on agriculture, the Swedish economy on agriculture and industry, and the Norwegian economy on the service sector (Bergman, 1999: 366; cf. Bergman et al., 1993: 509). Interest and inflation rates, however, were similar in the three countries, at least until the outbreak of World War I

(Bergman et al., 1993: 511; Bordo and Jonung, 2003: 54; Henriksen and Kærgård, 1995: 106–107). These economic discrepancies were exacerbated with the outbreak of war. There is a considerable debate about whether the union facilitated trade, but it apparently did not accelerate it (Bergman, 1999: 366, 375; Einaudi, 2000b: 98; Henriksen and Kærgård, 1995: 101–105; Øksendal, 2007: 133). The monetary union, however, did facilitate short-term financial flows and financial integration (Jonung, 2007: 93; Øksendal, 2007: 147, 148). The SMU did not create the extensive usage of a single currency, but it did encourage the circulation of coinage, especially smaller coins, which was mutually accepted (Jonung, 2007: 79–80; Ryan and Loughlin, 2018: 712–713) (Table 5.1, Column 3, Row 2). Gold coins, the basis of the currency union, however, rarely circulated (Bergman et al., 1993: 508; Jonung, 2007: 80).

These countries had similar political histories, but the monetary union did not create any specific political institutions (Table 5.1, Column 3, Row 3). The three nations were formally united under the Kalmar Union from 1397 until 1524, with a single monarch ruling over the three (Gustafsson, 2006: 207, 212). After Denmark lost control of Norway after the Napoleonic wars, Norway was forced into political union with Sweden in 1814 (Elgenius, 2011: 398). Beyond these general shared political institutions, however, no central bank was established (Bergman et al., 1993: 514; Jonung, 2007: 80). The monetary union agreement specified no central coordination of monetary or any other policies (Jonung, 2007: 80). Once the gold standard was dropped, the coordination among these central banks was weakened (Einaudi, 2000b: 98). There was also no customs union (Einaudi, 2000b: 98). The Scandinavian countries, however, had similar cultures and histories (Einaudi, 2000b: 98; Ryan and Loughlin, 2018: 712) (Table 5.1, Column 3, Row 4). A common currency unit was seen as an important symbol of Scandinavism that would bring the countries together (Jonung, 2007: 79).

In sum, the SMU, in addition to providing economic benefits, was primarily driven by cultural similarities. The SMU was established in a territory, and this changed little over time (Table 5.1, Column 3, Row 5).

5.3.4 *Austro-Hungarian Monetary Union (AHMU)*

5.3.4.1 *The History of the AHMU*

Seeking to avoid Hungarian secession, the Habsburg empire granted Hungary partial independence in a dualist arrangement in 1867 (Berend, 2005: 42; 2006: 4; Kann, 1974: 333). Though united under the same crown, each entity had separate governments and managed its own fiscal policy, including national budgets and debts (Berend, 2005: 42; Eddie, 1989: 845; Flandreau, 2003: 119; Ryan and Loughlin, 2018: 713). Important government functions were governed centrally, including military and foreign affairs, funded by both Austria and Hungary (Flandreau, 2003: 119; 2006: 6; Kann, 1974: 333). As part of the compromise, renegotiated every ten years, both entities enacted trade laws to uphold the customs union formed in 1850 and operated under a common monetary policy (Eddie, 1989: 815; Nautz, 2018: 71; Pammer, 2010: 139; Schulze and Wolf, 2012: 655).

Inherited from the Austrian Empire, the Austrian National Bank operated as a central bank, authorized by law to issue banknotes for the entire monetary union. The Hungarians, however, were not satisfied with this arrangement, so it was converted into the Austro-Hungarian Bank in 1878, which gave greater control to Hungarians (Flandreau, 2003: 124; 2006:7). The shared currency consisted of silver florins as well as inconvertible paper florins issued by the central bank and the Habsburg imperial authority, both backed by a silver monetary standard (Flandreau, 2003: 123). Following the declining value of silver, as well as wide exchange-rate fluctuation of the currency in exchange markets, a gold crown was introduced in 1892 to replace the silver florin (Flandreau, 2003: 125; 2006: 8; Flandreau and Komlos, 2002: 295; Kann, 1974: 422). Starting in 1896, in the absence of a formal political agreement between Austria and Hungary over gold convertibility, the Austro-Hungarian Bank operated a shadow gold standard that successfully stabilized the currency (Flandreau, 2003: 127; Flandreau and Komlos, 2002: 308).

World War I eliminated this shadow gold standard, as the government ended gold convertibility to pay for the war (Garber and Spencer, 1992: 4). With Austria-Hungary defeated in 1918, the monetary union and empire was dissolved into six countries—Czechoslovakia, Romania, Poland, Yugoslavia, Austria, and Hungary (Berend and Ránki, 1969: 169; Garber and Spencer, 1992: 7–8). Beginning in 1919, these successor states attempted to establish their own national currencies, and the assets of the Austro-Hungarian Bank were finally liquidated and divided between them by 1924 (Garber and Spencer, 1992: 8–17, 22–25; Gross and Gummer, 2013: 254–255; Helleiner, 2003: 145).

5.3.4.2 Institutions, Territories, Regions, and the AHMU

Although the AHMU was not an OCA, it did have some aspects of one, despite Austria and Hungary's economic differences (Table 5.1, Column 4, Row 1). Hungarian exports to the more-industrialized Austria were primarily agricultural, but in return Hungary received a large share of Austrian capital exports, conferring mutual economic benefits (Berend and Ranki, 1974: 40, 65–66; Bruckmüller and Sandgruber, 2003: 161; Good, 1984: 139; Sked, [2001] 2013: 204–205, 206). There was substantial labor market integration (Boulet, 1999: 118). Trade was also well-integrated; the two regions were each other's main trading partners. Between 1885 and 1910, around 73% of Hungary's exports went to Austria, and 70% of Hungarian imports came from Austria (Boulet, 1999: 140, 294–295; Eddie, 1977: 334–336). Monetary coordination increased trade (Flandreau and Maurel, 2005: 144). The evidence regarding capital mobility is more mixed—while the Hungarian countryside was initially only weakly integrated with the main economic centers, it became increasingly integrated in the early twentieth century (Berend and Ranki, 1974: 34; Boulet, 1999: 267–269).

Usage networks for the single currency in Austria and Hungary were established during the Austrian Empire preceding the Dual Monarchy (Table 5.1, Column 4, Row 2). Florin banknotes were first printed in 1762 (Botiş, 2016: 65). Paper bills became a permanent feature of the Austrian monetary system due to the apparent demand for them, though inflation became a significant problem due to the Napoleonic wars (Jobst and Stix, 2016: 96). In 1816, a new florin was issued by the newly created Austrian National Bank, depreciated florins were withdrawn from circulation, and the silver monetary standard was set (Jobst and Stix, 2016: 96). The branches of the bank, which expanded to Hungary in 1854 and 1855, formed a network that increased the supply of cash and credit and supported the creation of internal markets throughout Europe (Botiş, 2016: 71).

Except during Hungary's war of independence in 1848, the Austrian National Bank served as the sole bank of issue for both regions until it was converted to the Austro-Hungarian Bank after the establishment of the Dual Monarchy (Botiş, 2018: 76–78). Following the Compromise of 1867, Austria and Hungary had largely independent governments, and centralized political capacity was significantly reduced in accordance with the demands of the Hungarians (Haslinger, 2012: 112–113; Pammer, 2010: 139). While the AHMU did have a common debt, inherited from the pre-compromise period, payments were made by Austria (with contributions from Hungary in accordance with the quota) rather than a centralized fiscal authority (Flandreau, 2003: 119; Pammer, 2010: 141, 143, 151–152). Any common deficits were to be paid out of national budgets. The monarch did, however, play an important role when they could not reach agreement on their division of collective expenditures, as he was seen as a legitimate authority for mediating disputes between ethnic groups (Haslinger, 2012: 113). Though the AHMU had diminished centralized political institutions for managing fiscal policy, it did have a strong central bank that acted in the capacity of a lender of last resort in response to multiple economic crises (Jobst and Rieder, 2016: 140–141, Table 5.1, Column 4, Row 3).

Like the Austrian Empire before it, the AHMU was not culturally homogeneous, uniting many different lands under monarchic rule (Haslinger, 2012: 112) (Table 5.1, Column 4, Row 4). The union's main economic beneficiaries were the Hungarian agricultural business elites and Austrian leaders in industry, dominated by Magyar and German ethnic groups respectively (Bruckmüller and Sandgruber, 2003: 159; Schulze and Wolf, 2012: 656). In both regions, non-dominant ethnic groups struggled to advance themselves economically and politically throughout the nineteenth and early twentieth centuries, at times resulting in nationalist mobilization (Berend and Ránki, 1967: 180–186; Bruckmüller and Sandgruber, 2003: 160; Nautz, 2018: 82). There is debate as to whether nationalism or World War I undermined the union (Good, 1984: 98; Jászi, 1929: 4; Sked, [2001] 2013: 235–236; Wank, 1997: 45–46; see reviews in Schulze and Wolf, 2012: 654–655; Storm and Van Ginderachter, 2019: 748–749). There were certainly limits to how much the economic union acted as a unifying force, however. Throughout the union's existence, nationalist conflicts tested economic, not to mention cultural, unity (Good, 1984: 98).

In sum, the AHMU, in addition to facilitating trade, had at least some shared political institutions and instituted a territory of shared use. The AHMU was established in a territory that split into separate regions over time (Table 5.1, Column 4, Row 5).

5.3.5 European Monetary Union (EMU)

5.3.5.1 The History of the EMU

European interest in monetary integration grew out of dissatisfaction with the instability of the Bretton Woods international monetary agreement in the late 1960s and early 1970s. In 1972, the European Community (EC) created a fixed exchange rate regime called the Snake that limited fluctuations in exchange rates to 2.25% (Apel, 1998: 32, 36–37; Eichengreen and Frieden, 2001: 2; McNamara, 1998: 104–105, 107–109). Many states, however, left the Snake because the costs of maintaining their currencies within the band was too high (Apel, 1998: 42; McNamara, 1998: 107).

The Snake was replaced by the European Monetary System (EMS) in 1979 (Eichengreen and Frieden, 2001: 2; McNamara, 1998: 20–21). Like the Snake, the EMS allowed exchange rates to vary within bands of 2.25%. Though tumultuous in its initial four years, a period of relative stability followed (Apel, 1998: 57–58; Gros and Thygesen, 1998: 68–69). EC states increasingly committed themselves to low-inflationary economic policies, making them better able to remain within the system and encouraging others to join (Eichengreen and Frieden, 2001: 3; Gros and Thygesen, 1998: 87; McNamara, 1998: 129–143).

The Maastricht Treaty outlined the EC's three steps towards monetary union. Stage 1 entailed the removal of capital controls, increased convergence of inflation and interest rates, and improved policy coordination among member states (Apel, 1998: 104–105; Eichengreen, 1993: 1326; Eichengreen and Frieden, 2001: 4; McNamara, 1998: 163). In Stage 2, member states were required to increase the independence of their national banks, laying the groundwork for Stage 3, the creation of the European Central Bank (ECB) (Apel, 1998: 119, 124; Eichengreen, 1993: 1326; McNamara, 1998: 163–164). In 1997, the Sustainability and Growth Pact (SGP), which set limits on member states' public debt and budget deficits and subjected violators to sanctions, facilitated the economic convergence of the EMU (Lane, 2006: 48; Verdun, 2007: 204). The ECB and the single currency, the euro, were launched in 1999.

While the euro operated under relatively smooth economic conditions during its first decade, this came to an abrupt halt during the Eurozone crisis in 2009 (Frieden and Walter, 2017: 372). The crisis began when Greek officials revealed that their sovereign debt levels were significantly higher than previous estimates, far exceeding the limits set by the SGP (Copelovitch et al.,

2016: 814). The news prompted a downgrade of Greece's credit rating and caused borrowing costs to soar in peripheral EU states, increasing risks of default (Copelovitch et al., 2016: 815, 823).

One of the most significant responses to the Eurozone crisis was the creation of an EU bailout fund, first consisting of temporary facilities, followed by the establishment of the European Stability Mechanism in 2012, a permanent institution (Atik, 2016: 1217,1219; Gocaj and Meunier, 2013: 240). These institutions provided bailouts to Italy, Spain, Cyprus, Greece, Ireland, and Portugal, with conditions that imposed austere economic policies. The crisis also spurred efforts towards creating a banking union that could address the fragmentation of the European financial market (Howarth and Quaglia, 2013: 104–106). The Single Supervisory Mechanism was created in 2012, granting the ECB supervisory authority over all euro area banks (Howarth and Quaglia, 2013: 103). In 2014, the Single Resolution Mechanism was established to manage bank failure (Quaglia, 2019: n.p.). Originally, it was agreed that only national funds would be used to fund ailing banks, as opposed to a common fund or backstop (Quaglia, 2019: n.p.; Schild, 2018: 112). In 2020, however, it was agreed that the European Stability Mechanism be reformed to provide a common backstop (Eurogroup, 2020: n.p.). The main element rendering the EU's banking union incomplete is a common EU deposit insurance scheme, which remains controversial (Howarth and Quaglia, 2013: 110; Schild, 2018: 110).

The outbreak of the coronavirus pandemic in 2020 resulted in an economic crisis and a centralized EU response. EU officials enacted the General Escape Clause of the SGP, allowing member states to increase national spending and deficits above the EU targets (European Commission, 2020a: 1–2). In addition, a 750-billion-euro EU Recovery Fund was created, which allowed the EU to borrow collectively on financial markets to provide loans and grants to member states for economic recovery (European Council, 2020: 2–3).

5.3.5.2 Institutions, Territories, Regions, and the EMU

The euro has increasingly integrated the territory, eliminating exchange rate uncertainty and reducing transaction costs between and within EMU states (Baldwin, 2006: 36–48). Nevertheless, the EMU falls short of being an OCA, given the countries' different growth rates, inflation, deficits, and institutional configurations (Celi et al., 2018: 35; Frieden and Walter, 2017: 373–374; Höpner and Lutter, 2018: 90) (Table 5.1, Column 5, Row 1). There is relatively low factor mobility in the EU, reducing its capacity to adjust to asymmetric shocks (Jager and Hafner, 2013: 319). The EMU also lacks an institutionalized means of making transfer payments to ailing member states affected by asymmetric shocks, further making adjustment difficult (the Recovery Fund of 2020 is a one-time agreement).

Significant planning accompanied the shift from national currencies to the euro, establishing new social networks of use (Table 5.1, Column 5, Row 2). The currency was provided to retailers and banks throughout the EMU, and even outside the euro area, several months before the official introduction of the currency (ECB, 2002a: 13, 18–19). Citizens were encouraged to reduce their holdings of national currencies prior to the euro's physical introduction to reduce the need for future exchanges (ECB, 2002a: 11–14, 23). Starting on January 1, 2002, euros could temporarily be exchanged alongside national currencies (ECB, n.d.: n.p.; European Commission, 2002: n.p.). All ATMs, the biggest distributors of banknotes, were fully repurposed to distribute the euro by January 4 (ECB, 2002a: 35). National currencies ceased to be legal tender on March 1, 2002 (ECB, 2002a: 40; 2002b: n.p.).

Throughout its existence, the EMU has built up many of the political institutions necessary for the stability of monetary union, including a lender of last resort and most aspects of a banking union (Table 5.1, Column 5, Row 3). The EMU also has a democratically elected Parliament, though it has not always been included in important decision-making procedures, straining the democratic legitimacy of the political union (Schmidt, 2015: 90–91). The main political institutions lacking are those dedicated to centralized fiscal redistribution. During the financial crisis caused by the coronavirus pandemic, the Recovery Fund was created, but it was a one-time measure rather than a permanent institution with the authority to intervene in future crises. In addition, proposals to pool debt through euro-bonds or corona-bonds were never successful (European Commission, 2020b: 15; Matthijs and McNamara, 2015: 235). Without established centralized mechanisms for redistributing resources, the union must undergo ad-hoc processes to respond to crisis (Jones et al., 2016: 1027).

With the goal of constructing the EMU as a united political entity, EU officials promoted positive public attitudes toward the euro, using a wide variety of cultural media, to help ensure its success (European Commission, 1998: n.p.; Shore, 2000: 103–104) (Table 5.1, Column 5, Row 4). The design of the euro itself promotes a particular image of Europe as the governing entity. While the coins display national symbols local to the country where they are produced, banknotes do not depict specific historical figures or locations. Thus, the currency introduces the EU as a pan-European authority that does not conflict with existing nationalism, while increasing its sense of realness as an entity (Kaelberer, 2004: 170; McNamara, 2015b: 47, 55–57; Risse, 2003: 489, 493).

In sum, the institutions that support monetary union have been explicitly created for the EMU, including shared social networks, political institutions, and culture. The EMU, though originally a territory (of regions), has been turned into a well-defined region (Table 5.1, Column 5, Row 5).

5.4 Conclusions

We use Table 5.1 to compare our five cases, having established their basic characteristics in the historical reviews above. As we showed, in all five unions discussed here, trade flourished, and the economic benefits of monetary union assisted the member states. To a large extent, the nineteenth-century monetary unions simply could not survive the monetary disruptions of

World War I. Not surprisingly, none of these monetary unions were OCAs, but monetary unions supported trade and other economic activities (Table 5.1, Row 1).

The social, political, cultural, and spatial characteristics of these monetary unions, however, were quite different. The GMU, the AHMU, and the EMU created more social networks for the use of money, and in the case of the EMU, the use of the euro was mandated. In contrast, the SMU, and in particular, the LMU, created fewer shared networks of use (Table 5.1, Row 2). The EMU, though perhaps still lacking key political institutions, and the GMU created the most political institutions to support monetary functions. In contrast, the SMU, the AHMU, and the LMU had fewer (Table 5.1, Row 3). The SMU, the GMU, and the EMU had—or deliberately tried to create—a shared culture, while the LMU and the AHMU comprised quite distinct ones (Table 5.1, Row 4). Finally, the GMU and the EMU formed, eventually, single regions of use. The SMU and the LMU remained territories, while the AHMU split into distinct regions (Table 5.1, Row 5).

In sum, then, the LMU had few social, political, or cultural institutions to facilitate regional use of money, and few developed over time. It remained firmly based in a territory of distinct regions (Table 5.1, Column 2). The SMU had few shared political or social institutions but was instead held together to a large extent by a shared culture—a territorial Scandinavism. This territory, however, never developed into a region (Table 5.1, Column 3). The AHMU was in some sense the opposite of the SMU. It had stronger political institutions and social networks of use than the SMU but was composed of quite distinct cultures. The territory of the AHMU eventually split into quite separate monetary regions (Table 5.1, Column 4). The GMU, based to a large extent on a shared culture, developed an overarching political union as well as a shared currency, and eventually a single region of use, growing outwards from the Prussian Empire (Table 5.1, Column 1). Finally, the EMU formed in a different historical period, in which nation states were the regions comprising the initial EMU territory. However, a supra-region was quite deliberately created for the euro by mandating shared networks of use, establishing political institutions, and forging elements of a shared culture (Table 5.1, Column 5).

As this comparison shows, the GMU and the EMU regionalized—that is, they formed distinct monetary regions from territories, within which monetary use was bounded spatially. Furthermore, this monetary region coincided, at least to a large degree, with social, political, and cultural institutions. Of course, the EMU is an ongoing experiment, and a quite short one, especially in comparison to the GMU, and it was created under quite different historical circumstances. While we cannot point to definitive causes of successes or failures of monetary unions, our comparison does suggest that monetary unions do not depend on economic, social, political, or cultural institutions separately, nor do they fail because they lack any particular set of them. Instead, they depend on the conjoint regionalization of these institutions. With respect to the future of the EMU, we suggest that it will depend on the continued movement to regionalize economic, social, political, and cultural institutions. Whether other types of currencies, such as cryptocurrencies that succeed because of social networks of use that are not territorial or regional, can survive is, of course, another question.

We are certainly not the first to note that monetary unions are not only economic organizations but also depend on cultural, political, and social institutions (cf. Davis, 2017: 1; Dodd, 1994: x, 88–89, 122; Ingham, 2004: 12; Karatzas, 2007: xiii; Krämer, 1970: 11; McNamara, 2016: 353). Nor are we the first to compare historical European monetary unions to the EMU (e.g., Bordo and Jonung, 2003: 49–54; Einaudi, 2000b: 91–98; Krämer, 1970: 7–10; Pentecost and van Poeck, 2001: 3–4; Ryan and Loughlin, 2018: 710–716). However, we have systematically compared, more explicitly than others, these five European monetary unions with respect to all these conditions. Furthermore, while these previous comparisons were mostly directed towards a specific purpose of considering whether the monetary union was successful or not, or whether the historical monetary union provided any "lessons" for the EMU, we have instead considered how these factors work to create a region of monetary use out of a territory of use. In particular, we have shown that monetary unions corresponded with bounded regions when political, social, and cultural institutions intersected to reinforce monetary arrangements.

Acknowledgments

We would like to thank Corey O'Malley and Jasmine Vatani for their research assistance and the members of Emigh's working group for their comments. This research was supported by grants from the UCLA Faculty Senate and the UCLA CERS.

References

Apel, Emmanuel. 1998. *European Monetary Integration: 1958–2002*. London: Routledge.

Atik, Jeffery. 2016. "From 'No Bailout' to the European Stability Mechanism." *Fordham International Law Journal* 39, no. 5: 1201–1224.

Baldwin, Richard. 2006. "The Euro's Trade Effects." *Proceedings of June 2005 Workshop on What Effects is EMU Having on the Euro Area and Its Member Countries?* Working Paper Series, no. 594, Central European Bank. Frankfurt: European Central Bank.

Berend, I. T., and G. Ranki. 1974. *Hungary: A Century of Economic Development*. Newton Abbot, UK: David & Charles.

Berend, Ivan, and György Ránki. 1969. "Economic Problems of the Danube Region after the Break-up of the Austro-Hungarian Monarchy." *Journal of Contemporary History* 4, no. 3: 169–185.

Berend, Ivan T. 2005. "Democracy and Ethnic Diversity: The Case of Central and Eastern Europe." In *Political Democracy and Ethnic Diversity in Modern European History*, edited by André W. M. Gerrits, and Dirk Jan Wolffram, 32–48. Stanford: Stanford University Press.

Berend, Iván T., and György Ránki. 1967. "Economic Factors in Nationalism: The Example of Hungary at the Beginning of the Twentieth Century." *Austrian History Yearbook* 3, no. 3: 163–186.

Bergman, Michael, Stefan Gerlach, and Lars Jonung. 1993. "The Rise and Fall of the Scandinavian Currency Union 1873–1920." *European Economic Review* 37, no. 2–3: 507–517.

Bergman, U. Michael. 1999. "Do Monetary Unions Make Economic Sense? Evidence from the Scandinavian Currency Union, 1873–1913." *Scandinavian Journal of Economics* 101, no. 3: 363–377.

Bordo, Michael D., and Lars Jonung. 2003. "The Future of EMU: What Does the History of Monetary Unions Tell Us?" In *Monetary Unions: Theory, History, Public Choice*, edited by Forrest H. Capie, and Geoffrey E. Wood, 42–69. London: Routledge.

Botiş, Sorina. 2016. "Early Banknotes of the Habsburg Empire and Their Influence on Money Circulation in Transylvania." *Transylvanian Review* 25, no. 3: 63–78.

Botiş, Sorina. 2018. "Early Banknotes of the Habsburg Empire and Their Circulation in Transylvania until the Establishment of the Austro-Hungarian Dualism." *Transylvanian Review* 27, no. 4: 74–86.

Boulet, Jon M. 1999. "Austria-Hungary as an Optimal Currency Area: A Study of Austria-Hungary, 1867–1914." Ph.D. Dissertation, Clark University, Worcester, MA.

Bruckmüller, Ernst, and Roman Sandgruber. 2003. "Concepts of Economic Integration in Austria during the Twentieth Century." In *Nation, State and the Economy in History*, edited by Alice Teichova, and Herbert Matis, 159–180. Cambridge: Cambridge University Press.

Celi, Giuseppe, Andrea Ginzburg, Dario Guarascio, and Annamaria Simonazzi. 2018. *Crisis in the European Monetary Union: A Core-Periphery Perspective*. London: Routledge.

Copelovitch, Mark, Jeffry Frieden, and Stefanie Walter. 2016. "The Political Economy of the Euro Crisis." *Comparative Political Studies* 49, no. 7: 811–840.

Davis, Ann E. 2017. *Money as a Social Institution: The Institutional Development of Capitalism*. London: Routledge.

De Cecco, Marcello. 2007. "The Latin Monetary Union Revisited Once Again." In *From the Athenian Tetradrachm to the Euro: Studies in European Monetary Integration*, edited by Philip L. Cottrell, Gérassimos Notaras, Gabriel Tortella, and coedited by Monika Pohle Fraser, and Iain L. Fraser, 59–75. Aldershot, UK: Ashgate.

De Grauwe, Paul. 2014. *Economics of Monetary Union*. 10th ed. Oxford: Oxford University Press.

Dodd, Nigel. 1994. *The Sociology of Money: Economics, Reason & Contemporary Society*. New York: Continuum.

Dyson, Kenneth, and Ivo Maes. 2016. "Intellectuals as Policy-Makers: The Value of Biography in the History of the European Monetary Union." In *Architects of the Euro: Intellectuals in the Making of the European Monetary Union*, edited by Kenneth Dyson, and Ivo Maes, 1–29. Oxford: Oxford University Press.

Eddie, Scott M. 1977. "The Terms and Patterns of Hungarian Foreign Trade, 1882–1913." *The Journal of Economic History* 37, no. 2: 329–358.

Eddie, Scott M. 1989. "Economic Policy and Economic Development in Austria–Hungary, 1867–1913." In *The Cambridge Economic History of Europe. Vol. 8, The Industrial Economies: The Development of Economic and Social Policies*, edited by Peter Mathias, and Sidney Pollard, 814–886. Cambridge: Cambridge University Press.

Eichengreen, Barry. 1993. "European Monetary Unification." *Journal of Economic Literature* 31, no. 3: 1321–1357.

Eichengreen, Barry, and Jeffry A. Frieden. 2001. "The Political Economy of European Monetary Unification: An Analytic Introduction." In *The Political Economy of European Monetary Unification* 2nd ed., edited by Barry Eichengreen, and Jeffry A. Frieden, 1–21. Boulder, CO: Westview Press.

Einaudi, Luca L. 2000a. "From the Franc to the 'Europe': The Attempted Transformation of the Latin Monetary Union into a European Monetary Union, 1865–1873." *Economic History Review* 53, no. 2: 284–308.

Einaudi, Luca L. 2000b. "The Generous Utopia of Yesterday Can Become the Practical Achievement of Tomorrow: 1000 Years of Monetary Union in Europe." *National Institute Economic Review* 172 (April): 90–104.

Einaudi, Luca. 2001. *Money and Politics: European Monetary Unification and the International Gold Standard (1865–1873)*. Oxford: Oxford University Press.

Elgenius, Gabriella. 2011. "The Politics of Recognition: Symbols, Nation Building and Rival Nationalisms." *Nations and Nationalism* 17, no. 2: 396–418.

Eurogroup. 2020. "Statement of the Eurogroup in Inclusive Format on the ESM Reform and the Early Introduction of the Backstop to the Single Resolution Fund." Press Release no. 839/20, November 11, 2020. www.consilium.europa.eu/en/press/press-releases/2020/11/30/statement-of-the-eurogroup-in-inclusive-format-on-the-esm-reform-and-the-early-introduction-of-the-backstop-to-the-single-resolution-fund/

European Central Bank (ECB). 2002a. *Evaluation of the 2002 Cash Changeover*. Frankfurt am Main: European Central Bank.

European Central Bank (ECB). 2002b. "The Euro Becomes the Sole Legal Tender in All Euro Area Countries." Press Release, February 28, 2002. www.ecb.europa.eu/press/pr/date/2002/html/pr020228.en.html

European Central Bank (ECB). n.d. "Initial Changeover (2002)." www.ecb.europa.eu/euro/changeover/2002/html/index.en.html#:~:text=The%20cash%20changeover%20in%202002,of%20population%20of%20308%20million

European Commission. 1998. "Commission Communication on the Information Strategy for the Euro." European Commission, Directorate General for Economic and Financial Affairs, *Euro Papers* 16 (January). https://ec.europa.eu/economy_finance/publications/pages/publication_summary1278_en.htm

European Commission. 2002. "Review of the Introduction of Euro Notes and Coins." Commission Communication to the European Council. https://eur-lex.europa.eu/legal-content/EN/TXT/HTML/?uri=LEGISSUM:l25064

European Commission. 2020a. "Communication from the Commission to the Council on the Activation of the General Escape Clause of the Stability and Growth Pact." Brussels, February 3. https://eur-lex.europa.eu/legal-content/EN/TXT/HTML/?uri=CELEX:52020DC0123&from=EN

European Commission. 2020b. "Minutes of the 2334th meeting of the Commission held in Brussels (Berlaymont) on Wednesday 22 April 2020 (Morning)." PV(2020) 2334 final, Brussels, May 27. https://ec.europa.eu/transparency/documents-register/detail?ref=PV(2020)2334&lang=en

European Council. 2020. "Special Meeting of the European Council (17, 18, 19, 20 and 21 July 2020)–Conclusions." EUCO 10/20, CO EUR 8, CONCL 4, Brussels, July 21. www.consilium.europa.eu/media/45109/210720-euco-final-conclusions-en.pdf

Flandreau, Marc. 2000. "The Economics and Politics of Monetary Unions: A Reassessment of the Latin Monetary Union, 1865–71." *Financial History Review* 7, no. 1: 25–43.

Flandreau, Marc. 2003. "The Bank, the States, and the Market: An Austro-Hungarian Tale for Euroland, 1867–1914." In *Monetary Unions: Theory, History, Public Choice*, edited by Forrest H. Capie, and Geoffrey E. Wood, 111–141. London: Routledge.

Flandreau, Marc. 2006. "The Logic of Compromise: Monetary Bargaining in Austria-Hungary, 1867–1913." *European Review of Economic History* 10, no. 1: 3–33.

Flandreau, Marc, and John Komlos. 2002. "Core or Periphery? The Credibility of the Austro-Hungarian Currency 1867–1913." *Journal of European Economic History* 31, no. 2: 293–320.

Flandreau, Marc, and Mathilde Maurel. 2005. "Monetary Union, Trade Integration, and Business Cycles in 19th Century Europe." *Open Economies Review* 16, no. 2: 135–152.

Fleming, J. Marcus. 1971. "On Exchange Rate Unification." *The Economic Journal* 81, no. 323: 467–488.

Frankel, Jeffrey A., and Andrew K. Rose. 1997. "Is EMU More Justifiable Ex Post Than Ex Ante?" *European Economic Review* 41, no. 3–5: 753–760.

Frieden, Jeffry, and Stefanie Walter. 2017. "Understanding the Political Economy of the Eurozone Crisis." *Annual Review of Political Science* 20: 371–390.

Garber, Peter M., and Michael G. Spencer. 1992. "The Dissolution of the Austro-Hungarian Empire: Lessons for Currency Reform." IMF Working Paper, 92/66, International Monetary Fund.

Gocaj, Ledina, and Sophie Meunier. 2013. "Time Will Tell: The EFSF, the ESM, and the Euro Crisis." *Journal of European Integration* 35, no. 3: 239–253.

Good, David F. 1984. *The Economic Rise of the Habsburg Empire, 1750–1914*. Berkeley: University of California Press.

Gros, Daniel, and Niels Thygesen. 1998. *European Monetary Integration* 2nd ed. Harlow, UK: Longman.

Gross, Stephen G., and S. Chase Gummer. 2013. "Ghosts of the Habsburg Empire: Collapsing Currency Unions and Lessons for the Eurozone." *East European Politics and Societies* 28, no. 1: 252–265.

Gustafsson, Harald. 2006. "A State That Failed? On the Union of Kalmar, Especially Its Dissolution." *Scandinavian Journal of History* 31, no. 3–4: 205–220.

Haslinger, Peter. 2012. "How to Run a Multilingual Society: Statehood, Administration and Regional Dynamics in Austria-Hungary, 1867–1914." In *Region and State in Nineteenth-Century Europe: Nation-Building, Regional Identities and Separatism*, edited by Joost Augusteijn, and Eric Storm, 111–128. Basingstoke, UK: Palgrave Macmillan.

Helleiner, Eric. 2003. *The Making of National Money: Territorial Currencies in Historical Perspective*. Ithaca, NY: Cornell University Press.

Henderson, W. O. 1939. *The Zollverein*. London: Cambridge University Press.

Henriksen, Ingrid, and Niels Kærgård. 1995. "The Scandinavian Currency Union 1875–1914." In *International Monetary Systems in Historical Perspective*, edited by Jaime Reis, 91–112. Basingstoke, UK: Macmillan Press.

Holtfrerich, Carl-Ludwig. 1993. "Did Monetary Unification Precede or Follow Political Unification of Germany in the 19th Century?" *European Economic Review* 37, no. 2–3: 518–524.

Höpner, Martin, and Mark Lutter. 2018. "The Diversity of Wage Regimes: Why the Eurozone Is Too Heterogeneous for the Euro." *European Political Science Review* 10, no. 1: 71–96.

Howarth, David, and Lucia Quaglia. 2013. "Banking Union as Holy Grail: Rebuilding the Single Market in Financial Services, Stabilizing Europe's Banks and 'Completing' Economic and Monetary Union." *Journal of Common Market Studies* 51, no. S1: 103–123.

Ingham, Geoffrey. 2004. *The Nature of Money*. Cambridge: Polity.

Jager, Jennifer, and Kurt Hafner. 2013. "The Optimum Currency Area Theory and the EMU: An Assessment in the Context of the Eurozone Crisis." *Intereconomics* 48, no. 5: 315–322.

Jászi, Oscar. 1929. *Dissolution of the Habsburg Monarchy*. Chicago: University of Chicago Press.

Jobst, Clemens, and Kilian Rieder. 2016. "Principles, Circumstances, and Constraints: The Nationalbank as Lender of Last Resort from 1816 to 1931." *Monetary Policy and the Economy* Q3–4/16 (September): 140–162.

Jobst, Clemens, and Helmut Stix. 2016. "Florin, Crown, Schilling and Euro: An Overview of 200 Years of Cash in Austria." *Monetary Policy and the Economy* Q3–4/16 (September): 94–119.

Jones, Erik, R. Daniel Kelemen, and Sophie Meunier. 2016. "Failing Forward? The Euro Crisis and the Incomplete Nature of European Integration." *Comparative Political Studies* 49, no. 7: 1010–1034.

Jonung, Lars. 2007. "The Scandinavian Monetary Union 1873–1924." In *From the Athenian Tetradrachm to the Euro: Studies in European Monetary Integration*, edited by Philip L. Cottrell, Gérassimos Notaras, Gabriel Tortella, and coedited by Monika Pohle Fraser, and Iain L. Fraser, 76–95. Aldershot, UK: Ashgate.

Kaelberer, Matthias. 2004. "The Euro and European Identity: Symbols, Power and the Politics of European Monetary Union." *Review of International Studies* 30, no. 2: 161–178.

Kann, Robert A. 1974. *A History of the Habsburg Empire, 1526–1918*. Berkeley: University of California Press.

Karatzas, Theodoros B. 2007. "From the Athenian Tetradrachm to the Euro." In *From the Athenian Tetradrachm to the Euro: Studies in European Monetary Integration*, edited by Philip L. Cottrell, Gérassimos Notaras, Gabriel Tortella, and coedited by Monika Pohle Fraser, and Iain L. Fraser, xii–xv. Aldershot, UK: Ashgate.

Keller, Wolfgang, and Carol H. Shiue. 2014. "Endogenous Formation of Free Trade Agreements: Evidence from the *Zollverein*'s Impact on Market Integration." *The Journal of Economic History* 74, no. 4: 1168–1204.

Kenen, Peter B. 1969. "The Theory of Optimum Currency Areas: An Eclectic View." In *Monetary Problems of the International Economy*, edited by Robert A. Mundell, and Alexander K. Swoboda, 41–60. Chicago: University of Chicago Press.

Krämer, Hans R. 1970. "Experience with Historical Monetary Unions." Working Paper, *Kieler Diskussionsbeiträge*, no. 5, Leibniz Information Centre for Economics.

Lachmann, Richard. 2000. *Capitalists in Spite of Themselves: Elite Conflict and Economic Transitions in Early Modern Europe*. New York: Oxford University Press.

Lane, Philip R. 2006. "The Real Effects of European Monetary Union." *Journal of Economic Perspectives* 20, no. 4: 47–66.

Libæk, Ivar, and Øivind Stenersen. 1991. *History of Norway: From the Ice Age to the Oil Age*. Translated by Joan Fuglesang, and Virgina Siger. Oslo: Grøndahl & Søn Forlag A/S.

Marcussen, Martin. 2000. *Ideas and Elites: The Social Construction of Economic and Monetary Union.* Aalborg, Denmark: Aalborg University Press.

Masson, Paul R., and Mark P. Taylor. 1993. "Currency Unions: A Survey of the Issues." In *Policy Issues in the Operation of Currency Unions*, edited by Paul R. Masson, and Mark P. Taylor, 3–51. Cambridge: Cambridge University Press.

Matthijs, Matthias, and Kathleen McNamara. 2015. "The Euro Crisis' Theory Effect: Northern Saints, Southern Sinners, and the Demise of the Eurobond." *Journal of European Integration* 37, no. 2: 229–245.

Maurer, Bill. 2017. "Blockchains Are a Diamond's Best Friend: Zelizer for the Bitcoin Moment." In *Money Talks: Explaining How Money Really Works*, edited by Nina Bandelj, Frederick F. Wherry, and Viviana A. Zelizer, 215–229. Princeton: Princeton University Press.

McCallum, Bennet T. 2003. "Theoretical Issues Pertaining to Monetary Unions." In *Monetary Unions: Theory, History, Public Choice*, edited by Forrest H. Capie, and Geoffrey E. Wood, 7–25. London: Routledge.

McKinnon, Ronald I. 1963. "Optimum Currency Areas." *The American Economic Review* 53, no. 4: 717–725.

McNamara, Kathleen R. 1998. *The Currency of Ideas: Monetary Politics in the European Union.* Ithaca, NY: Cornell University Press.

McNamara, Kathleen R. 2015a. "The Forgotten Problem of Embeddedness: History Lessons for the Euro." In *The Future of the Euro*, edited by Matthias Matthijs, and Mark Blyth, 21–43. Oxford: Oxford University Press.

McNamara, Kathleen R. 2015b. *The Politics of Everyday Europe: Constructing Authority in the European Union.* Oxford: Oxford University Press.

McNamara, Kathleen R. 2016. "Regional Monetary and Financial Governance." In *The Oxford Handbook of Comparative Regionalism*, edited by Tanja A. Börzel, and Thomas Risse, 351–373. Oxford: Oxford University Press.

Minford, Patrick. 2004. "Britain, the Euro, and the Five Tests." *Cato Journal* 24, no. 1–2: 75–87.

Mundell, Robert A. 1961. "A Theory of Optimum Currency Areas." *The American Economic Review* 51, no. 4: 657–665.

Murphy, David T. 1991. "Prussian Aims for the Zollverein, 1828–1833." *The Historian* 53, no. 2: 285–302.

Nautz, Jürgen. 2018. "Ethnic Conflicts and Monetary Integration Lessons from the Austro-Hungarian Monetary Union?" In *Construction and Deconstruction of Monetary Unions: Lessons from the Past*, edited by Nathalie Champroux, Georges Depeyrot, Aykiz Doğan, and Jürgen Nautz, 69–87. Wetteren, Belgium: Moneta.

Nipperdey, Thomas. [1983] 1996. *Germany from Napoleon to Bismarck 1800–1866.* Translated by Daniel Nolan. Goldenbridge, Ireland: Gill & Macmillan.

Øksendal, Lars Fredrik. 2007. "The Impact of the Scandinavian Monetary Union on Financial Market Integration." *Financial History Review* 14, no. 2: 125–148.

Otero-Iglesias, Miguel. 2015. "Stateless Euro: The Euro Crisis and the Revenge of the Chartalist Theory of Money." *Journal of Common Market Studies* 53, no. 2: 349–364.

Pammer, Michael. 2010. "Public Finance in Austria-Hungary, 1820–1913." In *Paying for the Liberal State: The Rise of Public Finance in Nineteenth-Century Europe*, edited by José Luís Cardoso, and Pedro Lains, 132–161. Cambridge: Cambridge University Press.

Pentecost, Eric J., and André van Poeck. 2001. "The Historical Background to European Monetary Union." In *European Monetary Integration: Past, Present and*

Future, edited by Eric J. Pentecost, and André van Poeck, 1–11. Cheltenham, UK: Edward Elgar.

Ploeckl, Florian. 2021. "A Novel Institution: The Zollverein and the Origins of the Customs Union." *Journal of Institutional Economics* 17 no. 2: 305–319.

Price, Arnold H. 1949. *The Evolution of the Zollverein: A Study of the Ideas and Institutions Leading to German Economic Unification between 1815 and 1833*. Ann Arbor, MI: University of Michigan Press

Quaglia, Lucia. 2019. "The Banking Union in Europe." *Oxford Research Encyclopedia, Politics*. Oxford: Oxford University Press. https://doi.org/10.1093/acrefore/9780190228637.013.1474

Redish, Angela. 1993. "The Latin Monetary Union and the Emergence of the International Gold Standard." In *Monetary Regimes in Transition*, edited by Michael D. Bordo, and Forrest Capie, 68–85. Cambridge: Cambridge University Press.

Risse, Thomas. 2003. "The Euro between National and European Identity." *Journal of European Public Policy* 10, no. 4: 487–505.

Ryan, John, and John Loughlin. 2018. "Lessons from Historical Monetary Unions: Is the European Monetary Union Making the Same Mistakes?" *International Economics and Economic Policy* 15, no. 4: 709–725.

Schild, Joachim. 2018. "Germany and France at Cross Purposes: The Case of Banking Union." *Journal of Economic Policy Reform* 21, no. 2: 102–117.

Schmidt, Vivien A. 2015. "The Forgotten Problem of Democratic Legitimacy: 'Governing by the Rules' and 'Ruling by the Numbers.'" In *The Future of the Euro*, edited by Matthias Matthijs, and Mark Blyth, 90–114. Oxford: Oxford University Press.

Schmitt, Hans A. 1983. "Germany without Prussia: A Closer Look at the Confederation of the Rhine." *German Studies Review* 6, no. 1: 9–39.

Schulze, Max-Stephan, and Nikolaus Wolf. 2012. "Economic Nationalism and Economic Integration: The Austro-Hungarian Empire in the Late Nineteenth Century." *The Economic History Review* 65, no. 2: 652–673.

Sheehan, James J. 1989. *German History 1770–1866*. Oxford: Clarendon Press.

Shore, Cris. 2000. *Building Europe: The Cultural Politics of European Integration*. London: Routledge.

Sked, Alan. [2001] 2013. *The Decline and Fall of the Habsburg Empire, 1815–1918*. Abingdon, UK: Routledge.

Storm, Eric, and Maarten Van Ginderachter. 2019. "Questioning the Wilsonian Moment: The Role of Ethnicity and Nationalism in the Dissolution of European Empires from the Belle Époque through the First World War." *European Review of History: Revue européenne d'histoire* 26, no. 5: 747–756.

Talia, Krim. 2004. "The Scandinavian Currency Union 1873–1924: Studies in Monetary Integration and Disintegration." Ph.D. Dissertation, Stockholm School of Economics, Stockholm, Sweden.

Tilly, Richard H. 2007. "On the History of German Monetary Union." In *From the Athenian Tetradrachm to the Euro: Studies in European Monetary Integration*, edited by Philip L. Cottrell, Gérassimos Notaras, Gabriel Tortella, and coedited by Monika Pohle Fraser, and Iain L. Fraser, 42–58. Aldershot, UK: Ashgate.

Verdun, Amy. 2007. "A Historical Institutionalist Analysis of the Road to Economic and Monetary Union: A Journey with Many Crossroads." In *The State of the European Union*. Vol. 8, *Making History: European Integration and Institutional Change at Fifty*, edited by Sophie Meunier, and Kathleen R. McNamara, 195–209. Oxford: Oxford University Press.

Wank, Solomon. 1997. "The Habsburg Empire." In *After Empire: Multiethnic Societies and Nation-Building: The Soviet Union and the Russian, Ottoman, and Habsburg Empires*, edited by Karen Barkey, and Mark Von Hagen, 45–57. Boulder, CO: Westview Press.

Weber, Max. 1978. *Economy and Society: An Outline of Interpretive Sociology*. Edited by Guenther Roth, and Claus Wittich. Berkeley, CA: University of California Press.

Whaley, Joachim. 2012. *Germany and the Holy Roman Empire*. Vol. 2, *From the Peace of Westphalia to the Dissolution of the Reich 1648–1806*. Oxford: Oxford University Press.

Willis, Henry Parker. 1901. *A History of the Latin Monetary Union: A Study of International Monetary Action*. Chicago: University of Chicago Press.

Zelizer, Viviana A. 1989. "The Social Meaning of Money: 'Special Monies.'" *American Journal of Sociology* 95, no. 2: 342–377.

6 Brussels Under Pressure

Compliance, the Single Market, and National Purpose in the EU

Francesco Duina and Hermione Xiaoqing Zhou

6.1 Introduction

The European Union's (EU) establishment of a single market has relied on the implementation of its laws across the member states. Such laws have reached far beyond the strict confines of the economy. They concern public health, consumer rights, the environment, gender equality, and many other areas. This is because harmonization, or at least some degree of regulatory alignment, in those areas has been deemed necessary for the creation of an even and fair marketplace.

Considered mandatory and superior to national law, EU law has thus promoted certain values and principles. As a result, the member states have experienced what scholars have called 'Europeanization': extensive convergence in their economic, social, political, and even cultural dimensions (Radaelli 2003, 30). Countries in Europe increasingly look 'alike', as the forces of market integration push them toward homogeneity. Yet, the implementation of EU law has not always been a smooth affair. Significant variation in compliance exists. A body of research has thus developed to explain the observable differences.

While varied and insightful, those works have subscribed to one conceptually limiting stance: a view of the member states as the 'targets' of compliance pressures. The member states are inevitably depicted as being on the 'receiving ends' of EU law, sometimes willing to ensure compliance and at other times resisting it or simply not doing enough (Saurugger and Radaelli 2008, 214–215). Compliance pressures are thus not seen as originating from the member states. This stance is certainly logical: the EU promulgates laws deemed mandatory and superior to national laws, and the member states are the legislative and administrative units responsible for implementation. These laws include 'primary' law (the EU treaties) and 'secondary' law (directives, regulations, and decisions promulgated on the basis of primary law) (Nugent and Rhinard 2015). There is no question that pressures for compliance are aimed at the member states. But this may not be the full picture.

Inspired by new spatially informed research on EU institutions and processes that questions standard understandings of power (Jessop 2016; Mueller and Hechter 2019), this chapter explores an alternative perspective. Drawing

DOI: 10.4324/9781003191049-8

from works in international relations and sociology on state behavior, it advances the following proposition: that the member states can act as initiators of compliance pressures aimed at a disinterested, even resistant, EU. The idea may seem counterintuitive at first. However, it should be taken seriously as a possibility.

We suggest that this can happen when particular actors (such as manufacturers, for instance, or financial services providers) are seen as not respecting EU law in a given country, and that country intends, but is unable for various reasons, to ensure compliance. The EU may not always be inclined to redress matters. In such instances, the member state may actually pressure the EU for support. When it does so, it 'reverses' the 'normal' flow of compliance pressures. When this occurs, the member states themselves function as active promoters of Europeanization and social change through the compliance process – in line with the idea that compliance can empower member states in various ways (Saurugger and Radaelli 2008, 215).

Recent events around food quality standards offer initial empirical support for this possibility. The case concerned transnational corporations (TNCs) suspected by a group of Central and Eastern European countries (CEECs) – mostly the Visegrád countries (the Czech Republic, Hungary, Poland, and Slovakia) but also Croatia, Bulgaria, and Romania – of violating the principle of non-discrimination based on nationality (as set out in article 18 of the Lisbon Treaty) by selling food products of inferior standards in their markets. The CEECs, arguing that EU secondary law offered the TNCs a loophole, mobilized and asked the EU Commission for legal and administrative help to redress the situation. The Commission – the EU's executive body uniquely endowed with the power to initiate legislation – responded with indifference and even outright resistance. Starting in 2009, the CEECs embarked on a multi-pronged campaign to secure the Commission's attention, and legal and other resources. The efforts succeeded.

The dynamics speak directly to the questions at the core of this volume. Several countries traditionally seen as being at the periphery of an international and highly institutionalized marketplace, and in fact as prone to violating EU law (Falkner and Treib 2008; Sedelmeier 2008, 806), reversed traditional power balances vis-à-vis the center and TNCs from wealthier countries. More analytically put, the chapter examines the relationship between a formal supranational institution (law) and economic processes, in the context of cooperation and competition between territories as well as center-periphery relationships. Thus, attention also goes to vertical considerations, as we consider how traditionally weaker actors (in the periphery) managed to reverse pressure flows toward the normally stronger center (the EU).

The chapter is organized as follows. The first section reviews the existing literature on compliance. The second questions the underlying assumptions of that work and offers an alternative perspective. The third turns to the dual-quality food case in the CEECs for empirical evidence. The chapter closes by reflecting on the findings' implications for future research and power in the EU.

6.2 Standard Compliance Accounts

The EU compliance literature has grown significantly over the last two decades (Treib 2014). Scholars have focused on one key question: given that the member states are expected to ensure compliance in their territories, why don't they all do so all the time? That is, what can explain why sometimes we see compliance breaches in the member states?

Three main explanatory frameworks have been put forth. The earliest, and still invoked today, emphasizes the 'goodness of fit' between specific demands of any given EU law and a country's domestic institutional and regulatory realities (Börzel and Buzogány 2019; Duina 1999). When the law introduces principles in line with existing laws and practices, the member states ensure compliance. When EU law poses significant departures, they resist it. A number of factors, of course, influence 'fit' itself, not least the 'quality' and 'specificity' of the law in the first instance (Kaeding 2008).

The second school shifts the spotlight onto countries' abilities to manage compliance requirements. Rather than stressing questions of fit, scholars highlight different national administrative capacities, the number of domestic veto players, and political economic resources (Ademmer 2018; Börzel et al. 2010; Knill and Tosun 2009). The cross-national differences have prompted some (Falkner and Treib 2008) to propose that the member states belong to different 'worlds of compliance'. One group is able to regularly comply, two (including the CEECs) have mixed records, and a fourth is often in breach.

The third approach, intended largely as a corrective to the fairly deterministic tendencies of the first two, stresses the importance of domestic political preferences and cost-benefit calculations (Downs et al. 1996; Mastenbroek and Kaeding 2006). Here, institutional misalignments or capacity limitations can be overcome if national governments see positive results from compliance. This may happen, for instance, when complying may generate significant economic benefits or valuable international reputational outcomes.

These three perspectives share one common denominator. They assume that pressure for compliance comes from the EU and targets the member states. Recently, scholars have begun to ask whether the EU should be seen as the sole or even primary source of such pressure. They point to 'bottom-up' or 'horizontal' actors as potentially relevant. These include interest groups, the public, and civil society organizations (Hofmann 2018), and trans-governmental networks (Maggetti and Gilardi 2011) in fields ranging from social policy to telecommunications. Elements of the member states themselves may also act as sources of pressures. After all, states are not unitary entities.

Yet, while adding valuable sophistication around the sources of pressures, these additional perspectives still see the member states as inevitably the targets. Alternative scenarios where member states might play a rather different role are not entertained. Indeed, if anything, compliance scholars have spent considerable new efforts documenting situations where the member states *resist* such pressures. They often describe member states as 'obstinate' (Börzel et al. 2010) or the 'problem' (Thomann 2015). Their 'deviant' or 'resistant' behavior would persist unless 'corrected' by others (Mathieu and Bauer 2018).

The next section offers an alternative viewpoint of how compliance pressures may flow in the EU space. The focus is on the possibility of a reversed perspective which, if correct, also identifies new – and unexpected – pathways for how Europeanization (and thus convergence) may happen across the member states and traditional hierarchical relationships are challenged.

6.3 Roles Reversal: Member States Pressure the EU

This section starts by recognizing the possibility that the EU Commission may be disinterested in, or even resistant to, addressing a compliance issue. It then draws from international relations and sociological research on state behavior to outline a three-part framework for how the pressure may follow a reversed flow than typically described. Three distinct analytical steps are involved: the reasons prompting member states to increase pressure on the EU for compliance, the measures this entails, and the possible outcomes.

To be clear, the suggested framework does not assume that only the member states can be originators of pressures directed at an indifferent or oppositional EU. Transnational networks, for instance, or local interest groups may also do so. Nor does it presume that the member states become initiators of compliance pressures directed at the EU only when the latter seems uninvolved: they may do the same when the EU is actively engaged. These possibilities should certainly inform a broader articulation of the proposed framework.

6.3.1 A Disinterested or Resistant EU

The Commission is the EU institution with the most power and responsibility over legal implementation. It oversees compliance performance. It launches investigations and takes legal action against member states. It can refer cases to the Court of Justice, which in turn has the power to impose financial penalties.

Yet, importantly, scholars also stress that the Commission, as a complex and large bureaucracy, faces a multiplicity of demands and expectations (Nugent and Rhinard 2015, 347–348; Kassim and Laffan 2019). The Commission may therefore not always worry about compliance. Resources are limited, warnings from pertinent constituents may be absent, the available information may be inaccurate, and powerful actors may encourage the Commission to dismiss concerns. Given this, it seems reasonable to expect the Commission's interest in compliance to vary significantly, even when indications exist of a member state facing difficulties.

6.3.2 Reasons for Pressuring the EU

Liberal and rationalist international relations theorists see market and economic resources as powerful incentives for state behavior (Moravcsik 1997). This includes actions and choices around legal commitments

(Goodin and Moravcsik. 2011). It follows that, if officials in a country come to believe that breaches of EU law mean forgoing those resources, incentives exist to change course (Simmons 2000). For instance, new funds or markets might be available to those countries and firms that comply with environmental regulations in a given industry. In such cases, member states may have reasons to raise concerns with the EU when they face compliance challenges and the EU itself appears disinclined to take action.

Geopolitical developments may also drive member states to raise compliance issues. States participating in international organizations are often motivated by power and security calculations (Stone 2011). Their observance of international commitments reflects those calculations (Chayes and Chayes 1993). Non-compliance may have few consequences at one point but, as regional and global dynamics change, matters may be different later on. Such developments could serve as incentives for member states to turn to the EU law for support. Consider, for instance, the possibility that a rogue actor (for instance, Russia or a terrorist organization) is stealing national security data from an EU member state because some of its companies are not complying with EU digital market law. That member state may pressure the EU for additional legal or administrative resources.

We can expect 'softer' reasons – in line with more sociological perspectives – to also matter. Constructivists argue that countries care about their international reputation and moral standing in the world. They respond to comparative data and indicators of their status vis-à-vis their peers (Ikenberry 2015) and derive their identities from international interactions (Checkel 2005). Evidence comes, for instance, from research on OECD's country rankings (Niemann and Martens 2018), studies of isomorphism in world society (Meyer et al. 1997), and the EU's own soft law (Brøgger 2016) and enlargement processes (Sedelmeier 2008, 806). It follows that countries may be particularly driven to comply and pressure a disinterested or resistant EU if they believe that negative social or moral consequences may result from non-compliance.

Any of the above reasons can matter on its own or in combination. Changes in geopolitical conditions can generate different sorts of economic calculations, for instance, while social standing losses can have economic repercussions. In addition, other factors – such as changes in domestic political leaderships (Grieco et al. 2009) – can matter as well.

6.3.3 *Pressuring the EU*

Pressuring the EU for compliance support can take many forms. The relevant initiatives can potentially be complementary. International relations and sociological research on state behavior in international settings offers insights into what those initiatives might be.

First, the member states can approach the EU directly for recognition of the problem and support in addressing it. Research on compliance with international regulatory frameworks, such as those produced by the IMF and World Bank, indicates that countries facing implementation difficulties regularly make such requests to the very organizations that have produced the regulations in question (Bal Gündüz and Crystallin 2018). The requests can be monetary for additional personnel, equipment, and infrastructure. They can also focus on additional regulatory and administrative resources.

Second, member states can coordinate with each other to exchange information and data (Kahn-Nisser 2015; Maggetti and Gilardi 2016). Working in strategic alliances, they can define the nature of the perceived problems and compare possible solutions. This can help them approach the EU as well as other relevant actors that can in turn influence the EU – not unlike what happens in EU lobbying more generally (Panke 2012).

Third, concerned member states can leverage the EU institutional system to secure the Commission's attention. Commission officials face pressures from social, economic, and political actors (Broscheid and Coen 2003; Panke 2012). Other EU institutions may therefore serve as useful 'gateways' or 'bridges' to them (Christiansen 2001). The European Parliament (EP) is perhaps the most important venue. As the only institution with directly elected members, it can act with authority and established procedures vis-à-vis the Commission. Something similar may be said about the Council of Ministers. Importantly, the EP and Council are the two EU bodies empowered to vote on EU legislative proposals submitted by the Commission.

Fourth, concerned member states can craft effective or convincing language – what we may call impactful discursive practices. Language choices in the EU shape how problems, values, practices, and solutions are defined (Carstensen and Schmidt 2016). They can heighten sensitivities and awareness, and infuse issues with emotional, normative, and other content. Highly technical language can increase the apparent merits of a position. Certain discursive practices resonate better than others in particular environments. We can expect concerned member states to use such tools strategically to achieve their ends.

6.3.4 Outcomes of National Pressures

Member states' pressures on the EU for compliance can succeed or fail. Positive outcomes include an increased awareness on the part of the EU that problems exist, and the EU allocating resources to member states for improved compliance capacity. Rather than in binary terms, outcomes may be measured in degrees. Member states may succeed fully, in part, or not at all – much like actual compliance itself (Duina 1999).

Figure 6.1 summarizes the discussion so far. Compliance pressures can flow 'upwards' from the member states toward the EU, with possible positive results:

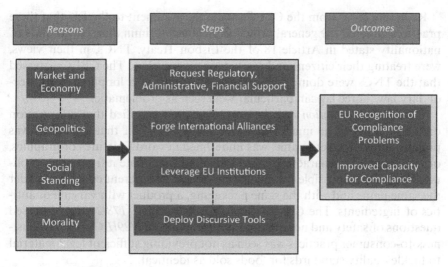

Figure 6.1 Member states pressure the EU for compliance.

6.4 The Case of Food Quality Standards in the CEECs

The recent and highly publicized food quality incident involving CEECs offers empirical evidence in favor of this alternative perspective. It serves as a valuable single case study that invites further assessments (Rueschemeyer 2003). The evidence presented here draws from academic research, news sources, governments' websites, EU official documents (such as EP reports and opinion documents, Council of Europe notes, and committees' reports), websites of Members of the European Parliament (MEPs), the 2018 one-hour Parliament debate on this issue, and correspondence with a senior EU Commission official from the Justice and Consumers Directorate General (henceforth CEU 2021) who was actively involved in the case.

6.4.1 The CEECs Raise Compliance Concerns

In the late 2000s, the CEECs began suspecting TNCs from Western countries of selling in their markets food products with the same name and packaging as those sold in the other member states even when the former were, in practice, of lower quality (Niculescu 2009). Products that appeared identical were hence different. Studies showed that fish fingers sold in the Czech Republic, for instance, contained less fish meat than those sold in Germany or Austria (Johnstone 2015). Coca-Cola's soft drinks sold in Slovakia, Hungary, Bulgaria, and Romania contained a cheaper sweetener than that used in countries like Germany and Austria (Euractiv 2013). Nutella in Hungary was found to be less creamy than its counterpart in Austria (Jancarikova 2017). Studies of other food products painted similar pictures (Gotev 2017; Minarechová 2017; Young 2016).

Representatives from the CEECs made the argument to the EU that these practices violated the general principle of non-discrimination on grounds of nationality stated in Article 18 of the Lisbon Treaty. TNCs, in their views, were treating their citizens and markets as second class. The CEECs stressed that the TNCs were doing so by taking advantage of a loophole in EU secondary law. Three laws in particular were seen as problematic.

The *Food Information Regulation (1196/2011)* specified that information on the quantity of an ingredient was mandatory only if that ingredient was present in the product's name, was underlined in words, pictures or graphics, or was essential to characterize the food and differentiate it from other products. It was thus possible for a company to sell in different countries, under the same name and with the same packaging, a product with varying quantities of ingredients. The *General Food Law Regulation (178/2002)* addressed questions of safety and not quality. And *Directive 2005/29/EC* on unfair business-to-consumer practices was seen as not providing sufficient legal material to tackle quality standards for foods sold as identical.

Given this, the CEECs asked that the Commission take explicit steps to help redress the situation. A fundamental principle of EU law was, in their view, being violated.

6.4.2 *A Resistant EU*

The Commission and other EU entities responded by expressing no interest. For several years, they in fact dismissed the CEECs by arguing that the evidence lacked generalizability and future studies might produce better data (European Parliament 2011a), EU resources were already being spent on issues potentially related to food (European Parliament 2013a), existing measures to increase transparency for consumers might prove helpful (European Parliament 2011a, 2012, 2013b), local tastes may be driving minor differences in food ingredients (European Commission 2009), and that existing EU laws sufficed and the real problem was the limited enforcement capabilities of the CEECs along with some populist politicians' desire to secure electoral votes at home by seeming to defend domestic consumers as well as producers (CEU 2021). The CEECs could have let the matter go. Instead, they embarked on a decade-long campaign aimed at the EU.

6.4.3 *What Prompted the CEECs to Seek EU Help?*

Social and moral considerations motivated the CEECs. Resentful of being treated unequally vis-à-vis the older and richer EU member states, the CEECs grew increasingly upset at what they considered to be discriminatory practices by Western TNCs. They were hence moved by reputational and fairness worries – though these stances, according to some in the Commission, connected to matters of market and economy as well, since CEECs officials could be seen favorably by acting to protect local food producers from external competition (CEU 2021).

The words of national leaders made this clear. Bulgarian Prime Minister Boyko Borissov stated in 2017, for instance, that the dual standards practice was "unacceptable and insulting. Maybe this is a remnant of apartheid – for some, food should be of higher quality, and for others, in Eastern Europe, of lower quality" (Boffey 2017). Czech Agriculture Minister Marian Jurecka stated in the same year that CEECs' citizens were tired of being "Europe's garbage can," while the top aide to Hungarian Prime Minister Viktor Orban called dual food standards "the biggest scandal of the recent past" (Jancarikova 2017). These sentiments were echoed in the EP by CEECs' nationals, such as Czech MEP Olga Sehnalová, who described the problem as a "highly political and symbolic issue, which is very relevant for the equality, justice and fairness of the European internal market" (European Parliament 2018a).

These moral concerns were articulated particularly around consumer rights. How could CEECs' consumers be treated differently than those in the rest of the EU? Government reports on lower quality and higher prices prompted widespread consternation among the public. Surveys (Kopřiva 2016; Šajn 2017, 3) showed that consumers felt "very puzzled" and "quite angry" (Michail 2018). CEECs' officials accordingly began to describe the problem as one of discrimination against Eastern European consumers. Hungary and the Czech Republic provided data to the Council of European Union in 2017 and stressed the need "to protect consumers and their trust in the quality of foodstuffs throughout the whole of the European Union" (Council of the European Union 2017).

Slovakian MEP Edit Bauer, in turn, wrote to the Commission in 2013 that "this practice clearly discriminates against consumers in some Member States – primarily those which joined the EU after 2004 – who can only acquire lower-quality products at what are often higher prices ... widens the gap and has a negative impact on social cohesion" (Bauer 2013). Other MEPs, such as Croatia's Dubravka Šuica, argued that "it is unacceptable to discriminate between consumers ... to do so goes against Europe's core policies and principles" (Šuica 2014). And Slovak MEP Ivan Štefanec stated that "we cannot accept this type of different treatment of European consumers" (EPP Group 2018).

Indeed, at stake was the very idea of equality at the foundations of the single market. As MEP Sehnalová argued, "the issue of dual quality is directly related to the essence of the functioning of the single market" (Sehnalová 2018a). Any differential treatment, she reasoned, would "undermine citizens' confidence in the fair functioning of the EU internal market" (Nadkarni 2018).

6.4.4 *Pressuring the EU*

The CEECs took four related sets of initiatives to pressure the EU for support. First, they contacted the EU directly to raise its awareness and request a more supportive regulatory framework and resources for enforcement (European Parliament 2013c; Šuica et al. 2017; Tarabella 2013), a "clear and uniform consumer protection system" (Bauer 2013), and a "code of fair principles on the marketing of branded products in the EU Single Market"

(General Secretariat of the Council of the EU 2017). These measures would give them, they claimed, the tools needed to end unlawful corporate behaviors (European Parliament 2018a).

As part of these requests, CEECs officials and their MEPs asked the EU for more tests, evidence collection, and surveys (European Parliament 2011b). They also requested that the Commission develop a "common testing methodology" across the EU (Sehnalová 2018b; The Slovak Spectator 2017). They asked for funding for large-scale analyses, a data-sharing system for test results, and an additional EU supervisory agency specialized in dual-quality issues (Sehnalová 2018a, 2018b).

Seeing limited results by mobilizing as single countries, the second set of initiatives entailed international coalition building. As already noted, one of the Commission's reasons for initially defending TNCs was the possibility that their practices reflected the availability of different ingredients and national tastes (European Commission 2009). EU officials argued that they could not "impose [a] 'one-recipe-for-all'" as long as corporations were not intentionally misleading consumers (European Parliament 2011b). The Commission also refused to carry out large-scale studies by saying that they would be "extremely complex, resource intensive and with little added value" (European Commission 2013; European Parliament 2016).

To reject this logic, starting in 2017 the CEECs formed alliances to exchange data and ideas, and to approach the Commission together. This contrasted to their earlier uncoordinated tests and campaigns (dTest 2016; Euractiv 2013; Gotev 2017; Kopřiva 2016; Minarechová 2017; MTI 2017). In March 2017, for instance, a group of CEECs' MEPs issued a major interpellation asking for a Commission proposal for supportive legislation (Šuica et al. 2017). Later that year, *The Summit for Equal Quality of Products for All* was held in Bratislava: present were the prime ministers of the Visegrád countries, representatives of all other member states, several EU Commissioners, and industry stakeholders (General Secretariat of the Council of the EU 2017). Led by Slovakia, the CEECs put forth requests for legislative measures and a common testing methodology (The Slovak Spectator 2017).

A third set of initiatives focused on leveraging strategically the EU institutional environment. They used the EP in particular as a platform to air grievances, hold debates, develop proposals, and call on the Commission for a "meaningful investigation" of compliance problems (European Parliament 2013c; Sehnalová 2016). MEPs from the CEECs submitted pointed questions to the Commission (e.g., Bauer 2013; Šuica 2014; Tarabella 2013; Vigenin 2011). In June 2018, the EP's Committee on Internal Market and Consumer Protection voted for a report calling for stronger measures at the EU and national levels to tackle non-compliance. They demanded that the Commission publish a common testing methodology and disclose data "no later by end of this year" (Nadkarni 2018). These efforts culminated into a one-hour parliament debate in September 2018, during which over 35 representatives directed forceful statements at the Commission (European Parliament 2018a).

Finally, the fourth set of initiatives involved using provocative and ethically charged language throughout all efforts. As already seen in the previous section, senior politicians and MEPs used vivid images to drive home their views (European Parliament 2011b; Šuica 2014). The 2018 hour-long EP debate offers many examples. "People shouldn't feel like second- or third-class citizens when buying a product," Slovak MEP Ivan Štefanec for instance implored. They rejected as disingenuous counterarguments that perhaps food companies were simply catering to local tastes. As Bulgarian MEP Momchil Nekov quipped, for example: "this is really unreasonable; how can you have a 3-month baby have local preferences?" MEP Sehnalová joined by saying that "to speak of food preferences of baby food, you must be joking." MEPs from other parts of the EU saw merit in these arguments. French MEP Eric Andrieu, for instance, agreed that "there is only one Europe. One single internal market and the same rules apply to everybody. There's no first- or second-class Europeans" (European Parliament 2018a).

6.4.5 Outcome: EU Recognition and Improved Compliance Capacity

The CEECs' persistence began yielding results in 2017. The Commission accepted that the principle of non-discrimination was being breached. The recognition came from the highest levels. Commission President Juncker unequivocally stated, for instance, in his State of the Union speech, that:

> I will not accept that in some parts of Europe, in Central and Eastern Europe, people are sold food of lower quality than in other countries, despite the packaging and branding being identical. Slovaks do not deserve less fish in their fish fingers. Hungarians less meat in their meals. Czechs less cacao in their chocolate. EU law outlaws such practices already. And we must now equip national authorities with stronger powers to cut out any illegal practices wherever they exist.
>
> (European Parliament 2018b, 7)

Senior commissioners voiced their support. Věra Jourová, Commissioner for Justice, Consumers and Gender Equality, joined, for instance, by promising that "I am determined to put an end to this practice, prohibited under EU law and make sure that all consumers are treated equally" (Voin and Wigand 2017).

In practice, this translated into significant initiatives. The EU set out to produce new specific legislation by 2019 (Visegrad Group 2017). Ahead of that, it published guidance lists and explanations of the requirements of the *EU Food Information Regulation* and *Unfair Commercial Practices Directive*. The Commission also promised €1 million to its Joint Research Centre (JRC) to develop a common methodology for comparative testing and enforcement actions (Voin and Wigand 2017). Later that year, the Commission sent representatives to the *Summit for Equal Quality of Products for All* in Bratislava (General Secretariat of the Council of the EU 2017).

In June 2018 the common testing methodology was published (European Commission 2018a) and Commissioner Jourová "encourage[d] all national authorities to use it in the coming months, so that we can put an end to this practice" (Voin and Wigand 2017). The Commission's legislative proposal, titled *New Deal for Consumer,* was published the same year. The aim was:

> to tackle dual quality of products by amending Article 6 of Directive 2005/29/EC to designate as a misleading commercial practice the marketing of a product as being identical to the same product marketed in several other Member States, when those products have a different composition or characteristics.
>
> (European Commission 2018b)

A year later, the results of the first major study by the Commission's JRC involving 1,400 food products in 19 EU countries showed important compositional differences, despite identical front-of-pack information, in 9% of the cases (Ulberth 2019). Another 22% of the products had different compositions despite similar front-of-pack information. While there was no conclusive East-West pattern, the results were sufficiently troubling. As Commissioner Jourová put it when commenting on them:

> There will be no double standards in Europe's single market. With the new laws penalizing the dual quality and strengthening the hands of the consumer authorities, we have the tools at hand to put an end to this practice. European consumers will be able to do their shopping in full trust that they buy what they see.
>
> (EU Science Hub 2019)

In November 2019, the EU adopted Directive 2019/2161. Article 3 amended Directive 2005/29 on unfair business-to-consumer practices by prohibiting explicitly "any marketing of a good, in one Member State, as being identical to a good marketed in other Member States, while that good has significantly different composition or characteristics, unless justified by legitimate and objective factors." This essentially closed any loophole in secondary legislation that TNCs might have exploited.

The CEECs' effort had thus borne fruit – even if some representatives from the CEECs still felt it was not enough (CEU 2021). The resulting Commission's regulatory, administrative, and financial measures empowered the member states to enforce EU food quality standards.

6.5 Conclusion

The dominant literature on EU law and the single market has depicted the member states as the 'targets' of compliance pressures stemming from the 'center.' This chapter proposed that the 'flow' of that pressure may in fact be 'reversed.' If correct, this offers a corrective to the primarily top-down view of compliance, the establishment of a single market, and Europeanization more generally.

These dynamics have special relevance for the CEECs, given claims – sometimes disputed to be sure (Bachtler et al. 2014; Ugur 2013) – that they represent a world of 'dead letters' prone to resisting EU law after the incentives of accession were gone, and their common depiction as weaker members of the EU. 'Peripheral' countries in the European market space appear capable of mobilizing, pressuring the 'center' to take action, and ultimately contributing in meaningful ways to the course of European economic integration.

The analysis thus speaks directly to this volume's core concerns with questions of territory, institutions, and economy. National actors, located at a perceived lesser place in the EU's geographical and political landscape, asserted themselves through the legal system in the broader process of economic integration. This was a multifaceted and subtle process – with those actors at once being able to obtain new standards and norms around compliance while, in so doing, managing to delegitimize the central authority and source of those very standards and norms. There is every reason to think that in the ever-evolving EU space, and also in other contexts where institutionalized realities are less than fully settled, we are likely to observe similar shifts and changes.

Future research should consider additional case studies, ideally across policy sectors, for further evidence. How frequent are cases of EU member states reversing pressures? Do their reasons for doing so vary across cases? Are there additional strategic steps they take beyond the ones identified here? In addition, the Commission has traditionally been depicted as the watchdog of potentially deviant member states. The possibility of a more equal relationship, in terms of interests and orientations, should be considered. Are there ways in which the Commission can prove more receptive to member states' concerns – capturing and sharing, more systematically the information they generate? Similar questions could be asked about other economic blocs and international organizations.

References

Ademmer, Esther. 2018. "Capitalist Diversity and Compliance: Economic Reforms in Central and Eastern Europe After EU Accession." *Journal of European Public Policy* 25, no. 5: 670–689.

Bachtler, John, Carlos Mendez, and Hildegard Oraže. 2014. "From Conditionality to Europeanization in Central and Eastern Europe: Administrative Performance and Capacity in Cohesion Policy." *European Planning Studies* 22, no. 4: 735–757.

Bal Gündüz, Yasemin, and Masyita Crystallin. 2018. "Do IMF Programs Catalyze Donor Assistance to Low-Income Countries?" *The Review of International Organizations* 13, no. 3: 359–393.

Bauer, Edit. 2013. "Parliamentary Questions: Consumer Protection and Discrimination." European Parliament E-001209-13.

Boffey, Daniel. 2017. "Multinationals Fobbing Us Off with Inferior Food, Says Bulgarian Minister." The Guardian, May 29, 2017. www.theguardian.com/world/2017/may/29/bulgaria-accuses-food-companies-cold-war-on-flavour-quality

Börzel, Tanja A., and Aron Buzogány. 2019. "Compliance with EU Environmental Law. The Iceberg Is Melting." *Environmental Politics* 28, no. 2: 315–341.

Börzel, Tanja A., Tobias Hofmann, Diana Panke, and Carina Sprungk. 2010. "Obstinate and Inefficient: Why Member States Do Not Comply with European Law." *Comparative Political Studies* 43, no. 11: 1363–1390.

Brøgger, Katja. 2016. "The Rule of Mimetic Desire in Higher Education: Governing Through Naming, Shaming and Faming." *British Journal of Sociology of Education* 37, no. 1: 72–91.

Broscheid, Andreas, and David Coen. 2003. "Insider and Outsider Lobbying of the European Commission: An Informational Model of Forum Politics." *European Union Politics* 4, no. 2: 165–189.

Carstensen, Martin B., and Vivien A. Schmidt. 2016. "Power Through, Over and In Ideas: Conceptualizing Ideational Power in Discursive Institutionalism." *Journal of European Public Policy* 23, no. 3: 318–337.

Chayes, Abram, and Antonia Handler Chayes. 1993. "On Compliance." *International Organization* 47, no. 2: 175–205.

Checkel, Jeffrey T. 2005. "International Institutions and Socialization in Europe: Introduction and Framework." *International Organization* 59, no. 4: 801–826.

Christiansen, Thomas. 2001. "Intra-Institutional Politics and Inter-Institutional Relations in the EU: Towards Coherent Governance?" *Journal of European Public Policy* 8, no. 5: 747–769.

Council of the European Union. 2017. "Experience of Certain EU Member States with Dual Quality of Foodstuffs in Free Movement Within the EU." Agriculture and Fisheries Council 6716/17.

Downs, George W., David M. Rocke, and Peter N. Barsoom. 1996. "Is the Good News About Compliance Good News About Cooperation?" *International Organization* 50, no. 3: 379–406.

dTest. 2016. "Different Food Quality in the EU." July 2016. www.dtest.cz/clanek-5156/rozdilna-kvalita-potravin-v-eu

Duina, Francesco. 1999. *Harmonizing Europe: Nation-States Within the Common Market*. Albany, NY: SUNY Press.

EPP Group. 2018. "Dual Quality of Products: We Do Not Accept Second-Class Consumers in the EU." Group of the European People's Party, July 12, 2018. www.eppgroup.eu/newsroom/news/dual-quality-of-products

EU Science Hub. 2019. "Dual Food Quality: Commission Releases Study Assessing Differences in the Composition of EU Food Products." The European Commission's Science and Knowledge Service, June 24, 2019. https://ec.europa.eu/jrc/en/news/dual-food-quality-commission-releases-study-assessing-differences-composition-eu-food-products

Euractiv. 2013. "Food Products: 'Lower Quality' in Eastern EU?" Euractiv, April 13, 2013. www.euractiv.com/section/health-consumers/news/food-products-lower-quality-in-eastern-eu/

European Commission. 2009. "Parliamentary Questions: Answer Given by Ms Kuneva on Behalf of the Commission." E-4962/2009.

———. 2013. "Parliamentary Questions: Answer Given by Mr Mimica on Behalf of the Commission." E-007154/2013(ASW).

———. 2018a. "EU Harmonised Methodology for Testing of Food Products." European Commission Directorate-General Joint Research Center, April 25, 2018. https://ec.europa.eu/knowledge4policy/publication/eu-harmonised-methodology-testing-food-products_en

———. 2018b. "Communication from the Commission to the European Parliament, the Council and the European Economic and Social Committee: A New Deal for Consumers." COM/2018/0183 final.

European Parliament. 2011a. "Parliamentary Questions: Joint Answer Given by Mr Dalli on Behalf of the Commission." E-004388/11, E-005055/11, E-005563/11, E-005756/11.

———. 2011b. "Some Products Are More Equal Than Others? MEPs Debate Food Quality Variations." European Parliament News, October 10, 2011. www.europarl. europa.eu/news/en/headlines/society/20111007STO28689/some-products-are-more-equal-than-others-meps-debate-food-quality-variations

———. 2012. "Parliamentary Questions: Answer Given by Mr Dalli on Behalf of the Commission." E-005217/2012.

———. 2013a. "Parliamentary Questions: Answer Given by Mr Mimica on Behalf of the Commission." E-007154/2013.

———. 2013b. "Parliamentary Questions: Answer Given by Mr Borg on Behalf of the Commission." E-001209/2013.

———. 2013c. "A New Agenda for European Consumer Policy." European Parliament P7_TA(2013)0239.

———. 2016. "Parliamentary Questions: Answer Given by Ms Jourová on Behalf of the Commission." E-000329/2016(ASW).

———. 2018a. "Dual Quality of Products in the Single Market Debate." EU Parliament, September 13, 2018. www.europarl.europa.eu/plenary/EN/vod. html?mode=chapter&vodLanguage=EN&startTime=20180913-09:20:30-531#

———. 2018b. *Quality Differences in Consumer Products in the EU Legislation Study.* European Parliament, Policy Department for Citizens' Rights and Constitutional Affairs. www.europarl.europa.eu/RegData/etudes/STUD/2018/608840/IPOL_STU %282018%29608840_EN.pdf

Falkner, Gerda, and Oliver Treib. 2008. "Three Worlds of Compliance or Four? The EU-15 Compared to New Member States." *Journal of Common Market Studies* 46, no. 2: 293–313.

General Secretariat of the Council of the EU. 2017. "Outcomes of the Summit for Equal Quality of Products for All." Bratislava, 13474/17.

Goodin, Robert E., and Andrew Moravcsik. 2011. *The New Liberalism.* Oxford: Oxford University Press.

Gotev, Georgi. 2017. "Lower Quality of Same Food Brands in Eastern Europe Raises Eyebrows." Euractiv, February 17, 2017. www.euractiv.com/section/health-consumers/news/lower-quality-of-same-food-brands-in-eastern-europe-raises-eyebrows/

Grieco, Joseph M., Christopher F. Gelpi, and T. Camber Warren. 2009. "When Preferences and Commitments Collide: The Effect of Relative Partisan Shifts on International Treaty Compliance." *International Organization* 63, no. 2: 341–355.

Hofmann, Tobias. 2018. "How Long to Compliance? Escalating Infringement Proceedings and the Diminishing Power of Special Interests." *Journal of European Integration* 40, no. 6: 785–801.

Ikenberry, G John. 2015. *Ranking the World: Grading States as a Tool of Global Governance.* Cambridge: Cambridge University Press.

Jancarikova, Tatiana. 2017. "East Europeans Decry "Double Standards" for Food, Seek Change to EU Law." Reuters, March 1, 2017. www.reuters.com/article/centraleurope-food-idUSL5N1GD4N4

Jessop, Bob. 2016. "Territory, Politics, Governance and Multispatial Metagovernance." *Territory, Politics, Governance* 4, no. 1: 8–32.

Johnstone, Chris. 2015. "Study Reveals Stark Differences in Same Food and Drink Sold on Czech and German Markets." Radio Prague International, July 8, 2015. www.radio.cz/en/section/curraffrs/study-reveals-stark-differences-in-same-food-and-drink-sold-on-czech-and-german-markets

Kaeding, Michael. 2008. "In Search of Better Quality of EU Regulations for Prompt Transposition: The Brussels Perspective." *European Law Journal* 14, no. 5: 583–603.

Kahn-Nisser, Sara. 2015. "The Hard Impact of Soft Co-Ordination: Emulation, Learning, and the Convergence of Collective Labour Standards in the EU." *Journal of European Public Policy* 22, no. 10: 1512–1530.

Kassim, Hussein, and Brigid Laffan. 2019. "The Juncker Presidency: The 'Political Commission' in Practice." *JCMS: Journal of Common Market Studies* 57, no. S1: 49–61.

Knill, Christoph, and Jale Tosun. 2009. "Post-Accession Transposition of EU Law in the New Member States: A Cross-Country Comparison." *European Integration online Papers (EIoP)* 13, no. 2: 1–18.

Kopřiva, Pavel. Februry 18, 2016. Research: Czech Consumers Want Food of the Same Quality as European Food. Czech Agriculture and Food Inspection Authority. www.szpi.gov.cz/clanek/tz-2016-vyzkum-cesky-spotrebitel-zada-stejne-kvalitni-potraviny-jako-evropsky.aspx

Maggetti, Martino, and Fabrizio Gilardi. 2011. "The Policy-Making Structure of European Regulatory Networks and the Domestic Adoption of Standards." *Journal of European Public Policy* 18, no. 6: 830–847.

Maggetti, Martino. 2016. "Problems (and Solutions) in the Measurement of Policy Diffusion Mechanisms." *Journal of Public Policy* 36, no. 1: 87–107.

Mastenbroek, Ellen, and Michael Kaeding. 2006. "Europeanization Beyond the Goodness of Fit: Domestic Politics in the Forefront." *Comparative European Politics* 4, no. 4: 331–354.

Mathieu, Emmanuelle, and Michael W. Bauer. 2018. "Domestic Resistance Against EU Policy Implementation: Member States Motives to Take the Commission to Court." *Journal of European Integration* 40, no. 6: 667–682.

Meyer, John W., John Boli, George M. Thomas, and Francisco O. Ramirez. 1997. "World Society and the Nation-State." *American Journal of Sociology* 103, no. 1: 144–181.

Michail, Niamh. 2018. "MEPs Push for a Dual Quality on Pack Logo." Food Navigator, September 13, 2018. www.foodnavigator.com/Article/2018/07/13/MEPs-push-for-a-dual-quality-on-pack-logo

Minarechová, Radka. 2017. "Some Food Really Is Better in Austria, Study Finds." The Slovak Spectator, February 16, 2017. https://spectator.sme.sk/c/20461101/some-food-really-is-better-in-austria-study-finds.html

Moravcsik, Andrew. 1997. "Taking Preferences Seriously: A Liberal Theory of International Politics." *International Organization* 51, no. 4: 513–553.

MTI. 2017. "There Is a Double Standard at Work on the European Food Market." March 29, 2017. www.kormany.hu/en/ministry-of-agriculture/news/there-is-a-double0-standard-at-work-on-the-european-food-market

Mueller, Sean, and Michael Hechter. 2019. "Centralization Through Decentralization? The Crystallization of Social Order in the European Union." *Territory, Politics, Governance* 9: 1–20.

Nadkarni, Isabel Teixeira. 2018. "Consumer Product Quality: MEPs Take Aim at Dual Standards." European Parliament, July 12, 2018. www.europarl.europa.eu/news/en/press-room/20180711IPR07740/

Niculescu, Rareş-Lucian. 2009. "Parliamentary Questions: Same Brand, Different Product Formulation." European Parliament E-4962/09.

Niemann, Dennis, and Kerstin Martens. 2018. "Soft Governance by Hard Fact? The OECD as a Knowledge Broker in Education Policy." *Global Social Policy* 18, no. 3: 267–283.

Nugent, Neill, and Mark Rhinard. 2015. *The European Commission (the European Union Series)*. Basingstoke, UK: Palgrave Macmillan.

Panke, Diana. 2012. "Lobbying Institutional Key Players: How States Seek to Influence the European Commission, the Council Presidency and the European Parliament." *JCMS: Journal of Common Market Studies* 50, no. 1: 129–150.

Radaelli, Claudio M. 2003. "The Europeanization of Public Policy." In *The Politics of Europeanization*, edited by Claudio M. Radaelli and Kevin Featherstone, 27–56. Oxford: OUP.

Rueschemeyer, Dietrich. 2003. "Can One or a Few Cases Yield Theoretical Gains?" In *Comparative Historical Analysis in the Social Sciences*, edited by James Mahoney and Dietrich Rueschemeyer, 305–336. Cambridge: Cambridge University Press.

Šajn, Nikolina. 2017. "Dual Quality of Branded Food Products Addressing a Possible East-West Divide." European Parliamentary Research Service PE 607.265.

Saurugger, Sabine, and Claudio M. Radaelli. 2008. "The Europeanization of Public Policies: Introduction." *Journal of Comparative Policy Analysis: Research and Practice* 10, no. 3: 213–219.

Sedelmeier, Ulrich. 2008. "After Conditionality: Post-Accession Compliance with EU Law in East Central Europe." *Journal of European Public Policy* 15, no. 6: 806–825.

Sehnalová, Olga. 2016. "Parliamentary Questions: Investigation of Differences in Products with the Same Brand and Packaging in the Internal Market." European Parliament E-000329-16.

———. 2018a. "Draft Report on Dual Quality of Products in the Single Market." European Parliament Committee on the Internal Market and Consumer Protection PR\1147015EN.docx.

———. 2018b. "Report on Dual Quality of Products in the Single Market." European Parliament Committee on the Internal Market and Consumer Protection P8_TA(2018)0357.

Simmons, Beth A. 2000. "International Law and State Behavior: Commitment and Compliance in International Monetary Affairs." *American Political Science Review* 94, no. 4: 819–835.

Stone, Randall W. 2011. *Controlling Institutions: International Organizations and the Global Economy*. Cambridge: Cambridge University Press.

Šuica, Dubravka. 2014. "Parliamentary Questions: Discrimination against Consumers in New Member States and Hence in Croatia." European Parliament E-001148-14.

Šuica, Dubravka, et al. 2017. "Parliamentary Questions: Major Interpellation – Differences in Declarations, Composition and Taste of Products in Central/Eastern and Western Markets of the EU." European Parliament O-00019/2017.

Tarabella, Marc. 2013. "Parliamentary Questions: Difference in the Quality of the Same Product in Different Countries." European Parliament E-007154-13.

The Slovak Spectator. 2017. "Bratislava Hosts Summit on Dual Quality of Food." October 12 2017. https://spectator.sme.sk/c/20671185/bratislava-hosts-summit-on-dual-quality-of-food.html

Thomann, Eva. 2015. "Customizing Europe: Transposition as Bottom-up Implementation." *Journal of European Public Policy* 22, no. 10: 1368–1387.

Treib, Oliver. 2014. "Implementing and Complying with EU Governance Outputs. Living Reviews in European Governance." *Living Reviews in European Governance* 1, no. 1: 5–47.

Ugur, Mehmet. 2013. "Europeanization, EU Conditionality, and Governance Quality: Empirical Evidence on Central and Eastern European Countries." *International Studies Quarterly* 57, no. 1: 41–51.

Ulberth, Franz. June 2019. Results of an EU Wide Comparison of Quality Related Characteristics of Food Products. European Commission. https://ec.europa.eu/jrc/en/publication/results-eu-wide-comparison-quality-related-characteristics-food-products

Vigenin, Kristian. 2011. "Parliamentary Questions: Quality and Content Variations in Products Being Marketed in Different EU Member States." European Parliament E-005055/2011

Visegrad Group. 2017. "Minister: EU Can Have Law against Dual Food Quality by 2019." Visegrad Group, October 3, 2017. www.visegradgroup.eu/news/minister-eu-can-have-law

Voin, Melanie, and Christian Wigand. 2017. "Dual Quality Food Products: Commission Guides Member States to Better Tackle Unfair Practices." European Commission, September 26, 2017. https://europa.eu/rapid/press-release_IP-17-3403_en.htm

Young, Lisa F. 2016. "Rozdielna Kvalita Potravín V Eú." dTest, July 2016. www.dtest.sk/clanek-5156/rozdielna-kvalita-potravin-v-eu

7 Institutional Context and Territorial Policy

Analyzing the New Regional Policy and Regional Development Agencies in Turkey

Nuri Yavan, Şükrü Yılmaz, and Aykut Aniç

7.1 Introduction

Institutions have recently become the main agenda topic for development and institutional economics (e.g. Rodrik, Subramanian, and Trebbi 2004; Hodgson 2006; Acemoglu and Robinson 2012), on the one hand, political science and economic sociology (e.g. Streeck and Thelen 2005; Mahoney and Thelen 2010; Scott 2013) on the other, as well as economic geography (Amin 1999; Martin 2000; Rodríguez-Pose and Storper 2006; Gertler 2010, 2018; Farole, Rodríguez-Pose, and Storper 2011; Rodríguez-Pose 2013, 2020; Storper 2013; Bathelt and Glückler 2014; Pike et al. 2015; Glückler and Lenz 2016, 2018; Zukauskaite, Trippl, and Plechero 2017). There is a growing consensus, also reflected in the EU and OECD's local and regional development policies, that institutions play a critical role in the production, consumption, distribution, exchange, and regulation processes of growth and development, and that growth and development is a process requiring good, effective, and inclusive institutions enabling cooperation between governments, organizations, and social groups (Acemoglu and Robinson 2012; Barca, McCann, and Rodríguez-Pose 2012; OECD 2012).

Numerous studies in economic geography have been conducted to examine the impact of local institutions on regional development, as well as the role of institutional quality in developmental processes such as innovation, entrepreneurship, clustering, competitiveness, social inclusion, political participation, etc. (Gertler 2018; Glückler, Suddaby, and Lenz 2018; Rodríguez-Pose 2020). On the other hand, there are still several issues and gaps that need to be addressed in this emerging literature on institutions, such as (1) inability to determine what institutions are and how they work, (2) considering them as static factors, challenges associated with conceptualizing institutional change, and the role of the agency, (3) emphasizing formal institutions at the expense of informal ones, (4) lack of multiscalar and methodological approaches to institutions and institutional change (Gertler 2018; Glückler, Suddaby, and Lenz 2018; Rodríguez-Pose 2020).

Based on the gaps and problems in the aforementioned literature, some recent theoretical and empirical studies in economic geography have proposed a new and alternative framework from a relational perspective, differing from

DOI: 10.4324/9781003191049-9

previous studies both theoretically and analytically, in order to extend knowledge about institutions, institutional context, and institutional changes (Bathelt and Glückler 2014; Glückler and Bathelt 2017; Glückler and Lenz 2016, 2018; Glückler 2020). These authors treat the trinity of institutional context as regulations–organizations–institutions with an analytical distinction and conceptualize "institutions" as stabilizations of mutual expectations and correlated interaction that develop contingently according to rules and regulations but also constituted by the interrelations between them. In addition, they introduced a new framework of institutional change adapting the temporal change of institutions in relation to the institutional context and the different mechanisms created by this change to regional scale and geographical studies (Bathelt and Glückler 2014; Glückler and Lenz 2016, 2018).

In this chapter, we aim to reveal the contextual nature of the relationship between the new regional development policy introduced after the EU accession process in Turkey and the existing institutional context. We argue that it will be difficult for the new regional development policy, which aims to substitute bottom-up social practices based on decentralization and cooperation among actors, instead of centralization and top-down social practices in Turkey, to comply with the existing institutional context. Therefore, it would be interesting to explore institutions, as stable patterns of interactions, which moderate the effects of newly introduced regional policy on social practices. Here we ask whether the new regional development policy in Turkey, really transformed the stable patterns of existing institutional context, and which institutions have developed with or in response to the new regional policy. Uncovering these institutions allows us to explain the causal factors behind the institutional compliance challenges we predicted. To do this, we include the "institutional context" framework of Glückler and Lenz (2016), which is constituted by three pillars –regulations, organizations, and institutions – into our institutional analysis. This analytical framework provides an opportunity to examine the differences between formal rules and actual social practices in specific contexts and the possibility to observe the systematic nature of the multi-dimensional interactions between rules and institutions in seeking to reveal the contextual nature of a policy (un)success on social practices (Bathelt and Glückler 2014; Glückler and Lenz 2016). In addition to the analysis of 'institutional context', employing Mahoney and Thelen's (2010) "gradual institutional change" approach, which defines multiple institutional strategies in conjunction with existing and emerging institutional structures, we analyze which of the four different institutional change mechanisms (displacement, layering, drift, and conversion) and institutional actors (both individuals and organizations) follow within Turkey specific institutional context.

Another goal of the chapter is to contribute to the literature on institutional economic geography and to evaluate the validity of Bathelt and Glückler's aforementioned relational framework in the context of Turkey's new regional development strategy and regional development agency (RDA) experience. Although some studies have been conducted on Turkey's regional development policy (e.g., Dulupçu 2005; Yavan 2011; Ertugal 2018; Ersoy

2018) and RDAs (Lagendijk, Kayasu, and Yasar 2009; Ertugal 2017; Sadioğlu, Dede, and Göçoğlu 2020), none of these studies have analyzed Turkey's regional policy and RDAs within the framework of institutional context and dynamics of institutional change in terms of the relational institutionalist approach adopted by this study.

Since institutions, unlike rules, are reproduced over time with inter-agency relations, expectation, and social practices, we conducted semi-structured interviews to observe Turkey-specific social practices and expectations with 18 different institutional actors (see Appendix 7.1) from the RDAs and the ministry coordinating them. In order to ensure the anonymity of the interviewees, the identities and regions of the interviewees are given as pseudonyms. We began our interviews with questions about the institutional context of new-regional policy and about the beginning of a new regional policy. Then, we continued with the questions in which the actors compared the new regional policies and new instruments with the previous ones, determined the breaking points and impacts of the policy change, and discussed how they were affected by the changes in the Europeanization and de-Europeanisation processes at the national level. To consolidate our qualitative analysis on the stabilized patterns of interaction emerging after the regional policy change, we co-evaluated and solidified our inferences from the interview data with the other empirical literature, official papers, programming activity, and working reports conducted in previous years.

The overall organization of the study takes the form of five sections, including the introduction section. In Section 7.2, the concept of institution, the institutional context, and the institutional change strategies pursued by actors are detailed. In Section 7.3, the drivers, scope and differences of the new regional development policy in Turkey and the roles, functions and structure of RDAs are explained. In Section 7.4, we present the findings based on questions/arguments asked, e.g., which types of institutions the new regional policy interacts with, how these institutions shape local-regional practices, and which institutional change strategies are followed by institutional actors. The last section summarizes the main results of this study and includes a discussion of the implication of the findings for future research.

7.2 Conceptual and Analytical Framework

7.2.1 Concept of Institution

There is a growing body of economic geography literature, with the institutional turn, which emphasizes the institutions and the institutional environment have decisive roles and multidimensional and far-reaching effects in local and regional economic development (Farole, Rodríguez-Pose, and Storper 2011; Gertler 2010; Martin 2000; Rodríguez-Pose 2013; Pike et al. 2015; Zukauskaite, Trippl, and Plechero 2017); see also Chapter 1 in the book. On the other hand, due to the subjective and controversial nature of the concept of institutions (Rodríguez-Pose 2013), there is not yet a consensus on what institution is in the social sciences in general and in the economic geography in particular, making analyses of the impact of institutions on economic development difficult. In addition, although most studies have emphasized the crucial role of institutions in local and regional economic development, there is a need for more explanations on what institutions exactly are, their emergence, persistence, and change in different contexts, as well as which institutions are crucial, when they are decisive and exactly how they shape growth (Glückler and Lenz 2016; Farole, Rodríguez-Pose, and Storper 2011; Rodríguez-Pose 2020; Pike et al. 2015). Therefore, the basic and primary step for research on institutions and institutional change is to analytically clarify the "institutions" that they tackle as the object of study within various identifications on "institutions."

Although many significant studies are examining and trying to define institutions (North 1990; Nelson 1993; Williamson 2000; Helmke and Levitsky 2004; Thelen 2004; Hodgson 2006; Scott 2013), the most cited definition and classification in the economic geography literature belong to North. North (1990), one of the representatives of the new institutional economics, defines institutions as both informal (such as cultural habits, traditions, customs, beliefs and sanctions) and formal constraints (constitutions, laws, property rights) that human beings devise to shape political, economic, and social interaction, and describes them as "the rules of the game in a society" (North 1990: 3). Along with this widely accepted definition of institutions, the majority of studies on institutions in the economic geography literature (e.g. Gertler 2010, 2018; Pike et al. 2015; Rodríguez-Pose 2013, 2020; Storper 2013) have focused on the outputs of institutions on economic and social development and determining which institutions are good for development and growth (Glückler and Lenz 2016). However, these studies having an instrumental approach tackle more institutions that can be compared and operationalized, and meaningful conceptualization of institutions is sacrificed in favor of standardized comparable quantitative assessments (Glückler and Lenz 2016; Rodríguez-Pose 2020; Pike et al. 2015). Therefore, in most of these institutional studies, hard-formal-society rules such as the rule of law or property rights stand out prominently, while soft–informal–community rules such as values, culture, trust, or social capital lag significantly behind (Rodríguez-Pose 2020). As a result, although there have been many studies on institutions, institutional change, and the effects of institutions on growth, innovation, and regional development in the economic geography, the leading researchers of the discipline have revealed the missing aspects and main obstacles to be overcome in the field of institutional economic geography (Gertler 2010, 2018; Bathelt and Glückler 2014; Glückler and Lenz 2016; Farole, Rodríguez-Pose and Storper 2011; Rodríguez-Pose 2013, 2020).

7.2.2 *Relational Approach to Institutions and Three Pillars of the Institutional Context*

Based on the critical review of the above-mentioned literature in economic geography more comprehensive research is required that does not view institutions as static structures, examines the mechanisms underlying institutional change in depth, and considers the interactions and co-evolution of formal and informal institutional variables in various sub-national contexts, as well as their mechanisms of transfer to different contexts (Gertler 2018; Rodríguez-Pose 2020). We believe that the conceptualization and analytical framework developed by Bathelt and Glückler, who have recently reconceptualized institutions in a "relational" perspective, may be suitable to respond to this requirement and fill the research gap (Bathelt and Glückler 2014; Glückler and Bathelt 2017; Glückler 2020; Glückler and Lenz 2016, 2018).

According to the relational perspective having an analytical distinction between each part of the institutional context (regulations, organizations, and institutions), institutions that are neither regulations nor organizations are the stable patterns of mutual expectations and correlated interactions between actors, developing contingently according to rules and regulations (Bathelt and Glückler 2014) (Table 7.1). In other words, unlike the rules of the game (regulations) and players the game (organizations) in the conceptualization of North (1990), Bathelt and Glückler (2014) and Glückler and Lenz (2016) conceptualize the institutions as expressing how the game is played differently in different contexts, but consistently in repetitive situations. Although policies, laws, and other rules define the basic framework for economic interaction and implementation, all of them do not always produce the expected results and are not applied equally (Glückler 2020). The impact of the regulations on real social practices is driven by institutions, the last analytical component of the institutional context. Therefore, *rules are*

Table 7.1 Three pillars of the institutional context

Component	Definition	Examples
Regulations	Rules that can be legally enforced, written, codified, and read and interpreted with others	Constitutions, laws, level of centralization and decentralization, regional policies, regional planning
Organizations	Organizations as individuals or as a set of collective actors, whether public or private, influencing the economic course	Local governments, companies, NGOs, universities, development agencies, unions, unions, associations
Institutions	Repeated stable patterns of behavior under similar conditions based on legitimate normative expectations among stakeholders, arising from formal rules or informal norms and conventions.	Stable and mutual expectation types between actors; such as perceptions of cooperation, coordination, competition, irregularity or corruption.

Source: Glückler and Lenz 2016; Glückler 2020.

not institutions yet (Bathelt and Glückler 2014, 7). They leave a scope of action for agents, which leads to deviations from expected codified norms and rules (Glückler and Bathelt 2017). What differentiates regulations and institutions is the scope of actions for agents. Institutions emerge in different contexts, sometimes according to rule (with the rule), in response to or against them. Even if there is a risk of sanctions, individuals or collective actors may sometimes not comply with these rules (Glückler and Lenz 2016). Despite sanctions, the social practices generated against the rule may be widely accepted, reproduced, and disseminated, thus institutionalized by the actors over time (Glückler and Lenz 2016). What matters are the beliefs of the actors about what these rules really are and what kind of stable interaction these beliefs produce. The interactions that these beliefs maintain affects the implementations and real social outputs of regulations in a specific context (Glückler and Lenz 2018). Thus, they have a context-dependent nature, and cannot be easily imitated or copied, just like capital, labor, technology, and coded information (Glückler and Lenz 2016). Finally, *institutions are not organizations* (Bathelt and Glückler 2014, 7), such as local and regional governments, business firms and associations, unions, NGOs, universities, affecting the economy in a geographical setting. Thus, institutions and organizations are analytically distinct concepts and there is a need to distinguish them and analyze as separate entities but in a relational way (cf. Zukauskaite, Trippl, and Plechero 2017). On the other hand, organizations are those who *"benefit from institutions"* in their interactions with others, whether they are individual or collective, public or private (Glückler and Bathelt 2017; Glückler and Lenz 2016). Whereas institutions are "the stability or correlation of interactions between the individual or collective actors in economic and social processes" (Bathelt and Glückler 2014, 346) and as Zukauskaite, Trippl, and Plechero (2017) stated, if organizations can be successful, they also have the potential to change existing institutional conditions.

Relational perspective to institutions–organizations–regulations (see Table 7.1), thus, provides us with the opportunity to examine the differences between formal regulations and actual interaction practices in specific contexts (Glückler and Bathelt 2017). The key point to be considered is that institutions cannot be observed empirically like regulations and organizations, rather they can be known through observing patterns of interaction in similar situations and are therefore perceived from practice, i.e., observed from actual practice. At this point, it should be noted here that it is problematic to transfer findings from one level of aggregation to another (Bathelt and Glückler 2014): "Therefore, interpretations of macro-scale trends in economic and societal development should be based on, at least, some micro-scale evidence related to practices of economic action and the social relations through which these practices are channeled" (Bathelt and Glückler 2014, 14). As a result, interactions between actors are the most appropriate units of analysis to understand the foundation of institutions in a given context (Glückler and Lenz 2016).

7.2.3 Interaction Types between Regulations and Institutions

The first important step in the analysis of an institutional context is to define each of its three components (regulation, organization, institution) making up the institutional context. Next comes the assessment of the quality and nature of the interaction between these components (Glückler 2020). At this point, we defined the first step in the previous section, and now, as the second step, we can focus on the type and nature of the interaction and relationship between institutions and regulations. In this context, Glückler and Lenz (2016, 265–270) defined four different interaction patterns that emerged between the institution and the regulations: The first type of interaction is mutual *reinforcement*, in which effective enforced regulation and institutions working towards the same goal support each other. The second type is *circumvention* in which effective rule is made against an existing institution or when a divergent institution emerges in response to an enforced rule. The third type is *substitution* in which aiming for the same goal, institutions develop in response to an ineffective or absent rule. The last one is *competition* in which institutions and ineffective regulation are at odds with each other, i.e., they differ with regard to their incentives and outcomes.

As can be seen, they identified four types of interaction depending on whether the aims of the rules and institutions are convergent or divergent, and whether the rules are enforced or non-enforced. These forms of interaction show us that the formal rules and policies applied may not always produce outcomes that are fit for the purpose for which they were produced and that positive or negative outcomes can be directed by the stable interaction patterns developing with or against the rules.

7.2.4 Institutional Change and Interactions among Organizations and Regulations

After revealing the components of the institutional context and the nature and types of the relationship between these components, it is necessary to analyze the institutional change in order to understand how the regulations, institutions, and organizations evolve and change over time. In this context, various typologies have been developed in the literature explaining institutional change and its mechanisms or modes. For example, Glückler and Lenz (2018) present an alternative perspective on institutional change by identifying four types of change typologies, namely statis, drift, morphosis and transformation, and distinguishing the change dynamics between institutional form and institutional function in an institutional context. However, the most appropriate approach for our study and context is the analytical framework put forward by Mahoney and Thelen (2010), which has been used frequently in the literature recently. Mahoney and Thelen (2010) introduced a new theoretical framework for the analysis of institutional change by defining four types of change modes (Layering, Drift, Conversion and Displacement) that actors pursue towards rules or organizations.

The institutional change approach and typology of Mahoney and Thelen (2010) does not represent the type of "radical institutional change" that is abrupt, disruptive, and external shocks based on the shaking and deterioration of established structures and institutions (Glückler and Lenz 2018). However, institutional change can occur gradually and in subtle ways to changes in the context based on internal dynamics, as a result of contradictions and tensions in the structure, or because of the power struggle between different interest groups or the enactment of a new rule, law, or policy. Here, the framework of change, which Mahoney and Thelen (2010) called "gradual institutional change," shows the direction, speed, characteristics, and mechanism of institutional change.

According to Mahoney and Thelen (2010), institutions[1] are the object of an ongoing conflict and they evolve and change gradually after they are established. That is, institutions do not emerge or decline overnight, but rather develop slowly and subtly (Zukauskaite, Trippl, and Plechero 2017). Institutions, when established, are not universally adopted and embraced; on the contrary, agents struggle over the form that they should take and the functions they should fulfill (Thelen 2004).

The struggle of the agents over the form and functions of institutions actually stems from the fact that institutions are distributional instruments laden with power implications (Mahoney and Thelen 2010). That is, institutions have distributional consequences for resources, many institutions aim to allocate resources to specific actors, not others (Mahoney and Thelen 2010). Actors with different resources tend to create different types of institutions according to their well-defined institutional preferences, an emerging institution is actually a composite of the relative, sometimes conflicting contributions and expectations of these differentially motivated actors (Mahoney and Thelen 2010). Disadvantaged actors by one institution may use their already advantageous status in other institutions to enact change (Mahoney and Thelen 2010). Struggles over the meaning, application, and enforcement of institutional rules are inextricably intertwined with the resource allocations they entail (Mahoney and Thelen 2010, 11). However, even if the rules are formally codified, the expectations of the rules are often vague, open to interpretation and discussion (Mahoney and Thelen 2010). Therefore, incremental change in institutions is largely due to gaps between the interpretation and application of the rule (Mahoney and Thelen 2010; Thelen 2004).

Mahoney and Thelen (2010) identified four forms of institutional change based on their relevance to the political context and rules (see Table 7.2). The first form of institutional change is *displacement*, which is the removal of existing rules and their rapid or slow replacement with new ones in the political context shaped by weak veto possibilities (Mahoney and Thelen 2010). Displacement occurs as new models emerge and spread that challenge previously accepted organizational forms and practices (Streeck and Thelen 2005). New institutions are often introduced by the losing actors of the pre-change system (Mahoney and Thelen 2010). The second is *layering*, which occurs when existing institutions or rules remain the same in a political context with

Table 7.2 The four types of institutional change

		Characteristics of the targeted institution	
		Low level of discretion in interpretation/ enforcement	High level of discretion in interpretation/ enforcement
Characteristics of the political context	Strong veto possibilities	**Layering** Removal of old rules: No Neglect of old rules: No Changed impact/ enactment of old rule: No Introduction of new rules: Yes	**Drift** Removal of old rules: No Neglect of old rules: Yes Changed impact/ enactment of old rule: Yes Introduction of new rules: No
	Weak veto possibilities	**Displacement** Removal of old rules: Yes Neglect of old rules: – Changed impact/ enactment of old rule: – Introduction of new rules: Yes	**Conversion** Removal of old rules: No Neglect of old rules: No Changed impact/ enactment of old rule: Yes Introduction of new rules: No

Source: Adapted from Mahoney and Thelen 2010.

strong veto possibilities but amendments, revisions, or additions to existing ones occur (Mahoney and Thelen 2010). In a political context shaped by strong veto possibilities, institutional challengers lack the power to remove existing rules, but defenders of the existing institutional structure also lack the power to prevent changes and additions to the rules (Mahoney and Thelen 2010). In the case of *drift*, the third type of institutional change occurs when the formal rules embodied in institutions remain constant, but the outcomes of these rules change. The trigger for change is context discontinuity, that is, the emergence of shifts in external conditions that existing institutions cannot sufficiently adapt themselves to cope with (Hacker, Pierson, and Thelen 2015). Namely, institutional change results from the failure to adapt and update an institution (Mahoney and Thelen 2010). The last type of change mode is *conversion*. Conversion occurs when the rules remain formally the same but when actors interpret these rules to fit their own interests, directing institutions towards goals beyond their original intentions. In the conversion type, not involved actors in the creation of the rules try to lead them to new goals, and the trigger of change is actor discontinuity, not context discontinuity as in the case of drift (Hacker, Pierson, and Thelen 2015).

As a result, in general, gradual institutional changes are seen as less important than external and abrupt ones but they can be equally consequential in terms of their outcomes. As can be seen, in this approach, each of the institutional change mechanisms characterizes a specific endogenous process that either transforms existing institutions in response to a new context (such as layering and conversion) or leads to gradual erosion (for example, drift) and eventual collapse (for example, displacement) (Glückler and Lenz 2018).

7.3 Background and Case Study: New Regional Development Policy and RDAs in Turkey

Focusing on the general view of the new regional development policy developed with the EU harmonization process in Turkey, this section primarily addresses the aspects, scope, and objectives of the new regional policy differing from the traditional regional development policy in the country, following by an explanation of the roles, structure, and functions of the RDAs as the main implementers of the new regional development policy of Turkey.

Turkey's unitary governance structure, despite the high population and chronically deep regional disparities, has a politically organizational gap at the regional level. Additionally, it has been characterized by the high inequality of political and administrative power between the central and local levels, and by the positional strength of the bureaucracy in making decisions regarding the allocation of authority and resources weakening the power of the political elite has become a stable institution (Luca 2017).

Turkey's traditional governance structure, on the other hand, has been reformed through the comprehensive arrangements initiated in 2004 with the aim of transforming the central and hierarchical administrative system into a decentralized, participatory, transparent, sensitive, and accountable one (CEC 2004). Furthermore, reforms transforming the traditional regional development paradigm (see Table 7.3) have been launched in Turkey in order to fulfill the requirements of the EU harmonization acquis and benefit from EU structural funds. Although there was an organizational gap at the regional level in Turkey until the early 2000s, it had a regional policy tradition, in accordance with the characteristics of the old regional development paradigm, which was dominated by largely space-blinded policies focusing on promoting sectors determined with a one-size-fits-all approach and was carried out with a central planning framework by the State Planning Organization (DPT) established in 1960 (Dulupçu 2005; Yavan 2011; Sezgin 2018). The regional policy shaped by the planned development approach, moreover, was addressed on the basis of providing a balanced income distribution, eliminating the imbalances between regions, and enabling the underdeveloped regions to develop faster. On the other hand, between 1980 and 1999 there was no fundamental development or change in the regional development approach in the plans that continued in the planned development period until the start of the full membership process to the EU in 1999. As a result of the planned period between 1960 and 1999, with a top-down approach, within the scope of priority region for development (KÖY) policies (Yavan 2011) aimed at developing the less-developed regions, practices such as providing relatively wage advantages for public employees in underdeveloped regions, giving

Table 7.3 Old and new regional development approach in Turkey

	Traditional policy/paradigm	*New policy/The results show paradigm*
Approach	• Centralized, top down	• Participatory, bottom up
	• Passive institutions and individuals	• Subsidiarity, local ownership and acceptance
	• Micromanagement	• Enhancing and activating local partners
		• Multi-layer governance
Objectives	• Balancing regional development levels	• Improving regional competitiveness
	• National economic growth	• Supporting internal growth dynamics
	• Encouraging investments in less developed regions	• Enhancing social and human capital
	• Building main infrastructure	• Improving institutional capacity
Focus	• Less developed regions	• All regions
Organization	• Central government and its regional branches	• Central government, local governments, RDAs, BKI, YDO, business environment, NGOs
Tools	• Comprehensive regional planning	• Strategic regional planning
	• Large scale infrastructure investments	• Action plans/program
	• Sectoral subsidies	• Program based financial and technical support
	• Prioritized regions policies	• Regional subsidies
		• New support mechanisms (Das, CMDP, KOYDES, BELDES, SODES)

Source: Adapted from Kalkınma Bakanlığı (2012: 13).

more advantageous investment incentives to relatively less-developed regions, investing in these regions, and providing the basic infrastructure were implemented. Despite all these policies and interventions, it is seen that no significant contribution has been made to the elimination of regional development disparities, which date back much further (Kalkınma Bakanlığı 2014).

It is possible to consider the accession process of Turkey to the EU in the 2000s as an important threshold in terms of its regional development policy (Ertugal 2017). On the one hand, with the aim of implementing the above-mentioned EU Accession Partnership Documents and the recommendations in the EU Progress Reports, and on the other hand, with the aim of integrating the "new regionalism" approach based on endogenous growth, cooperation, and innovation (Dulupçu 2005; Sezgin 2018), which has emerged all over the world, into Turkey's regional development policy. The initial turning point is establishing the NUTS classification for regions. The more important ones are Law on the Establishment of RDAs at the regional level was enacted in 2006 (TBMM 2006) with a radical arrangement aiming to develop decentralized and place-based policymaking practices in regional development policy, and then the establishment of RDAs in 26 NUTS2 regions was completed in 2009.

With the establishment of RDAs as the main implementers of the new regional development policy, there has been a paradigm shift (Ertugal 2018; Sezgin 2018), in Turkey's regional development approach, which changes the almost 50-year traditional planning tradition shaped by a top-down approach and by sector-based incentives (DPT 2006; Kalkınma Bakanlığı 2013, 2014). As seen in Table 7.3, Turkey's new regional development policy adopting participatory and multi-level governance mechanisms aims to both increase the competitiveness of regions and reduce regional inequalities by mobilizing their own internal resources instead of balancing regional development levels and promoting investments in less-developed regions. Moreover, instead of large-scale infrastructure investments, sector-oriented incentive mechanisms, and priority region for development (KÖY) policies, it aims to provide program-based financial and technical support and new incentive support mechanisms to all regions within strategic regional planning (see Table 7.3).

RDAs, defined as policy implementation tools by bureaucrats working in the field of regional development (Ertugal 2017), have been established to increase the capacity for participatory planning, programming, project production, and implementation at the local level, supporting endogenous growth dynamics, improve social and human capital, and institutional capacity, and encourage each region to prepare its own regional plans in a place-based, bottom-up approach and participatory processes (Kalkınma Bakanlığı 2012, 2014). Today, 26 RDAs (see Figure 7.1), as a non-typical public organization that can act in a flexible and dynamic manner in terms of the tender, employment, dismissal, and wage policy and expenditure procedures (TBMM 2005), continue to operate in Turkey. The organizational structure of RDAs consists of the Development Board, as the advisory body, the management board, as a decision-making body, as well as the general secretariat, as an

Figure 7.1 Geographical distribution of 26 RDAs in Turkey.
Source: KAGM 2021.

executive body, and established with the same organizational form in all regions. As of 2020, RDAs employ 955 major service workers, including 26 general secretaries, 155-unit heads (including YDO coordinators), 11 legal advisors, 5 internal auditors, 520 experts, and 238 support personnel (KAGM 2021). RDAs are structured as organizations with flexible legislation based on public sector-private sector-civil society partnerships, with implementation efficiency. They create regional plans and strategies with participatory methods, work on improving the investment environment, organize capacity-building training for local and rural development, and project-based financial and technical incentives and support to both public organizations and private sector companies and NGOs in reaching regional priorities.

Additionally, steps toward institutionalization in the new regional policy have been continued with the establishment of three more Presidency of Regional Development Administrations (PRDA) in different regions (DAP PRDA, DOKAP PRDA, and KOP PRDA), and the establishment of the Regional Development High Council (BGYK) under the chairmanship of the Prime Minister as the highest decision-making unit on a regional scale and the Regional Development Committee (BGK). In 2014, the Regional Development National Strategy (BGUS), action plans of PRDAs, and regional plans prepared by RDAs for 26 NUTS-2 Regions were approved by BGYK and enforced.

Through these improvements, Turkey's new regional development policy is conducted with national sectoral strategies in accordance with the national development plan, and regional plans consistent with the priorities specified in the BGUS. This framework also has implementing organizations, which manage the goals, objectives, and actions of these plans and actions at the local and regional level, including 26 RDAs, 4 PRDAs, 81 Investment Support Offices (YDOs), provincial organizations of the central government, and other local organizations (KOSGEB, etc.) (see Figure 7.2).

On the other hand, 2013 is seen as an important breaking period in terms of regional development in Turkey. The new regional policy in Turkey

Figure 7.2 New regional development policy framework of Turkey.

Source: Prepared by authors based on (Kalkınma Bakanlığı 2014, 141).

initiated after the EU membership process and gained momentum especially with the establishment of RDAs in 2006, and also showed a serious change and development in terms of approach, purpose, and implementation tools. In contrast, this momentum and wind of change have become reversed in parallel with the decreasing importance of Europeanization intentions and targets in the national policy agenda, as the country became introverted and centralized over time since 2013. The loss of focus against the regional development policy has been sharpened by the 2013 Gezi Park Protests, the centralization tendencies that gradually intensified with the coup attempt of 15 July 2016, and the 2018 Presidential Government System, in which these tendencies were embodied on a legal and institutional basis. As a result, the tendency of Europeanization has decreased considerably, and a centralized and security-oriented governmental approach at all levels has again become extreme. Following the disappearance of the EU accession initiative, objectives such as activating the local potential, improving participatory planning, and involving local actors in resource use within the regional plans and regional innovation strategies (RIS3) implemented by RDAs and PRDAs, have been off the agenda.

This situation manifests itself most concretely in the abolition of the BGYK and BGK in 2018 and the closure of the Ministry of Development (MD) as well as in the 11th Development Plan. As a matter of fact, after the abolition of the MD, the Strategy and Budget Presidency (SBB), which is subordinate to the Presidency to a large extent, took over the duties of the Directorate General of Regional Development and Structural Adjustment (formerly responsible for both regional development policy and the coordination of RDAs). However, in the new government system, RDAs have been not affiliated to the SBB but to the General Directorate of RDAs within Ministry of Industry and Technology (STB), which did not have any regional development and planning experience, and there was an apparent decline in the new regional development policy. Also, the centrist approach is noticeable in the 11th Development Plan, which is still in effect today, and both the BGUS and the regional plans were not sufficiently taken into account and were not effective in determining the strategies (SBB 2019). In addition, the existing Development Boards, which were dissolved after the coup attempt (2016), have still not been re-established in 23 agencies, with the exception of the RDAs in the three metropolitan cities.

7.4 Empirical Findings: Institutional Context and Change of New Regional Policy and RDAs in Turkey

Adapting to comprehensive changes in regional policy in Turkey following the EU harmonization process is likely to be one of the greatest challenges for institutional actors with long-term old-regional policy practices. Examining adaption difficulties to this new regional policy, deviations in its projected results, and the various reactions of institutional actors to the new regional policy has become a significant concern for regional studies and policymakers. However, focusing only on the regulations – as in other studies – which is one of the building blocks of the institutional context, will not suffice to identify variances in the adaptation to this new regional policy and deviations from its projected outputs. Considering the requirement to address the interaction between the components of the institutional context to observe the different results of a policy in practice, in this section, we revealed Turkey-specific institutionalized interaction patterns developing with or in response to the new regional policy using the conceptualization of Glückler and Lenz (2016). In addition, inspired by Mahoney and Thelen (2010), who define multiple institutional strategies pursued by actors about existing and emerging institutional structures, we analyzed which institutional change strategies are pursued by national and regional actors under the observed stable patterns of interactions.

Concerning regulations, the first pillar of the institutional context, we included the regulations on authorities and resources for new regional development policy in Turkey, official governance mechanisms, and national and regional development policies and strategies into our analysis. With our regulation-level analysis, we found that regional governance structures have been rescaled with their main objectives: (1) the establishing policy development processes that enable multi-level governance between central and local levels in the formulation of regional policies, (2) the improving democratic participation culture, and (3) the enabling regional collaboration between public–private–civil society actors (DPT 2006; TBMM 2005; Kalkınma Bakanlığı 2014). Additionally, within this new regional policy, it is also foreseen that RDAs would act as a common platform for the actors in the regions, which is participatory, gives importance to effective representation, ensures cooperation in the decision-making and implementation stages, and thus reinforces local ownership (DPT 2006; Kalkınma Bakanlığı 2013; TBMM 2005).

As a result of our institutional analysis, we found four different stable patterns of interaction competing the new regional policy rules, affecting the social consequences of the new regional policy and the effectiveness of institutional policy-making: first, *centralism, low trust, and disbelief for participation*, as the stable pattern of interaction undermining the vertical and horizontal dimensions of multi-level governance in regional policy development. The second one is *the stable pattern of disinterest in regional policies*, especially among public organizations. This stable pattern radically affects the objectives of aligning dispersed regional development resources with integrated and synergetic effects through the possessiveness of regional plans and RDAs as one of the most important outputs expected by the new regional policy. The third is the *repeated interaction type to consider RDAs as organizations only distributing financial support*. Regional organizations generally interact with RDAs in order to get financial support, which weakens their role as mediator, organizer, and catalyst that they are expected to assume in the region. The last one is the *widespread tendency to ignore long-term regional interests driven by short-termism*, leading to deviations from the target of regional policy to distribute the financial resources and support of RDAs in line with the needs determined.

7.4.1 Centralism and Low Trust, Disbelief for Participation

The new regional policy has been enforced in a *competing* direction with the institutionalized stable pattern of centralism and low trust triggered by the power distance between the central and local actors stemming from long-term and established centralist practices (Glückler and Lenz 2016). Unsurprisingly, competing with the expectations of multi-level governance between the center and the local, this institutionalized pattern was frequently emphasized in the interviews with both central-level actors and regional-level actors. Moreover, we observed that this institutionalized stable pattern has led to continually exceeding the limits of the statutory proposed coordination function of the center (ministry). This stable pattern can be clearly seen in one interviewee expression that "centralism is a corrective action and decentralization is not a condition or a necessity in all cases in regional policy development" (MCHD1).[2] Furthermore, social practices types of this stable pattern can be seen in the ministry's interventions in the decisions of the Management Boards of DAs, from decisions of RDAs about increasing the number of personnel in investment support offices to the final decisions related to guided projects (R1DE; R2DE; R3DE; R4FSG). Therefore, it can be commended that such types of interventions directed by centrist beliefs undermine the autonomous character of DAs and give them the appearance of a hierarchical organization of the ministry (R1CSG; R2CSG; R3FSG).

These observed centralism and low trust towards are also typical interactions patterns found in previous empirical studies and reports. Our findings are consistent with findings of the evaluation note prepared by Köroğlu (2011) after the meeting with regional and central actors in 2011, which establishes that centralism prevents RDAs from playing an active role in local public policy options. Similarly, our findings are consistent with those obtained by the report published by the Presidency State Supervisory Board (DDK) in 2014 based on interviews with 217 personnel from RDAs, a survey conducted with 630 agency personnel, and interviews with governors. Supportively, DDK (2014) found that centralism negatively affects the main qualities and accountability of the board of directors as the decision-making body and the vitality of the flexible and participatory structure included in the founding philosophy of RDAs. Likewise, this finding was also observed in the Audit Reports of the Development Agencies prepared by the Court of Accounts (Sayıştay 2021). Moreover, a similar interaction pattern was observed in Ertugal's (2017) study based on 37 semi-structured interviews conducted between 2011 and 2014. Ertugal (2017) stated that 32 interviewees complained that "centralization is still the norm" and argued that the central interventions prevent eliminating the development traps in local power relations by exceeding the role of coordination.

With regards to participation, the rules regulating the horizontal dimension of Turkey's new regional governance has envisaged improving democratic participation and reinforcing the appropriation of regional decisions at the local level with the 100-person (from public sector, private sector, and NGOs) Development Boards (Kalkınma Bakanlığı 2014; TBMM 2005, 2006). Previous empirical findings (DDK 2014; Ertugal 2017; Sayıştay 2021) and our findings provide important insights into an institutionalized stable pattern that RDAs are dysfunctional and ineffective to carry out a continuous and interactive participatory process in regional policy.

"The high interest and ownership in the first years of the establishment of the RDAs" (R1FMDB) have decreased considerably over time; the participation in the meetings has decreased below 50 percent (R5DE). In spite of the high number and quality of participants in the Development Boards in the initial years of RDAs, we observed the prevalent perception that the Development Boards were ineffective and unauthorized boards in due course among both members of Management Boards and Development Boards (R1DE; R2FMDB; R5DE). Since the decisions of Development Boards are in the nature of advice and precatory in the regulation and policy documents has strengthened this perception (R2DE). With regards to social outputs, this stable interaction has led to the Development Board meeting being considered ordinary and redundant meetings by regional organizations and affected to gradually decline in the interest of the participation of members in the Development Board. Furthermore, states such as not sending participants from the organizations to the meetings or the attendance of whoever is appropriate at that time in the organizations in question to each meeting have become permanent (R3DE; R4DE).

Although, the aim of weakening the centralism patterns envisaged within the new regional policy has worked for a while (R3FSG), due to the uncertainties in the law and legislation regulating the coordination relationship between the ministries and RDAs, central (ministry) actors, who were not largely involved in the creation of the new regional policy, acted as veto actors and followed a "conversion"-type institutional change strategy in the process (Mahoney and Thelen 2010). For example, regarding the vertical dimension of multi-level governance, the Ministry has been going beyond the limits in the rules regulating the tutelage (supervision) relation among Ministry and RDAs and has been implementing the existing rules in a way that leads to a more intense and hierarchical control effect (R1DE; MFHD1; MFHD3). In fact, from time to time, with the orders and instructions sent to the RDAs, suggestions are made about how the regional transactions, programs, and actions should be done (Sayıştay 2021). Similarly, in the horizontal dimension of governance, the lack of the enforced dimension of the rules regarding Development Boards and the tendency of the actors in the management board of RDAs to consider Development Boards insignificant have converted the practical function of these boards (Mahoney and Thelen 2010).

Moreover, after the EU harmonization agenda lost its importance and then the increasing centralization due to the shifts in country conditions since 2013, this institutionalized belief that competes with the predictions of the new regional policy has led to the activation of some drift-type institutional change strategies for regional policy (Mahoney and Thelen 2010). A prominent example of this is that after the military coup attempt in 2016, all Development Boards in 26 regions were dissolved. Although the rules regarding the duties, powers, and establishments of the Development Boards in the RDAs Law remain the same, these rules have been neglected since 2016, and the Development Boards have been re-established in agencies consisting of a single province, on the other hand, they have not been established yet in the remaining 23 RDAs. One interviewee expressed this drift-type change strategy as follows:

> The aim was to bring the decision-making processes closer to the local, and we gave up this ability and purpose to a certain extent over time. Turkey turned to more domestic politics and tended to centralize due to various risks, and the central intervention in the locale had to increase gradually over time.
>
> (MCHD2)

7.4.2 Stabilized Disinterest about Regional Policies

Another institutional pattern is the disinterest in regional policies among regional organizations. Since the plans, programs, and activities of many public and private organizations being scattered and lacking in synergy in a way that does not create a holistic-cumulative effect, the new regional policy of Turkey aims to stabilize the effective aligning of dispersed development resources and expertise through cooperation among organizations (DPT 2006; TBMM 2006). In line with, in order to transform the weak cooperation culture and to ensure the target-oriented alignment of the resources in line with the regional plans, the RDAs has been expected to act as coordinators, organizers, and catalysts, and to bring together the actors, resources and local dynamics with a place-based leadership (DPT 2006; TBMM 2006).

However, observed widely in our interviews, as a stable institutional pattern, the widespread perception among (especially) public organizations and other stakeholders that RDAs are ineffective and dysfunctional structures, and prevents that "the regional plans prepared by the RDAs are taken into account by the regional stakeholders" (R1DE). Moreover, "the regional plans are perceived only as of the annual activity and implementation plan of the RDAs" (R5DE), and "regional plans are not adopted as a reference point for regional public organizations" (R4DE). Therefore, this stable interaction pattern has been hindering their capacity, as a coordinator, to bring together the organizations (municipalities, provincial units, and support organizations such as KOSGEB) in the regions. Thus, it has been resulting in a deviation from the expected output of new regional policy about efficient and effective use of human and financial resources (R1CSG; R2CSG). Similarly, Köroğlu (2011) and DDK (2014) established that this stabilized institutional pattern, as we observed in the interviews, weakened the coordination task of the RDAs in the regions and this increased the irrelevance of the financial supports distributed by the regional stakeholders to the regional plans.

We found that this institutional pattern, which competes with the projected outputs such as coordination, cooperation, and resource alignment for a regional common vision in the new regional policy, is triggered by the uncertainties and deficiencies in the regulations and policy strategies shaping the regional policy and planning (Glückler and Lenz 2016). Indeed, when we examine the regulations for planning in Turkey, it has been observed that there is no clear and specific legislative arrangement about regional plans' the definition, scope, content, preparation, implementation, parties, bindingness, sanction, relationship with other plans and place in the planning hierarchy (cf. Cumhurbaşkanlığı 2018; TBMM 2006). Additionally, in support of this finding, the point frequently emphasized in the interviews is that the uncertain bindingness of the regional plans and the lack of enforcement and sanction power on other regional public organizations are the main obstacles to bringing together the organizations in the region in line with the plan (R1DE; R1CSG; R2DE; R2CSG; R3FSG).

Since there is a lack of adequate legislation and authority regarding the bindingness of regional plans (R4DE; R5DE), the expectation in new regional policy that regional public organizations should carry out their activities according to "the priorities in the regional plans remains only at the level of a declaration of intent" (R2DE). In addition, "the absence of a mechanism monitoring the compliance of the sub-scale plans with the regional plans and prescribing any sanctions if they are not appropriate" (R5DE) serves as the challenging force behind that RDAs cannot mobilize other regional public organizations to align their human and financial resources in line with the requirements of the regional plans. One interviewee commented with regards to this stable pattern:

> In the plans we prepare, we determine actions that are compatible with the future of the region, specific organizations, and target dates for these actions, but these actions are often not followed, each public organization acts according to its own agenda and interests. If the plans are not complied with, we cannot do anything because we do not have operational and legal enforcement power.
>
> (R1DE)

Similarly, in Köroğlu (2011), DDK Report (2014), Ertugal (2017), and Court of Accounts (Sayıştay 2021), there is evidence that the lack of bindingness of regional plans on national investments and local plans reinforces the institutionalized perception among regional organizations that RDAs are ineffective and dysfunctional structures. On the other hand, the BGYK and the BGK, which are expected to ensure organizational integration, were established in 2011, in order to convert this perception. With these organizations, it has been expected that they make regional plans visible and significant for the provincial organizations of the center through taking them into account at central government level. However, these Council and Committee were convened only once until its abolition in 2018 by neglecting the relevant regulations and rules. This institutional 'drift' (Mahoney and Thelen 2010) has resulted in missed opportunities that can significantly affect integration and coordination of regional organizations.

Moreover, the failure of regional organizations adopting the targets of regional plans has also been reflected in the practices of the RDAs. Although there were important changes such as the completion of the "Tenth Development Plan" covering the years 2014–2018 and the transition to the Presidential system, none of the regional plans implemented by RDAs in 2014 were revised as of the first half of 2021. In addition, any study has not been conducted to determine whether there is a need for revision in regional plans or not (Sayıştay 2021). Today, although some of the agencies emphasize this need (R1DE; R5DE), regarding reasons for the non-revision of plans, regional actors said that there was no official request from the STB as well as that even if the regional plans were renewed, there would not be an effective result since regional plans could not be fully implemented.

In addition, the conflicts of plans at the ministry level, especially the STB (former MD) and the Ministry of Environment and Urbanization, as well as lacking sufficient consideration of the regional plans and their priorities by ministries, are other causal factors preventing the coordination of too many organizations responsible for the same expertise and public service areas in the regions. Regarding this situation, one regional actor said that:

> since ministries do not include regional plans in their strategies, provincial units of ministries at the regional and local level do not care about these plans and therefore RDAs, and they act largely with orders from the central level for the plans and actions we foresee.
>
> (R3DE)

Moreover, all interviewees agreed that the motivation, expectation, and political support towards the RDAs during the period they were founded decreased gradually and this was reflected in the interactions of the regional organizations with the RDAs. So much so that in the new presidential system, after the MD was abolished, becoming RDAs' hierarchically organizational part of STB has created uncertainty and conflict in the expectations of regional public organizations towards the duties of the RDAs. In our interviews, regional actors often underlined that RDAs carry out wide-ranging projects such as rural tourism, social projects, and rural development, but these are not compatible with the duties of the STB and these incompatibilities are questioned by both the central and local actors. This organizational conversion has contributed to decline in the possessiveness of the RDAs by regional public organizations (R2DE; R2CSG; R4DE; R5DE). In response to this problem, almost all regional actors stated that being directly affiliated to the Presidency instead of the STB would be an important step in the new government system in order to activate the organizations in the region and to increase the bindingness of the RDAs.

7.4.3 Expectation of Obtaining Financial Support

Deficiencies in regulations, practices, shifts in external conditions over the years have led to the ignoring of expectations, like being coordinator, improving decentralization, providing technical support, from the RDAs in the years of establishment of RDAs. We observed that in parallel with Köroğlu (2011) and the DDK Report (2014), there is a deviation between the missions of the RDAs in regional policies and strategies and the perceptions of the local stakeholders about RDAs' duties. RDAs have become the focus of expectations for financial resources and grants by actors in the regions (R1DE; R2DE; R2B; R3B; R4FSG; R5DE; R5B). In other words, there is a perception among regional actors that the main task of RDAs is to distribute financial resources (R1DE; R5DE). Both the beneficiaries and the RDAs personnel have repeatedly emphasized that the expectation of providing financial resources directs the interaction of regional stakeholders with the agencies (R1DE; R2DE; R3DE; R5DE). One interviewee clearly expressed this institutionalized perception:

> In the first years, it was thought that the grants should be used as a trump card in order for RDAs to be accepted locally, regionally, and nationally. This benefited the awareness of the RDAs but diverted the RDAs from their original purpose. Because agencies are known by many organizations, officials and citizens in regions as an organization providing only (financial) support.
>
> (R3FSG)

Showing the general interaction of local governments, another interviewee stated that: "municipalities interact with RDAs until they receive financial resources, they do not interact with the RDAs after receiving the resource" (R2CSG). This expectation, which emerged and stabilized a while after the establishment of the agencies, has led to a 'drift'-change of the organizational structure of the RDAs (Mahoney and Thelen 2010). Over time, the majority of the personnel employed in the RDAs concentrated on the support and monitoring units and carried out the task of monitoring only the projects financially supported by the RDAs rather than other projects in the region (DDK 2014; KAGM 2021; Sayıştay 2021).

7.4.4 *Ignoring Long-term Regional Interests Driven by Short-Termism*

In our interviews, we observed widespread organizational practices that financial resources are not distributed rationally, which competes from the regional policy foreseeing the distribution of finance in line with the needs determined as a result of comprehensive analysis of the region (DPT 2006; Kalkınma Bakanlığı 2013, 2014; R1B; R2DE; R3FSG). Ignoring long-term regional Interests driven by short-termism has become an institutionalized pattern. It was frequently emphasized by both the beneficiaries and the agency staff that the priorities and designs of the financial support programs were not determined according to objective criteria (R1B; R5B) and that the resources were not allocated in a way that would allow the selection of the right investments and in the most appropriate manner to the regional needs (R3B).

This stable pattern substantially stems from the direct or indirect interventions of Management Board members in the selection of projects from time to time, and sometimes from the pursuing provincial-focused approach rather than the long-terms regional approach and regional interest (MFHD1; R1DE; R1CSG; R5FSG). For example, the Chamber of Industry and Commerce in the Management Board wants to support a certain sector, which sometimes leads to the support of projects that are highly unrelated to the regional plan and the determined priorities (R2DE; R3DE, R5DE). Since the members of the Board of Directors are elected or appointed for a short time, they tend to direct the RDAs according to their own interests in this short period of time and make demands and interventions that do not comply with the RDA goals (R2DE; R3DE, R5DE). Similarly, municipalities in Management Boards exhibit populist attitudes and tend towards projects that can receive a positive response from the electorate in the short term (R5DE, R3DE). In fact, in our interviews, RDAs personnel emphasized that each different governor tends to invest in the sector that is compatible with his own province, and this causes the RDAs to support different sectors every year in a way that is not compatible with holistic and long-term strategies (R1DE; R4FSG). Similarly, Ertugal (2017) revealed that the governors of different provinces in the region tend to support the number of projects that receive grants in their own provinces, unlike the projects in other provinces. Furthermore, in her research, Ertugal (2017) found that the members of the Management Boards compete to influence the distribution of financial resources and from time to time put pressure on the general secretaries.

Similarly, the stabilization of local micro-nationalism (province-focused thinking) was observed in the interviews of this study. Regarding such a type of approach, an expert interviewer stated that they determined the priority sectors in the region with the Smart Specialization Strategy by conducting extensive studies, but the sectors in the governor's own province were encouraged, not the priority sectors determined after the change of the governor (R1DE).

7.5 Conclusion

The purpose of this study was to investigate Turkey's new regional development policy since the 2000s, as well as the activities and practices of RDAs, the main instruments of this policy, on a national scale using the analytical frameworks of the institutional context and institutional change in the economic geography. The "institutional context" consists of three components: regulation–institutions–organizations. The analysis of institutional context employed in this study based on Glückler and Lenz (2016) includes an assessment of each component of institutional context and their interrelationships. Furthermore, an analysis of "institutional change" based on Mahoney and Thelen (2010) was conducted in order to observe which of the four different institutional change mechanisms (displacement, layering, drift, and conversion) and institutional actors (both individuals and organizations) follow within these interrelationships between components of institutional context.

The introduction of the new regional development policy in Turkey and the establishment of RDAs in the regions have resulted in a serious shift and improvements in Turkey from the traditional top-down and largely space-blinded regional development policy to a more place-based regional development policy integrating the bottom-up approach. Although, examining whether these improvements have become a stable pattern of interaction, that is, institutionalized, is not within the scope of this study, our observations and analyses showed us that the RDAs have made significant contributions to the Turkish regions and regional development practice in the following five areas in the context of place-based regional policy: (1) *a project developing culture* has been burgeoned, compared to the past, in the organizations in the regions together with the support activities of the RDAs, (2) *involving a wide range of regional actors*, at least in preparing regional plans for each regions, has been stabilized, (3) regional organizations started to be able to *benefit from international funds* through the technical support of the RDAs, (4) RDAs have become *knowledge pools* with detailed knowledge of each region, and (5) a widespread consensus among national and local actors that *incentives should now be distributed locally* rather than nationally as in the past.

On the other hand, we observed four basic institutions that restrict and even sometimes hinder the effectiveness and applicability of burgeoning place-based regional development policy in Turkey, by considering the relations of convergence and divergence between institutions and regulations, as well as examining the ways in which organizations and regulations interact and change over time. The first one is *centralism, low trust, and disbelief for participation*, which hinders flexibility between the national and regional level and between regional agents, risks regional agents working together, and complicates aligning effectively regional resources for a shared vision. With these institutional patterns, conversion strategies have been pursued on uncertain provisions regulating the coordination among the ministry and RDAs, resulting in increasing power distance in favor of the center (ministry) in multi-level governance, and also governors as the center-connected agents. The second one is *disinterest (especially among public organizations) in regional policies*, which is supported by the widespread perception that RDAs are increasingly ineffective and dysfunctional structures. Generally, the weaknesses in the power of sanction regarding the relations of RDAs with other public organizations, the uncertainties in terms of the bindingness of their regional plans as well as particularly the decrease in interest in the new regional policy developed simultaneously with the post-2013 de-Europeanisation process have made integration and coordination among public organizations difficult, and a significant part of the organizer and catalyst role of RDAs has resulted in disappointment. The third one is the *perception among regional actors that the key task of RDAs is to distribute financial resources.* RDAs have highly become the focus of expectations for financial resources and funds by actors in the regions. Especially, this stable interaction has triggered the drift-type organizational change in the RDAs, and the staff employed in the RDAs were concentrated in the financial support and monitoring units in due course. Finally, as a stable institution competing with policy targets regarding investments and financial resources in the region, we have observed *ignoring long-term regional interests driven by short-termism.* It is a stable institution that is highly 'competing' with the expectation of the new regional policy's goals of supporting areas that are consistent, realistic, and complementary to regional needs, having the risk of a series of interventions inconsistent with regional plans.

As a result, this study demonstrates that the regional institutional context has not adequately responded to Turkey's new regional development policy change after 2000, and that, despite some promising and meaningful changes, there were serious difficulties and deficiencies in the overall realization of policy objectives. Analyzing the institutional context of Turkey's new regional development policy with the interrelationships of regulations, institutions, and organizations, this study extends our knowledge of stable patterns of interactions, called institutions, as causal factors behind the limited compatibility to institutional context and ineffectiveness of RDAs and policies. In this way, our findings about stable patterns of interaction have supported the approaches and analytical frameworks of Bathelt and Glückler (2014) and Glückler and Lenz (2016, 2018), who propose to analyze institutions from the perspective of relational economic geography.

Further research analyzing institutional context and institutional change should employ a multi-scaler (cf. Grillitsch and Rekers 2016) and multi-actor approach (cf. Hassink, Isaksen, and Trippl 2019), which has recently been emphasized in the institutional, relational, and evolutionary economic geography literature, as well as comparative case study methodology (cf. Gertler 2010, 2018). In this context, it is considered very important and critical to conduct research through a "regional lens" beyond the national scale, especially in the analysis of institutional change (Gertler 2018) because certain types of institutional changes can be observed in some regions and not in others (Zukauskaite, Trippl, and Plechero 2017). In this direction, it would be interesting to assess whether any of the institutionalized patterns obtained from this study vary in regions and whether they differ according to the development level of regions. Further studies can examine region-specific situations of each of the identified institutionalized patterns, by involving more types of actors (beneficiaries, NGO representatives, provincial organization representatives, etc.) through a longitudinal and comparative analysis method. The study of Glückler and Lenz (2018), which offers an alternative perspective on institutional change in order to analyze whether the function and form of an institution has changed, can be an inspiring example for further research. Thus, studies focusing on the institutional context and institutional change, which are still in their infancy in the economic geography literature, will make a meaningful contribution to the field.

Acknowledgments

We would like to thank all interview participants for their valuable insights and time in answering our questions.

Notes

1 Mahoney and Thelen (2010) do not define institutions in a triple classification as Glückler and Lenz (2016) do. Institutions in the insight of Mahoney and Thelen (2010) correspond largely to the concept of regulations and organizations within the framework of Glückler and Lenz (2016).
2 See Appendix 7.1 for more details on all the pseudonyms we used in this part.

Appendix 7.1 Actors and Organizations Interviewed in Depth for the Study

Region-centre	Pseudonym	Position of interviewee	Date of interview	Duration	Type of interview
Region 1	R1DE	Development Expert	02.09.2020	60 min	Zoom – online
Region 1	R1CSG	Current Secretary General	20.04.2021	60 min	Zoom – online
Region 1	R1FMDB	Former Member of the Development Board	10.04.2021	45 min	Zoom – online

(Continued)

Appendix 7.1 (Continued)

Region-centre	Pseudonym	Position of interviewee	Date of interview	Duration	Type of interview
Region 1	R1B	Beneficiary	05.05.2021	60 min	Zoom – online
Region 2	R2DE	Development Expert	10.09.2020	55 min	Zoom – online
Region 2	R2CSG	Current Secretary General	13.05.2021	45 min	Zoom –online
Region 3	R3DE	Development Expert	25.09.2020	60 min	Zoom – online
Region 3	R3B	Beneficiary	14.05.2021	60 min	Zoom – online
Region 3	R3FSG	Former Secretary General	20.05.2021	60 min	Zoom – online
Region 4	R4DE	Development Expert	28.09.2020	55 min	Zoom – online
Region 4	R4FSH	Former Secretary General	25.05.2021	60 min	Zoom – online
Region 5	R5DE	Development Expert	01.10.2020	60 min	Zoom – online
Region 5	R5B	Beneficiary	01.06.2021	45 min	Zoom – online
Ministry	MFHD1	Former Head of Department	01.11.2020	50 min	Zoom – online
Ministry	MCHD1	Current Head of Department	27.03.2021	60 min	Zoom – online
Ministry	MCHD2	Current Head of Department	01.04.2021	60 min	Zoom – online
Ministry	MFHD2	Former Head of Department	01.05.2021	50 min	Zoom – online
Ministry	MFHD3	Former Head of Department	17.05.2021	60 min	Zoom – online

Note: In order to ensure the anonymity of the interviewees, the identities and regions of the interviewees are given as pseudonyms.

References

Acemoglu, Daron, and James A. Robinson. 2012. *Why Nations Fail: The Origins of Power, Prosperity, and Poverty.* New York: Crown Business.

Amin, Ash. 1999. 'An Institutionalist Perspective on Regional Economic Development.' *International Journal of Urban and Regional Research* 23 (2): 365–378. https://doi.org/10.1111/1468-2427.00201

Barca, Fabrizio, Philip McCann, and Andrés Rodríguez-Pose. 2012. 'The Case for Regional Development Intervention: Place-Based Versus Place-Neutral Approaches.' *Journal of Regional Science* 52 (1): 134–152. https://doi.org/10.1111/j.1467-9787.2011.00756.x

Bathelt, Harald, and Johannes Glückler. 2014. 'Institutional Change in Economic Geography.' *Progress in Human Geography* 38 (3): 340–363. https://doi.org/ 10.1177/0309132513507823

CEC. 2004. '2004 Regular Report on Turkey's Progress towards Accession, 6.10.2004,SEC(2004) 1201.' Accessed June 21, 2021. www.ab.gov.tr/files/AB_ Iliskileri/Tur_En_Realitons/Progress/Turkey_Progress_Report_2004.pdf

Cumhurbaşkanlığı. 2018. 'Bakanlıklara Bağlı, İlgili, İlişkili Kurum Ve Kuruluşlar İle Diğer Kurum Ve Kuruluşların Teşkilatı Hakkında Cumhurbaşkanlığı Kararnamesi.' Accessed June 21, 2021. www.mevzuat.gov.tr/MevzuatMetin/19.5.1.pdf

DDK. 2014. 'Türkiye'nin Kalkınma Ajansları Uygulamasının Değerlendirilmesi Raporu.' Accessed June 21, 2021. www.memurlar.net/common/news/documents/ 456927/20140130-2014-03.pdf

DPT. 2006. *Dokuzuncu Kalkınma Planı (2007–2013)*. Ankara: DPT.

Dulupçu, Murat Ali. 2005. 'Regionalization for Turkey: An Illusion or a Cure?.' *European Urban and Regional Studies* 12 (2): 99–115. https://doi. org/10.1177/0969776405048496

Ersoy, Aksel. 2018. *Turkey: An Economic Geography*. London: I.B Tauris.

Ertugal, Ebru. 2017. 'Challenges For Regional Governance In Turkey: The Role of Development Agencies.' *METU Journal of the Faculty of Architecture* 34 (2): 203–224. https://doi.org/10.4305/METU.JFA.2017.2.7

———. 2018. 'Learning and Policy Transfer in Regional Development Policy in Turkey.' *Regional Studies* 52 (9): 1181–1190. https://doi.org/10.1080/00343404.2017 .1417582

Farole, Thomas, Andrés Rodríguez-Pose, and Michael Storper. 2011. 'Human Geography and the Institutions That Underlie Economic Growth.' *Progress in Human Geography* 35 (1): 58–80. https://doi.org/10.1177/0309132510372005.

Gertler, Meric S. 2010. 'Rules of the Game: The Place of Institutions in Regional Economic Change.' *Regional Studies* 44 (1): 1–15. https://doi.org/10.1080/ 00343400903389979

——— 2018. 'Institutions, Geography, and Economic Life.' In *The New Oxford Handbook of Economic Geography*, edited by Gordon L. Clark, Maryann P. Feldman, Meric S. Gertler, and Dariusz Wójcik, 230–242, Oxford: Oxford University Press. https://doi.org/10.1093/oxfordhb/9780198755609.013.12

Glückler, Johannes. 2020. 'Institutional Context and Place-based Policy: The Case of Coventry & Warwickshire.' *Growth and Change* 51 (1): 234–255. https://doi. org/10.1111/grow.12362

Glückler, Johannes, and Harald Bathelt. 2017. 'Institutional Context and Innovation.' In *The Elgar Companion to Innovation and Knowledge Creation*, edited by Harald Bathelt, Patrick Cohendet, Sebastian Henn, and Laurent Simon, 121–138. Cheltenham: Edward Elgar. https://doi.org/10.4337/9781782548522

Glückler, Johannes, and Regina Lenz. 2016. 'How Institutions Moderate the Effectiveness of Regional Policy: A Framework and Research Agenda.' *Investigaciones Regionales – Journal of Regional Research*, 36: 255–277

———. 2018. 'Drift and Morphosis in Institutional Change: Evidence from the "Walz" and Public Tendering in Germany.' In *Knowledge and Institutions*, edited by Johannes Glückler, Roy Suddaby, and Regina Lenz, 111–133. Cham: Springer.

Glückler, Johannes, Roy Suddaby, and Regina Lenz. 2018. 'On the Spatiality of Institutions and Knowledge.' In *Knowledge and Institutions*, edited by Johannes Glückler, Roy Suddaby, and Regina Lenz, 1–19. Cham: Springer. https://doi. org/10.1007/978-3-319-75328-7_1

Grillitsch, Markus, and Josephine V Rekers. 2016. 'How Does Multi-Scalar Institutional Change Affect Localized Learning Processes? A Case Study of the Med-Tech Sector in Southern Sweden.' *Environment and Planning A: Economy and Space* 48 (1): 154–171. https://doi.org/10.1177/0308518X15603986

Hacker, Jacob S., Paul Pierson, and Kathleen Thelen. 2015. 'Drift and Conversion: Hidden Faces of Institutional Change.' In *Advances in Comparative-Historical Analysis*, edited by James Mahoney, and Kathleen Thelen, 180–211. Cambridge: Cambridge University Press.

Hassink, Robert, Arne Isaksen, and Michaela Trippl. 2019. 'Towards a Comprehensive Understanding of New Regional Industrial Path Development.' *Regional Studies* 53 (11): 1636–1645. https://doi.org/10.1080/00343404.2019.1566704

Helmke, Gretchen, and Steven Levitsky. 2004. 'Informal Institutions and Comparative Politics: A Research Agenda.' *Perspectives on Politics* 2 (4): 725–740. https://doi.org/10.1017/S1537592704040472

Hodgson, Geoffrey M. 2006. 'What Are Institutions?' *Journal of Economic Issues* 40 (1): 1–25. https://doi.org/10.1080/00213624.2006.11506879

KAGM. 2021. 'Kalkınma Ajansları 2017, 2018, 2019 ve 2020 Yılı Genel Faaliyet Raporu.' Accessed June 21, 2021. www.sanayi.gov.tr/plan-program-raporlar-ve-yayinlar/faaliyet-raporlari

Kalkınma Bakanlığı. 2012. *Kalkınma Ajansları 2011 Yılı Genel Faaliyet Raporu.* Ankara: Kalkınma Bakanlığı.

———. 2013. *Onuncu Kalkınma Planı (2014–2018).* Ankara: Kalkınma Bakanlığı.

———. 2014. *Bölgesel Gelişme Ulusal Stratejisi 2014–2023.* Ankara: Kalkınma Bakanlığı.

Köroğlu, Tunga. 2011. 'Bölge Planlama ve Bölgesel Kalkınmada Yeni Açılımlar -1: Bölge Kalkınma Ajanslarının Planlama Deneyimleri ve Sorun Alanları. Değerlendirme Notu'. TEPAV.

Lagendijk, Arnoud, Serap Kayasu, and Suna Yasar. 2009. 'The Role of Regional Development Agencies in Turkey: From Implementing EU Directives To Supporting Regional Business Communities?' *European Urban and Regional Studies* 16 (4): 383–396. https://doi.org/10.1177/0969776409102188

Luca, Davide. 2017. 'Boon or Bane for Development? Turkey's Central State Bureaucracy and the Management of Public Investment.' *Environment and Planning C: Politics and Space* 35 (6): 939–957. https://doi.org/10.1177/0263774X16670666

Mahoney, James, and Kathleen Ann Thelen, eds. 2010. *Explaining Institutional Change: Ambiguity, Agency, and Power.* Cambridge: Cambridge University Press.

Martin, Ron. 2000. 'Institutional Approaches in Economic Geography.' In *A Companion to Economic Geography*, edited by Eric Sheppard, and J. Barnes Trevor, 77–94. Oxford: Blackwell.

Nelson, Richard R., ed. 1993. *National Innovation Systems: A Comparative Analysis.* New York: Oxford University Press.

North, Douglass C. 1990. *Institutions, Institutional Change, and Economic Performance.* Cambridge: Cambridge University Press.

OECD. 2012. *Promoting Growth in All Regions.* Paris: OECD.

Pike, A., D. Marlow, A. McCarthy, P. O'Brien, and J. Tomaney. 2015. 'Local Institutions and Local Economic Development: The Local Enterprise Partnerships in England, 2010.' *Cambridge Journal of Regions, Economy and Society* 8 (2): 185–204. https://doi.org/10.1093/cjres/rsu030

Rodríguez-Pose, Andrés. 2013. 'Do Institutions Matter for Regional Development?' *Regional Studies* 47 (7): 1034–1047. https://doi.org/10.1080/00343404.2012.748978

———. 2020. 'Institutions and the Fortunes of Territories.' *Regional Science Policy & Practice* 12 (3): 371–386. https://doi.org/10.1111/rsp3.12277

Rodríguez-Pose, Andrés, and Michael Storper. 2006. 'Better Rules or Stronger Communities? On the Social Foundations of Institutional Change and Its Economic Effects.' *Economic Geography* 82 (1): 1–25. https://doi.org/10.1111/j.1944-8287.2006. tb00286.x

Rodrik, Dani, Arvind Subramanian, and Francesco Trebbi. 2004. 'Institutions Rule: The Primacy of Institutions Over Geography and Integration in Economic Development.' *Journal of Economic Growth* 9 (2): 131–165. https://doi. org/10.1023/B:JOEG.0000031425.72248.85

Sadioğlu, Uğur, Kadir Dede, and Volkan Göçoğlu. 2020. 'Regional Development Agencies in Turkey on the Scope of Governance and Local Elites: An Evaluation after 10-Years-Experience.' *Lex Localis: Journal of Local Self-Government* 18 (2): 371–394. https://doi.org/10.4335/18.2.371-394(2020)

Sayıştay. 2021. 'Kalkınma Ajansları 2012, 2015, 2016, 2017 ve 2018 Yılı Genel Denetim Raporları.' Accessed June 21, 2021. https://sayistay.gov.tr/reports/ category/9-kalkinma-ajanslari-genel-denetim-raporlari

SBB. 2019. On Birinci Kalkınma Planı (2019–2023). Accessed June 21, 2021. www. sbb.gov.tr/wp-content/uploads/2019/07/OnbirinciKalkinmaPlani.pdf

Scott, William Richard. 2013. *Institutions and Organizations: Ideas, Interests, and Identities*. London: Sage.

Sezgin, Ervin. 2018. 'New Regionalism in Turkey: Questioning the "New" and the "Regional".' *European Planning Studies* 26 (4): 653–669. https://doi.org/10.1080/09 654313.2017.1403571

Storper, Michael. 2013. *Keys to the City*. Princeton: Princeton University Press.

Streeck, Wolfgang, and Kathleen Thelen. 2005. 'Introduction: Institutional Change in Advanced Political Economies.' In *Beyond Continuity: Institutional Change in Advanced Political Economies*, edited by Wolfgang Streeck, and Kathleen Thelen, 1–39. New York: Oxford University Press.

TBMM. 2005. 'Kalkınma Ajanslarının Kuruluşu, Koordinasyonu ve Görevleri Hakkında Kanun Tasarısı ve Avrupa Birliği Uyum Ile Plan ve Bütçe Komisyonları Raporları (1/950).' Accessed June 21, 2021. www.tbmm.gov.tr/tutanaklar/ TUTANAK/TBMM/d22/c091/tbmm22091124ss0920.pdf

———. 2006. 'Kalkınma Ajanslarının Kuruluşu, Koordinasyone ve Görevleri Hakkında Kanun.' Accessed June 21, 2021. www.resmigazete.gov.tr/eskiler/2006/02/20060208-1. htm

Thelen, Kathleen. 2004. *How Institutions Evolve: The Political Economy of Skills in Germany, Britain, the United States, and Japan*. Cambridge: Cambridge University Press. https://doi.org/10.1017/CBO9780511790997

Williamson, Oliver E. 2000. 'The New Institutional Economics: Taking Stock, Looking Ahead.' *Journal of Economic Literature* 38 (3): 595–613.

Yavan, Nuri. 2011. 'Yeni Yatırım Teşvik Sisteminin Bölgesel Kalkınma Politikaları Çerçevesinde Değerlendirilmesi.' In *5. Bölgesel Kalkınma ve Yönetişim Sempozyumu, Sanayi Politikasının Yönetişimi Bildiriler Kitabı*, 125–154. Ankara: TEPAV.

Zukauskaite, Elena, Michaela Trippl, and Monica Plechero. 2017. 'Institutional Thickness Revisited.' *Economic Geography* 93 (4): 325–345. https://doi.org/10.1080 /00130095.2017.1331703

Part III

Social Inequalities, Displacement and Conflicts between Social Groups

8 Bureaucrats, Local Elites, and Economic Development

Evidence from Chinese Counties

Ling Zhu and Xueguang Zhou

8.1 Research Issues

In the literature on governance practice in China, local elites have occupied drastically different positions before and after 1949, the year of the founding of the People's Republic of China. Historically, local elites—those in the gentry class and kinship authorities—were central figures in rural governance in traditional Chinese society (Hsiao 1960). Kinships, temples and local governance bodies formed the cultural nexus of power to integrate local communities (Duara 1988). The role of local elites has been a consistent theme as the organizational basis of Chinese society in the vast literature of China studies.

In contrast, the role of local elites has been conspicuously absent since the founding of the People's Republic of China in 1949, with good reasons. The land reform and collectivization in the early years of the People's Republic of China wiped out the gentry class and literati associated with the old regime, and in their place emerged those cadres and officials, at county, township, or even village level, as part of the bureaucratic apparatus of the state. The Party state has reached different corners and arenas, and thoroughly organized the Chinese society by the elaborate, multilayered Chinese bureaucracy, from cities, to counties, down to townships and villages. Even in the reform era, the bureaucratic state has continued its strengthening and expansion and played an active role in all walks of life. As a result, the stratum of local elites appeared to be missing in governance of the People's Republic of China. Or is it?

Research on grassroots social movements and collective actions has shown that those local elites have never disappeared from the social scenes; instead, they have taken on more subtle roles and appeared in different forms. We have glimpses of their figures and maneuvers in the research literature. They appeared as those stubborn figures that led popular resistance to injustice and abuse of power by local authorities (Ying 2001); they were active organizers of informal social institutions (Tsai 2007) who can effectively mobilize resources and get things done. Kinship organizations were suppressed but survived in the collectivization era and prospered again in the post-Mao era (Friedman et al. 1991).

DOI: 10.4324/9781003191049-11

Moreover, local elites found their places in government offices in townships and counties. Given the all-encompassing role of the Party state in society, there is little room beyond the government offices for local elites to play their roles. Those elites are absorbed into the Chinese bureaucracy through several channels, such as the recruitment into the lower rank of local offices and moving up the bureaucratic ladders, entering the civil service after college education, or a mixture of these channels. The defining characteristics of this group of local elites are: first of all, they stay in the same locality for their entire career and life course; second, as a result, they are embedded in rich networks of social relations including marriage, alum, kinship, among others. Many have been recruited into government offices and become bureaucrats at various levels, and some eventually were promoted to those offices at the county level. We label them office-based *local elites* to distinguish them from those local elites who were not incorporated into the government institutions. We will refer to them as local elites whenever the context does not cause confusion.

In this study, we examine and assess the role of office-based local elites in county governments and their effect on local economic development. There are two goals in this study: (1) We identify this distinct group of local elites and their distribution in the local government offices; (2) we assess their role in socioeconomic development within their localities. In so doing, we address the larger issue about the double-identity of the local officials as the agent of the Chinese state and as the representative of local interests, and their role in China's governance.

Empirically, we focus on those leadership positions in county governments broadly defined, including the Party HQs, government HQs, the county People's Congress (CPC), and the county Communist Party Political Consultative Conference (CPPCC)—the so-called *sitao banzi* 四套班子—for several reasons. Most of those local officials were promoted to the present position after their extensive work experience at lower-level positions in the administrative jurisdiction. A large proportion of them have worked in their locality for their entire careers. They play important roles in local economic development, tax collection, pollution inspection, and the maintenance of social stability (*weiwen*), and so on. Our chapter focuses on this group of local officials and examines their roles as local elites.

8.2 From Local Elites to Bureaucrats and Back: Theoretical Arguments

There is a sizable literature on career mobility in the study of Chinese bureaucracy, focusing on bureaucrats that occupy key positions and enjoy sponsored mobility, preferential treatment, and privileged access to opportunities (Walder 1995). In this picture, bureaucrats at different levels are either adhering to the Party line or incentivized in the tournament model to engage in economic development (Landry et al. 2018; Li and Zhou 2005; Zhou 2007).

We argue that there is a need to differentiate two types of bureaucrats in local governments: one consisted of outside officials and the other consisted of local officials respectively. These two groups, we submit, are distinctive in their career patterns, which cultivate different identities and political allegiance. Those local officials in the second group play the role of local elites in the traditional society. They experience their career and social life within the same administrative jurisdiction, and they are deeply embedded in the dense network of multiple, overlapping social relations. This is in sharp contrast to the first group, who are movers transient in and out of a locality in a short period of time. The distinction between the two helps us clarify and address several tensions in the literature, regarding strategic groups in local governments (Heberer and Schubert 2012) and collusive behaviors at local levels (Zhou 2010).

To identify and assess the role of local elites, one needs to appreciate the functions they play in traditional China and how these functions are served in local governments. One challenge to address this issue is that bureaucratization has hidden the trace of local elites. And local elites are now regulated by both the traditional local social institutions and the vertical, bureaucratic organization. It is between these two logics we find the voice and shape of local elites. In so doing, we also propose a corrective to the overwhelming image of formal organizations in terms of incentives and disciplines, such as the tournament model of competition (Landry 2008; Li and Zhou 2005). These images are more useful for a small fraction of the top officials, but less relevant to the majority of local officials who are not directly linked to such performance evaluation, and whose behaviors need to be explained by other logics of action.

The role of local officials in contemporary China is better understood in light of the historic transformation of political order in the Party state since 1949, to which we now turn.

8.2.1 *Local elites in traditional society*

That local elites played a central role in local governance in traditional China has been well documented in the literature. As clearly shown in Hsiao's (1960) authoritative, encyclopedic study of rural China in the late Qing dynasty, local elites—consisting of gentry class, literati, and kinship leaders—provided the infrastructure on which rural China was organized (see also Chu 1965). Kinship organizations, temples, and other local governance bodies formed "the cultural nexus of power" that integrated social groups in local areas effectively (Duara 1988). These patterns were a recurrent theme in more recent studies of kinships, regional variations in China (Faure 2007; Li 2005; Liu 1997).

The role of local elites in governance was best captured in Fei Xiaotong's (Fei 2012) "two-track politics" metaphor: The first one is the "formal track" of the bureaucratic hierarchy that carried out directives from the top-down process, which ended at the county level. Below the county level, the second "informal track" took over, which was based on local elites in organizing villagers, in local problem solving, and in interactions with governments in tax collection and other affairs. These two tracks were linked to each other through informal interactions between county officials and local elites, but at the same time were independent of each other as both kept their distinctive identities and separateness, characteristic of a loosely coupled system (Weick 1976). In this two-track institutional arrangement, local affairs were largely dealt with by local elites within local boundaries.

8.2.2 From the two-track system to the centralized bureaucratic control

Significant changes have taken place since the founding of the People's Republic of China. Through land reform and the collectivization campaign between the late 1940s and 1950s, the Communist Party state eliminated the social strata of the gentry class in social life and put all areas and arenas under the bureaucratic control of the Party state (Parish 1984). Workplace had been organized by the bureaucratic organizations of the Party state (Walder 1986). Local officials of the Chinese bureaucracy down to the township level played the organizing role at the grass-roots level. The decollectivization in the post-Mao era has led to a considerable relaxation of control of rural areas, but the bureaucratic structure has been largely intact.

One important consequence of the political order since 1949 is the conspicuous absence of the strata of local elites at grassroots level. The centralization of authority led to the bureaucratic organization of the society, eliminating the role of the traditional gentry class and local elites. On the other hand, research has shown that local institutions such as kinship survived political pressures in the collectivization era (Friedman et al. 1991). In the post-Mao era, informal institutions came to play a significant role in local problem solving (Tsai 2007; Zhou 2012). Many studies have recognized the importance of local elites in organizing social protests (Cai 2008; Ying 2001).

More importantly, the key roles of local elites were not eliminated but incorporated into local governing bodies of the Party state. That is, the two-track system has been transformed into one bureaucratic system into which both tracks are incorporated – one "formal" track and the other one "informal track" (Zhou 2019). The former is played by those outside officials as "movers," whereas the latter are local elites who are now recruited into local government offices and distributed in those local offices where their local knowledge contributes to local problem solving and development. The difference is that they are now incorporated into the local government offices and acquired double identity as the agent of the state and as the representative of local interests (Zhou 2016). The Chinese bureaucracy also becomes both an organizational hierarchy and as a web of social relations.

8.2.3 Stratified mobility and the remaking of local elites

Zhou (2016) proposed a model of stratified mobility to highlight the bifurcation of "movers" and "stayers" in career mobility in the Chinese bureaucracy. That is, at each level of the administrative jurisdiction, only a very small proportion of the top officials (<< 5%) act as "movers" who transit in and out of administrative jurisdictions, whereas the majority of the bureaucrats are the "stayers" who spend their career within the same locality. Those stayers at the key positions are likely to play the role equivalent to "local elites" in the traditional society. In the Chinese bureaucracy, positions are differentiated for movers and stayers. The top key positions in the four leadership offices, especially those in the Party HQs and the government HQs, tend to be reserved for outside officials, whereas the stayers occupy the majority of other positions.

The transformation of local elites into local officials is made possible because of the bottom-up recruitment and incorporation process through which local elites are identified and absorbed into the cadre system at lower levels and then gradually move upward through bureaucratic ladders. Some of them were tested through jobs at lower-levels such as township governments or low-rank staff jobs; some were recruited from college graduates and then promoted from the bottom up. They tend to stay in the same jurisdictions for an extended period of time, cultivate webs of dense social relations, and occupy key positions in such networks. In other words, they are characteristic of local elites in traditional China.

What are the implications of the local officials? One may find an analogy between the two-track system and the bureaucracy consisting of "outside vs. local" officials. The outside officials tend to occupy key leadership positions with political allegiance to the vertical authority, who controls their career opportunities beyond the administrative jurisdiction. In contrast, local officials will spend their entire career in the same locality, and they are fully aware of this prospect; as a result, they tend to act as local elites in interest articulation and in problem solving (Feng 2010). These considerations point to the significant role of local elites inside the Chinese bureaucracy and their double-identity as the state agent and as the representative of local interests. The image of fragmented authoritarianism and honeycomb model of cellular social structure are captured in the personnel flow patterns.

This recognition has important implications. First, we need to develop a more nuanced view of bureaucrats in the administrative jurisdiction and differentiate those who are outside officials and those local officials, with important implications for principal-agent relationships and incentive design. Second, the recognition of local elites also raises the issue about the role of local elites in economic development. Previous studies have attributed economic development to the top leaders of the administrative jurisdictions. Their roles as well as that of the local elites need to be carefully examined.

In the rest of our study, we turn to address these two issues. We will (1) develop theoretical arguments and propose ways to identify and analyze the location and distribution of local elites in local governments in China; and on this basis, (2) examine the role of local elites, relative to that of the chief Party/government heads, on local economic development.

We will examine these two issues in the context of over 100 counties (or county-equivalent cities or districts) in Jiangsu Province. Below we first briefly outline the empirical context before we turn to these two research tasks.

8.3 The Empirical Context: Counties in Jiangsu Province, China

We choose county-level for analyses because counties have been historically key administrative jurisdictions in China's bureaucratic system (Creel 1964). Counties are also the basic level at which GDP and other important socioeconomic statistics are collected and can be meaningfully interpreted. Previous studies examined local political leaders' economic influence mostly at the provincial level (e.g., Wang et al. 2009; Xu and Wang 2010; Zhang and Gao 2007), and findings from these studies are often inconsistent with one another. County-level jurisdictions are the basic units of jurisdictions with comprehensive administrative and financial power in the territory-based hierarchy of the Chinese bureaucracy. Landry et al. (2018) argued that economic performance plays a limited role in provincial leaders' promotions, but has a strong effect at the lower administrative level, which means that local officials care more about economic performance.

Our empirical examination of the economic influences of local leadership groups draws on the information from *counties in Jiangsu province* between 1992 and 2007. Jiangsu province is located in the eastern coastal area of China. Jiangsu province has the second-largest economy in China, and experienced spectacular economic growth since the 1990s. It is often cited as the example of the Yangtze-river delta development model (长三角发展模式) in China, which emphasizes the importance of local governments in leading economic growth. Thus, Jiangsu province is an appropriate site to analyze the influence of local leadership groups on regional development. We have comprehensive information on the organizational mobility among all chief officials at provincial, prefecture, and county levels that allow us to trace the origins and destinations of career mobility, thereby identifying whether they are local elites or not.

As mentioned earlier, at the county level, there are four highest offices of the county-level administrative jurisdiction (sitaobanzi 四套班子)—the

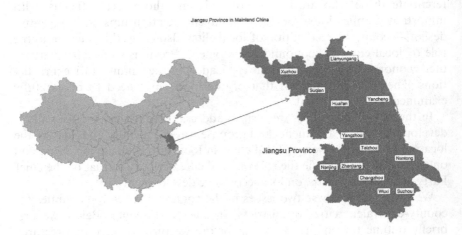

Figure 8.1 Map of China and Jiangsu Province.

Party HQs, the government HQs, the county People's Congress (CPC), and the county Chinese People's Political Consultative Conference (CPPCC). We will use the label of *"four leadership offices"* thereafter to refer to these four institutions. In the hierarchical system of the Chinese bureaucracy, all four offices are at the same "chu" rank (处级), but the Party HQs occupy a central and leadership position, dominating all other offices. We should note that, unlike the institutional design of the separation of power in the US, officials often have joint appointments across these offices. For example, the county Party secretary may also be the head of the CPC, and a vice county government head may be the vice head of the CPPCC. Therefore, it is appropriate to treat them as four leadership offices in different functional areas.

Our data covers all counties and the members of the four leadership offices, between 1990 and 2008. This extended period of time allows us to distinguish outside officials versus local ones.

8.4 Identifying Local Elites in the Chinese Bureaucracy: Positions and Distributions

We now turn to our first task: Identify office-based local elites in the Chinese bureaucracy, their trajectory of mobility and distribution in the Chinese bureaucracy. We propose that local elites are likely to reside in key offices at each level of the administrative jurisdiction. By examining patterns of organizational mobility, we are able to distinguish those movers and stayers associated with the key positions in an administrative jurisdiction. We propose that *there is a clear demarcation of stayers and movers associated with specific positions across local governments. Also, those in charge of economic activities are more likely to be promoted into the top leadership offices.*

Empirically, we focus on the four leadership offices of the county-level administrative jurisdiction (sitaobanzi 四套班子). The four leadership offices hold a set of positions for top officials in each administrative jurisdiction. For our research purpose, we distinguish two types of officials in these offices: In the first group are those *outside officials* who enter these positions from outside the jurisdiction. These officials are likely to be "movers" who usually exit the jurisdiction after a duration of a few years. The second group includes those *local officials* who experience internal mobility (including promotion) to these positions. They tend to be "stayers" who work in the jurisdiction for the entire career. Local elites consist of the second group. Obviously, the division of outside vs. local officials is not binary; rather, it may also be seen as a continuous variable that measures one's length of duration in the locality.

Because of the organizational regulation, the number of key positions in the four leadership offices are fixed; hence, the size of the leadership offices across counties and over time are relatively stable. In our data, for the Party HQs, only the Party secretary, the vice Party secretaries, and the standing members of the Party committees are included. Table 8.1 provides the number of personnel in each leadership office in a given year in our data.

Table 8.1 The composition of the sample

Year	Number of jurisdictions	Number of bureaucrats			
		Party HQs	Government HQs	CPCs	CPPCCs
1990	51	171	413	295	377
1991	58	228	448	280	341
1992	58	287	445	278	354
1993	58	313	443	544	747
1994	71	459	533	369	525
1995	71	430	507	377	501
1996	61	340	457	376	456
1997	95	598	735	558	671
1998	108	654	779	688	806
1999	103	627	734	586	670
2000	99	638	717	594	662
2001	108	702	805	726	791
2002	108	776	802	664	752
2003	106	781	790	656	737
2004	106	830	842	578	647
2005	105	787	808	606	664
2006	105	800	851	664	688
2007	105	781	877	750	789
2008	97	816	816	514	548

8.4.1 Positions associated with movers versus stayers

We examine the distribution of positions associated with movers and stayers in the bureaucracy. As we argued before, the key characteristic of local elites is that they stay in the same locality for an extended period of time, most of them for their entire career and that the movers and stayers occupy distinctive positions in the leadership offices. This recognition provides a key criterion to identify outside versus local officials associated with different positions in the leadership office, and a microscopic view of the local dynamics around outsider and insiders at different levels of local governments. In this conceptualization, the mover versus stayer is not a binary choice. Rather, it is a continuous variable. An official may enter and leave a locality over time. The longer he/she stays, the greater the extent of his/her role as a stayer. Similarly, we may identify a position as more likely to be occupied by a mover than a stayer. This set of analyses also provides information on the distribution of local elites in the local governments, based on the trajectory of their job movements across offices.

Specifically, our empirical strategy to identify *local vs. outside officials* in each jurisdiction is to differentiate those who were promoted internally *vs.* transferred from other jurisdictions at an earlier time. We define *local officials* of a jurisdiction as those whose origin jurisdictions are the same as their current work jurisdictions, and *outside officials* as those whose origin jurisdictions are different from their current ones. Since we only have the records of bureaucrats between 1990 and 2008, we do not have the complete work histories for some officials in our data. Operationally, we identify a bureaucrat's origin jurisdiction to be the one where he/she first appears in our dataset if the time that this bureaucrat first appears is later than the first year that we have records for that jurisdiction. For example, we have information for Hanjiang county starting in 1990. If a bureaucrat first appeared in our dataset in Hanjiang county in 1991, then we will assign Hanjiang county as this bureaucrat's origin jurisdiction. The liability of this identification strategy is that some officials appear in the dataset at the first year we have information for a jurisdiction, and thus we are unable to determine their origin jurisdictions because of the left censoring issue. We categorize these officials as *unknown* cases regarding their origin jurisdictions.

Table 8.2 provides information on a number of key positions in the four leadership offices and the statistics related to their characteristics as movers or stayers. As Table 8.2 shows, there is a clear demarcation of positions associated with movers and stayers. Take for example the "Party secretary" position—85% of the occupants are outside officials (85%), and 63% of the occupants exit the jurisdiction, with an average of 3.2 years of duration in that position. The government head position shows similar characteristics (72%, 40.9%, and 2.6 years of duration, respectively). Overall, we find that those positions in the Party sector tend to have a higher inflow rate, outside official percentage, and outflow rate than those positions in other sectors. The heads of CPC and CPPCC also have a relatively high volume of outside officials, among positions in local leadership offices, entering these positions but tend to have a low rate of moving to other outside positions, indicating that many may have retired from these positions. In addition, positions that are dominated by outsiders would have higher outflow rate as well. For example, in our data 63.1% of Party committee heads were eventually transferred out of the current jurisdictions, and their annual outflow rate is as high as 22.6%. For government heads, 40.9% were eventually transferred out of the current jurisdiction, and their annual outflow is 17.0%. Note that since our data ends in 2008, these statistics have adjusted for right-censoring issues.

Some local officials—those promoted within the four top leadership offices—may also be transferred to other jurisdictions, but their probability of such spatial mobility is noticeably lower than those outside officials. Figure 8.2 shows the annual outflow rates between local and outside officials, measured as the percentage of officials in each group that were transferred out of a jurisdiction in a given year. It is clear that the annual outflow rate of outside officials is always twice as high as that of local officials. On average, the annual outflow rate of outside officials is about 15%, whereas it is only about 5% for local officials. Consistent with our proposition, compared with local officials, outside officials are more likely to be "movers" with a higher probability to leave the present jurisdiction.

Table 8.2 Outside official percentage and spatial outflow rate across positions

Positions	# of person-positions	Outside official%	Outside official % (excluding left-censored unknown cases)	Average tenure (including left- and right-censored cases)	Average tenure (excluding left- and right-censored cases)	Outflow rate of person-positions (%)	# of person-years	Average annual inflow rate (%)	Average annual outflow rate (%)
Party HQ: Head	542	56.6	85.0	3.0	3.2	63.1	1,592	13.6	21.7
Government HQ: Head	601	51.4	72.0	2.6	2.6	40.9	1,524	14.8	16.1
Party HQ: Vice-head	1,533	26.3	37.2	3.0	2.9	35.2	4,480	8.0	12.1
CPC: Head	345	18.3	52.9	3.7	3.8	8.4	1,273	1.9	2.4
CPPCC: Head	412	15.3	33.0	3.9	4.1	5.1	1,568	2.3	1.4
Party HQ: Standing member	1,898	13.4	14.8	2.7	2.7	22.9	4,928	4.1	8.9
Government HQ: Vice-head	3,228	12.5	16.8	3.5	3.6	25.9	10,994	3.6	7.7
Government HQ: Head assistant	162	8.6	8.6	1.6	1.5	11.1	255	5.5	7.1

CPC: Vice head	1,891	5.4	9.2	4.6	4.8	4.5	8,552	0.9	1.1
CPPCC: Vice head	2,000	5.0	7.9	4.9	5.2	5.5	9,541	1.1	1.3
CPPCC: Secretary general	75	1.3	1.6	2.8	3.1	4.0	206	0.0	1.5
CPC: Standing member	301	0.3	0.3	1.3	1.3	5.3	352	0.0	4.5
CPPCC HQ: Standing member	208	0.0	0.0	1.3	1.3	5.3	243	0.0	4.5

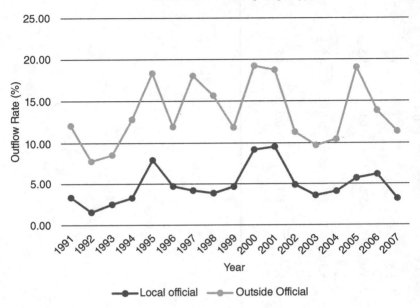

Figure 8.2 Annual outflow rates of local and outside officials in Jiangsu counties: 1990–2008.

8.4.2 Stability of the Leadership Offices

We now examine the stability of personnel in the four leadership offices, in terms of the types of external flow into or out of these offices in a given year. Specifically, we measure the inflow and outflow rates of the leadership offices and positions in those fixed key positions. The former refers to the percentage of bureaucrats that enter these leadership offices from outside origins (i.e., from the prefectural levels or lateral transfer) in a specific year. Outflow rate refers to the percentage of officials that leave the administrative jurisdiction to other destinations in a given year. Inflow and outflow events introduce changes in the leadership offices. Intuitively, the lower the inflow/outflow rate, the more stable the leadership offices as well as local elites in that locality.

Figure 8.3 shows inflow/outflow rates for the four leadership offices combined in a locality. Specifically, we differentiate their origin and destination jurisdictions. Interestingly, inflow rates (4%) are on average lower than outflow rates (7%). That is, a larger number of officials in a locality will be rotated to other jurisdictions in a given year. It is most likely that the vacancies in these positions will be filled by local elites promoted within the locality.

In terms of mobility origins and destinations, the inflow/outflow events occur mostly either vertically between the county and the prefecture to which it subordinates or lateral transfers among those counties within the same prefecture. Mobility events from (or to) other outside origins (destinations) (i.e.,

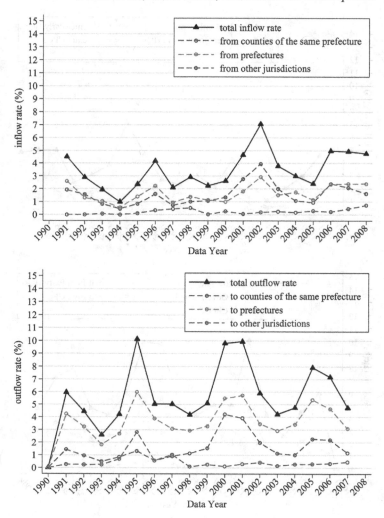

Figure 8.3 Inflow and outflow rates of bureaucrats in leadership offices of Jiangsu counties: by origin and destination jurisdictions, 1990–2008.

other prefectures or the province) are rare. These patterns are consistent with the distribution of authority across administrative jurisdictions and the conventional institutional practice in the Chinese bureaucracy.

Figure 8.4 reports the destination of these inflow/outflow rates for specific offices. It is obvious that most "movers" events pertain to the Party HQs and the government HQs, whereas mover events occur from and to CPC and CPPCC only occasionally, indicating the final destination of the local elites tend to concentrate in the latter two offices. Moreover, these patterns also show that, although personnel flow across localities takes place every year, the observed periodic rhythms are consistent with the term limits of chief officials (five years) in these localities.

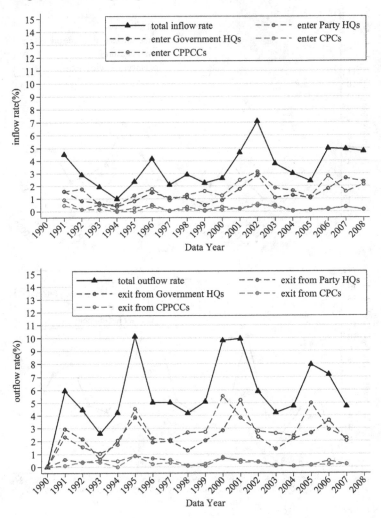

Figure 8.4 Inflow and outflow rates of bureaucrats in leadership offices of Jiangsu Counties: by origin and destination offices, 1990–2008.

Figure 8.5 further provides variations across counties of inflow and outflow rates of bureaucrats in the four local leadership offices. We use a three-year average to smooth the annual fluctuations and to capture the overall stability of leadership offices in the locality. Clearly, inflow and outflow rates not only fluctuate over time but also significantly vary across localities.

8.4.3 Local Knowledge Stock in the Leadership Offices

The distinction between movers and stayers indicates the important role of the latter group as local knowledge stock in attending to local welfare and problem solving. Much of the local knowledge in an organization resides in

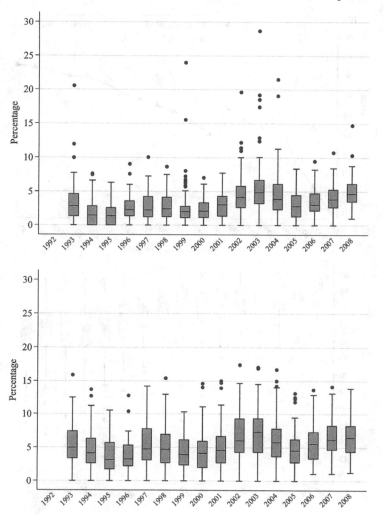

Figure 8.5 Inflow and outflow rates by counties and years: 3-year average, 1990–2008.

individual experience (Stinchcombe 1990); hence, the size of the stayers provides a measure of local knowledge stock. The larger the proportion of local officials in these offices, the larger the local knowledge stock in the locality. Given the variations in the stability of the leadership offices each year, the size of local knowledge stock varies accordingly.

We use the percentage of local officials in the four leadership offices, relative to the total size, as a measure of local knowledge stock. The higher the percentage, the larger the number of local officials, the greater the local knowledge stock. Figure 8.6 (first panel) shows the composition of bureaucrats in leadership offices by their origins, averaging across counties. We also show the category of those officials with "unknown origins" (i.e., with left-censored work history) because they consisted of a considerable part

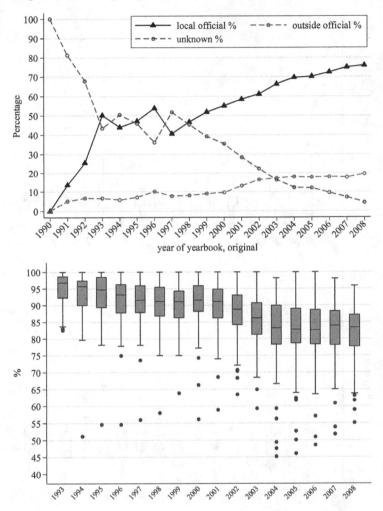

Figure 8.6 Origins of bureaucrats in local leadership offices of Jiangsu counties: 1990–2008.

of the leadership offices during this period. As one can see, the origins of most bureaucrats are not identifiable in the first several years (especially before 1993) because of the left censoring issue. With the decrease of unknown official percentage, the local official percentage increases significantly, indicating that the majority of officials with unknown origins should be local officials. In contrast, the average percentage of outside officials in the leadership offices remains relatively stable and is of small size over time. Even in the last several years (e.g., after 2000), when the majority of officials' origins are identifiable, the average percentage of outside officials across jurisdictions is still under 20%. Since the size of the leadership offices is around 25–30 bureaucrats per jurisdiction-year, it suggests that only five to six key leadership positions are occupied by outside officials.

Overall, the majority of the occupants in the four leadership offices are local officials.

Figure 8.6 (second panel) shows variations of local knowledge stock across jurisdictions, based on the assumption that officials with unknown origins are also local officials. We use a three-year average to smooth the annual fluctuations and to capture the cumulative effects of local knowledge in a locality. As is shown, the proportion is relatively stable over time, and most counties' local official percentage is well above 80%. But there are significant variations across counties. The noticeable downward trend after 2000 was due to changes in personnel policies on term limits and personnel exchanges for selected key positions.[1]

Given the two types of officials, outside and local, in the top leadership offices, a further question is: What is the distribution of these officials across the four leadership offices? In other words, how is local knowledge stocked in different leadership offices of county governments? Figure 8.7 shows the local leadership stock of different leadership offices and over time. Figure 8.7 (first panel) provides a clear contrast of local official percentage across the four leadership offices in local governments. It is noticeable that there is a significant decrease in local official percentage since the early 2000s. The Party HQs always have the lowest local official percentage, especially after 2000. It has stayed between 65% and 70% since 2002. The government sector has the next lowest local official percentage, ranging between 75% and 80% since 2002. In contrast, the local official percentage in CPCs and CPPCCs are always high, which is rarely below 90% between 1990 and 2008.

Given the low inflow and outflow rates for the CPC and CPPCC offices, local officials appear to be concentrated in these two offices. Moreover, the extremely low spatial inflow rates imply that the majority of outside officials in these two offices are those who previously served in the Party and government departments. Therefore, the length of stay of these outside officials should have had much longer stay in that jurisdiction than the average duration of those in the local Party committees and governments. Figure 8.7 (second panel) shows that this is indeed the case: on average, the length of stay of outside officials in the Party and government offices in a jurisdiction is about three years after 1994, while those in the CPC and CPPCCs can reach five or even six years. This pattern also indicates that these outside officials are most likely to retire from their positions in CPCs and CPPCCs. Given the length of their stay in these two offices and the likelihood that they stayed in other offices in the same locality before they entered the two offices, to a great extent these "outside" officials have acquired "local" identities.

To sum up, we can clearly identify two distinctive groups of officials in county governments: the movers and the stayers, or the outside officials and the local officials. These findings are not surprising, given the stratified mobility patterns discussed earlier and the known personnel management policies in the Chinese bureaucracy. The next question is: What are the implications of these patterns for local economic welfare and development?

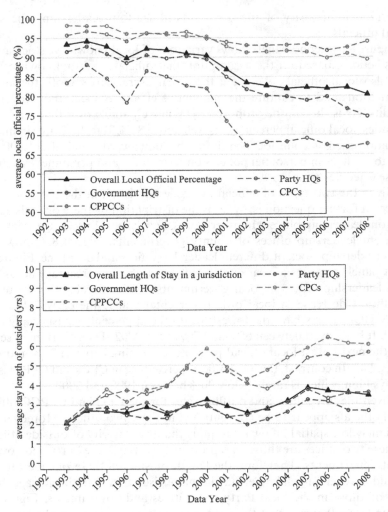

Figure 8.7 Local knowledge stock across local leadership offices in Jiangsu counties: 1990–2008.

8.5 Outside Officials, Local Elites, and Economic Development

We now turn to the second topic of inquiry in our study: the role of local elites in socioeconomic development. The literature on China's economic development has emphasized the role of local governments, especially the role of chief officials (Party secretaries, government heads) in promoting GDP growth, and their behaviors are seen as consistent with the tournament model of competition (Chen et al. 2005; Li and Zhou 2005; Zhou 2007). Studies that examined the effects on economic growth of these chief officials' tenure (e.g., Yao and Zhang 2015; Zhang and Gao 2007), industrial background (e.g., Zhang 2010), bureaucratic connections (e.g., Wang and Xu

2008), and replacement (e.g., Wang et al. 2009; Xu et al. 2007) have found some significant influence.

However, our previous discussion on the movers and stayers in the administrative jurisdiction raises serious questions about this proposition. For one thing, on average those in chief official positions tend to stay in their position for only about three years. These officials are fully aware of this mobility pattern. Therefore, they are likely incentivized to pursue short-term economic performance, often at the expense of long-term prosperity. Under this assumption, if we acknowledge that economic development depends on long-term strategies rather than short-term goals, then, we can hardly attribute economic performance in these regions and counties to those top officials as movers.

On the other hand, it is likely that local elites play a significant role in socioeconomic development within their own jurisdiction, for several reasons. First, because of their long career in the same locality, their interests are more aligned with long-term socioeconomic well-being there; second, they have better capacities in problem solving and information processing than outside officials. Therefore, we would expect that local elites play a larger role in socioeconomic development than those chief officials as "movers."

8.5.1 Variables

8.5:1.1 Measuring Economic Performance

We focus on the economic growth rate of a jurisdiction over time. Following the literature, we use two economic measures: GDP and annual government fiscal revenue.[2] This information was collected from various statistical yearbooks such as China County Statistical Yearbook and the annals of each county of interest. Thus, we have two dependent variables: (1) annual GDP growth rate and (2) annual government fiscal revenue growth rate.

Figure 8.8 shows the distribution of the two economic measures across counties and over time using box plots. The qualitative trend patterns of these two measures are largely consistent. Since Jiangsu counties are among those experiencing fastest economic growth in China, the average growth rates are high for both measures during this period.[3] However, there was significant variation of these growth rates across counties and over years as well.

8.5.1.2 Measuring Stability of Local Leadership Offices

We use a series of cumulative inflow measures to capture the stability of bureaucrats in local leadership offices. *inR_ma3* is the three-year moving average of spatial inflow rate of a jurisdiction for year t, t-1, and t-2. It indicates how often bureaucrats in local leadership offices are replaced by outside bureaucrats over these three years. Thus, this measure captures the overall stability of local leadership offices in a county.

As shown in the earlier sessions, the mobility patterns have distinct features in the four leadership offices. We can thus separately estimate the stability for

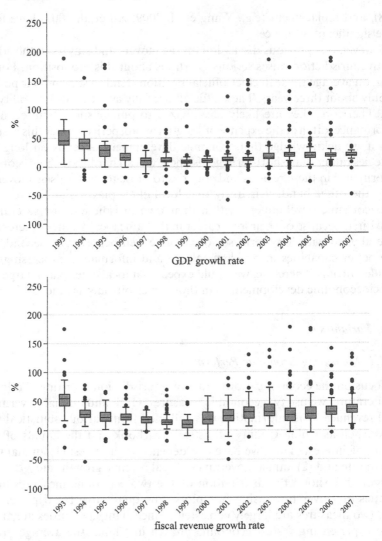

Figure 8.8 Distribution of economic growth rates over time and jurisdictions.

each of them. *inPartyR_ma3, inGovR_ma3, inCPCR_ma3,* and *inCPPCCR_ma3* are the three-year moving average of spatial inflow rates for years *t, t-1, and t-2* attributed to the four leadership offices: the Party HQs, the government HQs, CPCs, and CPPCCs, respectively. Previous descriptive patterns have illustrated that the destinations of transfer-in bureaucrats are mainly the Party and the government HQs, where bureaucrats play a more crucial role in formulating and executing economic policies. In contrast, spatial inflow rates of bureaucrats in CPCs and CPPCCs are very small.

In addition, descriptive analyses showed that the origins of spatial inflow of bureaucrats are mainly prefectures or other counties within the same prefecture. Bureaucrats are the carriers of human and social capital. On one hand, bureaucrats from prefectures may bring resources from prefectures and thus contribute to the local economy. On the other hand, however, bureaucrats from prefectures may be less experienced in local governance, and thus they may slow down regional development. Thus, we further differentiate spatial inflow rates with regard to the origin jurisdictions of the transfer-in bureaucrats. *fromCityR_ma3, fromInCountyR_ma3,* and *fromOtherR_ma3* are the three-year moving average spatial inflow rates for years *t, t-1, and t-2* with origin jurisdictions to be prefectures, counties within the same prefecture, and other jurisdictions (such as province and counties outside the same prefecture), respectively.

8.5.1.3 Measuring Local Knowledge Stock

As is mentioned in the earlier section, we use the percentage of local officials in the leadership offices as a measure of local knowledge stock. *localR_ma3* is the three-year moving average of this percentage in a jurisdiction over the past three years t, t-1, and t-2. It is the major indicator of the cumulative effects of local knowledge in a locality.

Considering the distinction of the four local leadership offices, we construct *localR_party_ma3, localR_gov_ma3, locaR_CPC_ma3,* and *localR_ CPPCC_ma3*, which are the measures of the moving average of local official percentage attributed to the Party HQs, government HQs, CPCs, and CPPCCs, respectively, over the past three years t, t-1, and t-2.

Moreover, as we have noted, the division of outside vs. local officials is not binary but may also be continuous given one's length of duration in a locality. Thus, we measure outside officials' length of stay in a locality as a continuous indicator of their local knowledge stock. Specifically, *stay_length_out_pty_ma3, stay_length_out_gov_ma3, stay_length_out_CPC_ ma3,* and *stay_length_out_CPPCC_ma3* are the three-year moving average of outside officials' length of stay in a jurisdiction over the past three years in a jurisdiction separately for the four local leadership offices: Party HQs, government HQs, CPCs, and CPPCCs.

8.5.2 Model

Following the model specification of prior studies that examine regional economic growth rates (Besley 2005; Besley and Case 2003; Wang et al. 2009; Zhang and Gao 2007), our primary model is specified in Formula (8.1):

$$g_{it} = \alpha_0 + \alpha_1 * y_{i(t-1)} + \beta' * X_{it} + YearFE_t + RegionFE_i + \tau_{it} \qquad (8.1)$$

where g_{it} is the growth rate of GDP/government fiscal revenue of county *i* in year *t*, $y_{i(t-1)}$ is the logarithm of GDP/government fiscal revenue of county *i* in year *t-1*, X_{it} represents a series of predictors and control variables, *YearFE_t* indicates the fixed-effects of year t, $RegionFE_i$ is the fixed-effects of county i, and τ_{it} is the error terms.

We have also collected complete information of each jurisdiction's area, jurisdiction restructure, urban population, and urban share to control for jurisdiction-specific characteristics that may affect regional economic growth. These variables are included in all regression analyses as controlled variables.

8.5.3 Analysis 1: Replicating Previous Studies in the Literature

Our first set of analysis is to replicate key findings on the role of chief officials, local Party secretaries and government heads, on economic growth rates. Most previous studies examined these effects at provincial level. These studies analyzed how provincial chief officials' personal characteristics, such as tenure (Yao and Zhang 2015; Zhang and Gao 2007), industrial background (Zhang 2010), bureaucratic connections (Wang and Xu 2008), origins (Xu et al. 2007), and replacement (Wang et al. 2009), affected local economic growth.

Since this study focuses on the differences of movers and stayers in a locality, we estimate the effects of two characteristics of Party secretaries and government heads on the local economy: their replacement and their origins (outside vs. local). The former evaluates the importance of their stability and the latter assesses the importance of their local knowledge stock. Regarding the replacement effect, Wang et al. (2009) showed that provincial chief officials' replacement would lead to a short-term shock in a locality and result in a significant short-term negative influence on the local economy. This is also consistent with the findings of Zhang and Gao (2007), which indicates that economic growth rates are usually lowest at the start of a leader's tenure. Regarding the origin effect, Xu et al. (2007) found that transferred-in Party secretaries and governors from other jurisdictions would have positive effects on local economic growth. They argued that it was because those transferred-in chief officials were more likely to respond to the political incentives set by the higher-level authorities rationally and actively. If the logic of these arguments holds, we should observe similar effects at the county level.

Accordingly, we estimated two models: the first model estimates the effects of chief officials' replacement, as well as whether they are outside officials or not, on the annual GDP growth rate, controlling for other covariates; the second model estimates these effects on the annual government fiscal revenue growth rate. Table 8.3 summarizes the regression results.

First, there are some immediate replacement effects of local chief officials on the economic growth rates. In M1 and M2, we can see that the replacement effects of chief officials are all negative. Particularly, the coefficients of M2 show that when both the Party secretary and the government head are replaced, there is a significantly negative influence on the fiscal revenue growth rate in that year. Although the replacement effects are not all

Table 8.3 Economic influences of chief officials: fixed-effects model[a]

	M1 GDP growth rate	M2 Fiscal Rev growth rate
Chief official replacement (*ref.* no replacement)		
Party secretary replace	−1.98	−1.48
	(2.23)	(2.11)
Gov head replace	−0.44	−1.42
	(1.93)	(1.82)
Both replace	−0.32	−2.99[+]
	(1.89)	(1.79)
Party secretary: outside official	−3.47[*]	−1.37
	(1.69)	(1.59)
Gov head: outside official	2.65	1.73
	(1.65)	(1.56)
Log(gdp)	−14.83[***]	
	(1.99)	
Log(fiscal Rev)		−17.69[***]
		(2.30)
Control variables[b]	Yes	Yes
Constant	53.94[***]	42.87[***]
	(11.56)	(10.70)
N	1,040	1,040
R^2	0.32	0.35

Notes
[+]$p < 0.10$, [*]$p < 0.05$, [**]$p < 0.01$, [***]$p < 0.001$.
a Standard errors in parentheses.
b Controlled variables include area, area change, urban population, urban share, the fixed-effects of years, and the fixed-effects of jurisdictions.

significant, the qualitative patterns of the finding are consistent with the results presented in the previous literature.

Moreover, the origin of local Party secretaries and government heads also matters for local economic growth, but the effects are not very consistent with the outcomes provided by Xu et al. (2007) based on the provincial data. Counties with government heads that are outside officials have higher economic growth rates, which is compatible with the pattern suggested by Xu et al. (2007). However, these effects are not statistically significant. In contrast, the coefficients in M1 show that jurisdictions with Party secretaries that are outside officials, compared to those with locally promoted ones, have significantly lower GDP growth rates, which are in a different direction as the effects proposed by Xu et al. (2007). Xu et al. (2007) did not estimate the origin effects of Party secretaries and government heads separately, and thus it is hard to tell whether they also found the difference between the two types of chief officials. However, the finding that outside officials have a negative impact on economic growth aligns with our key proposition that the local knowledge stock of bureaucrats in leadership offices contributes to the local economy.

8.5.4 Analysis 2: Stability of Local Leadership Offices and Economic Growth

Our next set of analyses examine how the stability of local leadership offices contributes to local economic growth, in addition to the influences of government heads and Party secretaries. We estimate three models for each outcome variable. The first model estimates the effect of the overall inflow rates of bureaucrats in local leadership offices on local economic growth, controlling for chief officials' characteristics and other covariates. The second model estimates the effects of inflow rates separately for the four leadership offices. The third model estimates the effects of inflow rates attributed to the different origin jurisdictions of the transferred-in bureaucrats. All inflow rates are measured in the form of a three-year moving average before the year of analysis. They capture the frequency of which members of the four leadership offices are replaced by outside officials. High inflow rate would undermine the stability of local leadership offices. If the stability of local leadership offices plays a positive role in the local economy, the inflow rate should have an adverse effect on the local economy. The findings are presented in Table 8.2.

The coefficients in Table 8.4 show that the effects of inflow rates of local leadership offices are mostly negative, which is consistent with our expectation. In particular, the effect of the inflow rate of the Party committee (*inPartyR_ma3*) is significantly negative on the fiscal revenue growth rate. That is, the more frequently that Party committee members are replaced by outside officials, the slower the fiscal revenue growth rate would be. Moreover, the origins of the newly transferred bureaucrats also matter in fiscal revenue growth rates. There is a significantly negative effect of *fromCityR_ma3*, suggesting that the inflow from the prefecture level has an adverse effect on the fiscal growth rate.

Interestingly, the significant effects of local leadership office stability only appear in models of the fiscal revenue growth rate and not in those of GDP growth rate. In contrast, the significant effects of chief officials' characteristics mostly appear in models of GDP growth rate. A plausible explanation is that local chief officials are specifically sensitive to GDP growth rate because it signals their performance and affects their career advancement, while local stable elites concern more with fiscal revenue collection as fiscal revenue is a crucial source of local budgets and fiscal power.

8.5.5 Analysis 3: Local knowledge stock and economic growth

We now examine the role of local knowledge stock, as measured by the share of local officials in a county's leadership office as well as the length of stay of outside officials, on local economic performance, controlling for other factors. We measure local knowledge stock in three ways. The first one is the overall, relative size of local officials in the four top offices combined in a county (*insiderR_ma3*). The second measure decomposes these local elites into the four specific offices, in order to evaluate their relative roles associated with these specific offices. The third measure captures the average length of

Table 8.4 Local stability and economic growth rate: fixed-effects model[a]

	GDP growth rate			Fiscal revenue growth rate		
	M1	M2	M3	M4	M5	M6
inR_ma3	−0.35			−0.33		
	(0.38)			(0.36)		
inPartyR_ma3		−0.40			−1.69**	
		(0.59)			(0.55)	
inGovR_ma3		−0.02			0.91	
		(0.66)			(0.62)	
inPeoConR_ma3		−0.88			−0.96	
		(1.40)			(1.32)	
inCPPCCR_ma3		−0.50			1.75	
		(1.28)			(1.20)	
fromCityR_ma3			−0.38			−1.21*
			(0.58)			(0.55)
fromInCtyR_ma3			−0.28			0.37
			(0.49)			(0.46)
fromOtherR_ma3			−1.02			−1.59
			(1.60)			(1.50)
Chief official replacement (*ref.* no replacement)						
Party secretary replace	−1.76	−1.76	−1.80	−1.31	−1.08	−1.14
	(2.24)	(2.25)	(2.25)	(2.12)	(2.11)	(2.12)
Gov head replace	−0.42	−0.48	−0.43	−1.35	−1.33	−1.30
	(1.93)	(1.93)	(1.93)	(1.82)	(1.82)	(1.82)
Both replace	−0.16	−0.25	−0.14	−2.90	−3.00+	−2.73
	(1.90)	(1.91)	(1.90)	(1.80)	(1.79)	(1.79)
Party secretary: outside official	−3.31+	−3.27+	−3.33+	−1.13	−0.07	−0.98
	(1.70)	(1.73)	(1.70)	(1.60)	(1.62)	(1.60)
Gov head: outside official	2.88+	2.73	2.90+	2.15	1.44	2.28
	(1.67)	(1.71)	(1.67)	(1.57)	(1.60)	(1.57)
Log(gdp)	−14.74***	−14.76***	−14.80***			
	(1.99)	(2.01)	(2.00)			
Log(fiscal Rev)				−17.19***	−16.95***	−17.51***
				(2.28)	(2.28)	(2.28)
Control variables[b]	Yes	Yes	Yes	Yes	Yes	Yes
Constant	55.72***	55.80***	56.10***	35.31***	33.72***	35.60***
	(11.72)	(11.75)	(11.77)	(7.52)	(7.50)	(7.50)
N	1040	1040	1040	1040	1040	1040
R^2	0.32	0.32	0.32	0.34	0.35	0.35

Notes

[+]$p < 0.10$, [*]$p < 0.05$, [**]$p < 0.01$, [***]$p < 0.001$.

a Standard errors in parentheses.

b Controlled variables include area, area change, urban population, urban share, the fixed-effects of years, and the fixed-effects of jurisdictions.

stay of outside officials in the four offices in that locality. All variables use three-year moving average measures to capture the cumulative effects of the local knowledge stock. Our hypothesis is that the greater volume of local knowledge stock, the more competent bureaucrats in local leadership offices in handling local issues, and the higher growth rates of the local economy.

The findings are presented in Table 8.5. Due to the multicollinearity concerns, we do not include measures of inflow rates in these models.[4] For both

Table 8.5 Local knowledge stock and economic growth rate: fixed-effects model[a]

	DV: GDP growth rate			DV: Fiscal revenue growth rate		
	M1	M2	M3	M4	M5	M6
insiderR_ma3	0.14			−0.18		
	(0.15)			(0.14)		
insiderR_party_ma3		0.18**	0.11		0.14*	0.11+
		(0.06)	(0.07)		(0.06)	(0.06)
insiderR_gov_ma3		0.03	0.04		−0.05	−0.11
		(0.09)	(0.11)		(0.09)	(0.10)
insiderR_CPC_ma3		−0.19+	0.07		−0.20*	−0.02
		(0.10)	(0.15)		(0.10)	(0.14)
insiderR_CPPCC_ma3		−0.10	−0.14		−0.33**	−0.27+
		(0.11)	(0.15)		(0.11)	(0.14)
stay_length_out_pty_ma3			−1.22+			−0.15
			(0.70)			(0.65)
stay_length_out_gov_ma3			0.17			−0.41
			(0.76)			(0.71)
stay_length_out_CPC_ma3			1.37*			1.06*
			(0.54)			(0.50)
stay_length_out_CPPCC_ma3			0.48			0.88
			(0.58)			(0.55)
Chief official replacement (*ref.* no replacement)						
Party secretary replace	−2.05	−2.43	−2.38	−1.39	−1.82	−1.63
	(2.23)	(2.22)	(2.22)	(2.11)	(2.09)	(2.09)
Gov head replace	−0.53	−0.73	−0.36	−1.31	−1.53	−1.37
	(1.93)	(1.93)	(1.93)	(1.82)	(1.82)	(1.82)
Both replace	−0.29	−0.24	−0.52	−3.03+	−2.92	−3.12+
	(1.89)	(1.88)	(1.88)	(1.79)	(1.77)	(1.77)
Party secretary: outside official	−3.11+	−1.07	−0.36	−1.83	0.40	0.69
	(1.73)	(1.85)	(1.88)	(1.63)	(1.74)	(1.77)
Gov head: outside official	2.91+	2.75	3.01+	1.38	1.33	1.59
	(1.67)	(1.77)	(1.78)	(1.58)	(1.66)	(1.67)
Log(gdp)	−14.74***	−15.85***	−15.85***			
	(1.99)	(2.05)	(2.06)			
Log(fiscal Rev)				−17.68***	−19.18***	−19.31***
				(2.30)	(2.32)	(2.32)
Control variables[b]	Yes	Yes	Yes	Yes	Yes	Yes
Constant	43.24**	64.55***	47.78*	56.71***	80.28***	62.57***
	(16.17)	(16.92)	(19.39)	(15.02)	(15.44)	(17.59)
N	1040	1040	1040	1040	1040	1040
R²	0.32	0.33	0.34	0.35	0.36	0.37

Notes
+$p < 0.10$, *$p < 0.05$, **$p < 0.01$, ***$p < 0.001$.
a Standard errors in parentheses.
b Controlled variables include area, area change, urban population, urban share, the fixed-effects of years, and the fixed-effects of jurisdictions.

GDP and fiscal revenue growth, the overall local official percentage has no significant effect on local economic growth rates (M1 and M4). However, when we decompose the overall outsider percentage into specific offices (M2 & M5), we find that the average local official percentage does have a significant, positive effect in both measures of economic growth. In particular, the local official percentage in the Party committee (*insiderR_party_ma3*) has a positive impact on GDP and fiscal revenue growth rates. But, somewhat surprisingly, the effect associated with the county government is not statistically significant and has a small effect size. This is likely because bureaucrats in the Party HQs usually play the most dominant role in local leadership offices.

A puzzling finding in the model for fiscal revenue is that the two variables of local officials in CPC and CPPCC (*insiderR_peopleCon_ma3, insiderR_CPPCC_ma3*) have negative effects on fiscal revenue growth rates, which is inconsistent with our theoretical prediction. They suggest that the larger share of local officials in CPC and CPPCC, the slower the local economic growth rate will be. That is, conversely, outside officials in these two offices contribute more to local development than local officials. We took a closer look at the careers of the officials in these two offices closely and found that outside officials tended to be those outside officials who previously worked in the Party and government offices and then moved to these two offices and who are likely to retire from these offices in that jurisdiction. As a result, these outside officials acquired local identities due to their extended length of stay in a locality. In Model 3 and 6 (M3 and M6), we further control outside officials' average length of stay in a jurisdiction over the past three years in these four sectors (i.e., *stay_length_out_pty_ma3, stay_length_out_gov_ma3, stay_length_out_CPC_ma3, and stay_length_out_CPPCC_ma3*) and evaluated how the inclusion of these measures would affect the patterns. The results show that the negative effects of the local official percentage in CPC and CPPCC can be explained by the length of stay of outside officials in a particular jurisdiction. That is, the length of stay of bureaucrats in people's congress and CPPCC have a significant, positive effect on economic development. And controlling for this, there is no significant effect of outside officials from these two offices.

To sum up, after controlling for the effect of top officials (i.e., the movers), we find evidence that local officials contribute positively to GDP growth and the growth of fiscal revenue. These observed patterns should be sharper if we have data for a longer period of time and more accurate indicators of local officials.

8.6 Discussion and conclusion

The rapid and sustained economic growth in China since the 1980s has led to many theoretical interpretations (Boisot and Child 1996; Qian and Weingast 1997; Xu 2011). It is widely acknowledged that the Chinese central government has decentralized administrative, fiscal, and urban planning power to local governments to boost local economy since the 1990s. Some studies emphasized the leading role of local governments, especially the role of incentive for the top officials in the jurisdiction for economic development (Landry et al. 2018; Li and Zhou 2005; Oi 1995). These studies have almost exclusively focused on the top officials in each administrative jurisdiction, i.e., the Party secretaries and governors at the county, prefecture, or provincial levels.

Our study shifts attention from those Party and government heads in the administrative jurisdictions to a larger group of bureaucrats in the leadership groups of local governments. We argue that there are two types of bureaucrats in local governments: one consists of outside officials and the other one consists of local officials respectively. These two groups are distinctive in their career patterns, which cultivate different identities and political allegiance. Those Party secretaries and governors, which are the focus of the previous literature, usually belong to the first group, and they are often movers transient in and out of a locality in a short period of time. In contrast, local officials who work in these localities for an extended period of time play the role of local elites in the traditional society. They experience their career and social life within the same administrative jurisdiction, and they are deeply embedded in the dense network of multiple, overlapping social relations. Consequently, their involvement in the organization and coordination in economic activities plays a substantial role in local development.

Using systematic data on the mobility patterns of top four offices in all counties in Jiangsu Province between 1990 and 2008, we examined the role of office-based local elites in county governments and their effect on local economic development. We differentiate *stayers* vs. *movers* in the four top local leadership offices by identifying local officials who are internally promoted vs. outside officials who are externally transferred into a locality. On this basis, we further (1) identify their relative distribution in the local government offices, and (2) assess their respective roles in socioeconomic development within their localities. In so doing, we address the larger issue about the double-identity of bureaucrats in local governments as agents of the Chinese state and as representatives of local interests, and their role in China's governance.

The findings reveal a clear demarcation of local vs. outside officials in local governments. The former are mostly stayers in one locality while the latter are mostly movers in and out of different localities. Moreover, outside officials usually occupy the most powerful positions in the local leadership offices, such as Party secretaries and government heads. Local officials, in contrast, are more likely to concentrate in other supporting positions in these top offices. We find some evidence on the positive role of local elites in local economic growth measured by GDP and fiscal revenue. Our finding shows that, in addition to the well-established effects of the chief officials in local development, the stability and knowledge stock of local officials, as measured by the proportion of local officials in the four leadership offices, also significantly contribute to economic growth.

Our findings contribute to the understanding of political incentives of local bureaucrats in regional development. While the current wisdom has emphasized the importance of economic-performance-based career incentives, we stress that local development is a collaborative outcome by outside and local officials, who have distinct career paths, incentives, and local embeddedness. On the one hand, outside officials have a higher degree of political allegiance to the higher-level authorities and they oversee local officials to avoid collusive behaviors. On the other hand, local officials assist top officials from outside to formulate and carry out local policies as well as forestalling harmful developmental projects that only aim for short-term gains. To some extent, these mutual checks and balances paradoxically alleviate the principal-agency problem in the long chain of command in the multilayered Chinese bureaucracy.

Our chapter contributes to this edited volume by revealing the nexuses between economies, institutions, and territories in the context of Chinese counties. While territory-based knowledge is indispensable for effective local governance and economic growth, local elites are incorporated in political institutions that face challenges of collusive behaviors and corruption. In the China context, local elites concentrate in supportive positions while outside officials dominate the supervisory positions in local governments. Local officials assist top supervisors to accomplish tasks, but they can hardly obtain those supervisory positions. Even if they are promoted to such positions, they are likely to be transferred to other jurisdictions. Obviously, local and outside officials are distinct social groups in local governments with different interests, incentives, as well as different capacities and knowledge. In an era when economic growth becomes a national goal set by the central government, these two groups may work together to contribute to the regional economy through organizational design and incentive provision.

Given the limited information and time span in our data, our findings are by no means conclusive. We hope that our study is a first step to bring the role of local elites back into the study of sources of economic development in China and that future studies, with richer data and more refined research designs, can explore this theme further.

Notes

1 In 1999, the Central Committee put forward the *Provisionary Regulations on the Exchange of Party and Government Cadres* (Dangzheng Lindao Ganbu Jiaoliu Gongzuo Zanxing Guiding). In this provisionary regulation, it is required that newly appointed government and Party leaders, as well as leaders in local people's courts, supervision departments, and financial departments must be outside officials. It also specifies that ten years should be a maximum term for bureaucrats in these positions. This provisionary regulation was revised and finalized in August 2006, featured by the effectuation of *Regulations on the Exchange of Party and Government Cadres* (Dangzheng Lindao Ganbu Jiaoliu Gongzuo Guiding).

2 For studies using GDP as the economic measure, see Li and Zhou 2005; Zhang and Gao 2007; Wang et al. 2009; Xu and Wang 2010. For studies using annual government fiscal revenue as the economic measures, see (Jiang 2018; Landry et al. 2018; Lü and Landry 2014).

3 There are some extreme outliers for the GDP and fiscal revenue growth rates in the original data, and some can go as high as more than 1000%. I checked the data and found that the extreme outliers whose growth rates are above 200% are all because of the change measurements of either GDP or fiscal revenues during certain years. Therefore, I exclude the extreme outliers in the regression models.

4 For example, the correlation between the inflow rate (*inR_ma3*) and the local official percentage (*insiderR_ma3*) is −0.54. This is expected because outside officials,

as we have argued, are frequent movers; hence, a considerable number of outside officials in a locality in a specific year are newly transferred into that locality in that year, which is captured in the inflow rate.

References

Besley, Timothy. 2005. "Political Selection." *Journal of Economic Perspectives*. 19, no.3 (Summer): 43–60.

Besley, Timothy, and Anne Case. 2003. "Political Institutions and Policy Choices: Evidence from the United States." *Journal of Economic Literature* 41, no.1 (March): 7–73.

Boisot, Max, and John Child. 1996. "From Fiefs to Clans and Network Capitalism: Explaining China's Emerging Economic Order." *Administrative Science Quarterly* 41, no.4 (December):600–628.

Cai, Yongshun. 2008. "Local Governments and the Suppression of Popular Resistance in China." *The China Quarterly* 193, (March):24–42.

Chen, Ye, Hongbin Li, and Li-An Zhou. 2005. "Relative Performance Evaluation and the Turnover of Provincial Leaders in China." *Economics Letters* 88, no.3 (May):421–425.

Chu, Tung-Tsu. 1965. *Law and Society in Traditional China*. Paris: Mouton & Co.

Creel, H. G. 1964. "The Beginnings of Bureaucracy in China: The Origin of the Hsien." *The Journal of Asian Studies* 23, no.2 (February):155–184.

Duara, Prasenjit. 1988. *Culture, Power, and the State: Rural North China, 1900–1942*. Stanford, CA: Stanford University Press.

Faure, David. 2007. *Emperor and Ancestor: State and Lineage in South China*. Stanford, CA: Stanford University Press.

Fei, Xiaotong. 2012. *Rebuild the Soil—The Development of Chinese Society (in Chinese)*. Changsha: Yuelu Press.

Feng, Junqi. 2010. *Cadres in the Mid-County*. Beijing: Unpublished dissertation, Department of Sociology, Peking University.

Friedman, Edward, Paul G. Pickowicz, and Mark Selden. 1991. *Chinese Village, Socialist State*. New Haven: Yale University Press.

Heberer, Thomas, and Gunter Schubert. 2012. "County and Township Cadres in China as a Strategic Group: A New Approach to Political Agency in China's Local State." *Journal of Chinese Political Science* 17, no.3 (August):221–249.

Hsiao, Kung-Chuan. 1960. *Rural China: Imperial Control in the Nineteenth Century*. Seattle: University of Washington Press.

Jiang, Junyan. 2018. "Making Bureaucracy Work: Patronage Networks, Performance Incentives, and Economic Development in China." *American Journal of Political Science*, 62, no.4 (October), 982–999.

Landry, Pierre F. 2008. *Decentralized Authoritarianism in China: The Communist Party's Control of Elites in the Post-Mao Era*. New York: Cambridge University Press.

Landry, Pierre F., Xiaobu Lü, and Haiyan Duan. 2018. "Does Performance Matter? Evaluating Political Selection Along the Chinese Bureaucratic Ladder." *Comparative Political Studies*, 51, no.8 (July):1074–1105.

Li, Hongbin, and Li-An Zhou. 2005. "Political Turnover and Economic Performance: The Incentive Role of Personnel Control in China." *Journal of Public Economics*, 89, no.9–10 (September):1743–1762.

Li, Huaiyin. 2005. *Village Governance in North China: 1875–1936*. Stanford, CA: Stanford University Press.

Liu, Zhiwei. 1997. *Between the State and Society: Study on the Basic Political Units, and Tax and Corvee System in Guangdong during Ming and Qing Dynasty (in Chinese)*. Guangzhou: Sun Yat-sen University.

Lü, Xiaobo, and Pierre F. Landry. 2014. "Show Me the Money: Interjurisdiction Political Competition and Fiscal Extraction in China." *American Political Science Review* 108, no.03 (July):706–722.

Oi, Jean C. 1995. "The Role of the Local State in China's Transitional Economy." *The China Quarterly* no.144 (December):1132–1149.

Parish, William L. 1984. "Destratification in China." In *Class and Social Stratification in Post-Revolution China*, edited by J. Watson, 84–120. New York: Cambridge University Press.

Qian, Yingyi, and Barry R. Weingast. 1997. "Federalism as a Commitment to Preserving Market Incentives." *Journal of Economic Perspectives* 11, no.4 (Autumn):83–92.

Stinchcombe, Arthur L. 1990. *Information and Organizations*. Berkeley: University of California Press.

Tsai, Lily L. 2007. "Solidary Groups, Informal Accountability, and Local Public Goods Provision in Rural China." *The American Political Science Review* 101, no.2 (May):355–373.

Walder, Andrew G. 1986. *Communist Neo-Traditionalism: Work and Authority in Chinese Industry*. Berkeley: University of California Press.

Walder, Andrew G. 1995. "Career Mobility and the Communist Political Order." *American Sociological Review* 60, no.3 (June):309–328.

Wang, Xianbin, and Xianxiang Xu. 2008. "The Origins, Destinations, Tenures of Local Leaders and Economic Growth: Evidence from Provincial Leaders." *Management World (in Chinese)* no.3 (March):16–26.

Wang, Xianbin, Xianxiang Xu, and Xu Li. 2009. "Provincial Governors' Turnovers and Economic Growth: Evidence from China." *China Economic Quarterly* 8, no.4 (April):1301–1328.

Weick, Karl E. 1976. "Educational Organizations as Loosely Coupled Systems." *Administrative Science Quarterly* 21, no.1 (March):1–19.

Xu, Chenggang. 2011. "The Fundamental Institutions of China's Reforms and Development." *Journal of Economic Literature* 49, no.4 (December):1076–1151.

Xu, Xianxiang, and Xianbin Wang. 2010. "Growth Behavior in the Appointment Economy." *China Economic Quarterly* 9, no.3:1447–1466.

Xu, Xianxiang, Xianbin Wang, and Yuan Shu. 2007. "Local Officials and Economic Growth." *Economic Research Journal* 42, no.9 (September):18–31.

Yao, Yang, and Muyang Zhang. 2015. "Subnational Leaders and Economic Growth: Evidence from Chinese Cities." *Journal of Economic Growth* 20, no.4 (May):405–436.

Ying, Xing. 2001. *Dahe Yimin Shangfang De Gushi [the Story of Immigration Petitions in Dahe]*. Beijing: Sanlian Chubanshe.

Zhang, Er-sheng. 2010. "Economic Growth and Local Governors' Entrepreneurial Background: Evidence from the Secretary of the CPC Provincial Committee and Provincial Governor." *China Industrial Economics* 3 (March):36–50.

Zhang, Jun, and Yuan Gao. 2007. "Term Limits and Rotation of Chinese Governors: Do They Matter to Economic Growth?" *Economic Research Journal (Jingji Yanjiu)* 11:91–103.

Zhou, Li-an. 2007. "Zhongguo Difang Guanyuan De Jinbiaosai Moshi Yanjiu [Governing China's Local Officials: An Analysis of the Promotion Tournament Model]." *Jingji Yanjiu [Economic Research]* 7:36–50.

Zhou, Xueguang. 2010. "The Institutional Logic of Collusion among Local Governments in China." *Modern China* 36, no.1 (November):47–78.

Zhou, Xueguang. 2012. "The Road to Collective Debt in Rural China: Bureaucracies, Social Institutions and Public Goods Provision." *Modern China* 38, no.3 (May):271–307.

Zhou, Xueguang. 2016. "The Separation of Officials from Local Staff: The Logic of the Empire and Personnel Management in the Chinese Bureaucracy." *Chinese Journal of Sociology* 2, no.2 (April):259–299.

Zhou, Xueguang. 2019. "Lun Zhongguo Guanliao Tizhizhong De Feizhengshi Zhidu (Informal Institutions in the Chinese Bureaucracy: An Essay)." *Qinghua Shehui Kexue (Tsinghua Social Sciences)* 1, no.1 (September):1–42.

9 Working at the Nexus of Global Markets and Gig Work

US Gig Workers, Credential Capitalization, and Wealthy International Clientele

Alexandrea J. Ravenelle and Ken Cai Kowalski

9.1 Introduction

This chapter examines how US-based elite gig workers navigate the institutionalization of a new international and territorial division of labor processes in the global gig economy. Typically associated with low-paid jobs like ridesharing and food delivery, the gig economy has also thrust freelance professionals into global competition for consulting, bookkeeping, marketing, and other white-collar work. Yet, while gig platforms such as Uber, DoorDash, and TaskRabbit have become household names, and their low wages and workplace exploitation has been well-documented (Ravenelle 2019), online platforms that enable remote work are often described in more positive terms as enabling workers to manage multiple gigs simultaneously and increase their income accordingly (Wood et al. 2019a). As the popularity of white-collar gig work continues to expand across the globe, US-based workers' understandings of these competitive pressures reflect and reinforce the territorialization of remote work within an international status hierarchy.

Previous research has documented long-standing concerns about labor outsourcing in high-income countries (Egger and Egger 2005), as well as mixed consequences from the expansion of gig work in developing nations (De Stefano 2015), but little is known about the experience of international competitive pressures within the gig economy, especially on elite platforms. While early research suggests that workers turn to these platforms after being pushed out of a job, to fill a gap in employment, to supplement a full-time job with additional income, or to pivot careers (Ravenelle et al. 2021), little is known about the experience of international competitive pressures within the gig economy. In this chapter, we ask, as online gig platforms reduce the barriers to obtaining international work, how do US-based elite gig workers view their increased international competition? Does the ability to digitally traverse borders connect US-based workers with an expanded pool of potential clients, strengthening their position in the workplace, or do workers perceive the lack of boundaries as possibly reducing their ability to receive a sufficient income? And how do US-based workers understand their competitive position within the global market for these professional services?

DOI: 10.4324/9781003191049-12

Drawing on demographic surveys and qualitative interviews with 35 workers with experience working on elite gig platforms such as Catalant, Graphite, and TopTal, this study examines how US-based gig workers perceive the global market and leverage the cachet of their US credentials to command higher wages. Although these high-status workers typically join online platforms to enjoy flexibility and a $1000/day minimum wage, they soon find themselves competing with much cheaper foreign labor. At the same time, Americans with prestigious degrees and employment histories find themselves in high demand for easy, lucrative, and sometimes legally dubious jobs offered by wealthy overseas clients. These workers come to see themselves and their output as especially skillful, deserving a wage premium over labor from developing countries. This perception justifies migration to platforms that cater to higher-end clients, including foreign governments and multinational firms, which enables them to better capitalize on their American credentials.

9.2 Literature Review

9.2.1 Globalization of the Gig Economy

As digital connectivity spread throughout the 1990s, it accelerated labor outsourcing and eventually enabled remote work to be completed anywhere in the world with access to adequate internet infrastructure (Graham et al. 2017a). Concerns about job loss and worker exploitation in offshored manufacturing industries soon expanded to include digital services, communication, and entertainment sectors facing the prospect of "twenty-first century offshoring" (Levy 2005). In fact, white-collar work is now as "offshorable" as manual labor: in the United States, up to 25% of all jobs can potentially be performed overseas, with those requiring higher education more likely to be offshorable (Blinder and Krueger 2013).

Although global labor sourcing has sometimes been credited with raising wages in impoverished countries and allowing corporations to reach new markets with cheaper goods (Friedman 2005), logistical challenges and lack of regulatory oversight have historically posed serious risks to workers and consumers alike (Dicken 2015). In wealthy countries like the United States, gig platforms purport to offer workers flexibility, while skirting labor regulations that apply to conventional employers (Katta et al. 2020). Likewise, a range of NGOs and initiatives in developing countries tout digital gig work as a pathway out of poverty, yet many workers experience long hours, low pay, and uncertain schedules (Graham et al. 2017b).

The global expansion of digital freelance work has unevenly incorporated different parts of the world. On platforms like Upwork, contracts are overwhelmingly offered by clients in the Global North to workers in the Global South (Horton et al. 2018). By 2013, employers from high-income countries outnumbered those from lower-income countries by 10 to 1 on the digital job platform oDesk (now known as UpWork), while 4.5 times as many workers hailed from lower-income countries compared to high-income countries (Agrawal et al. 2015). The United States represents a partial exception, supplying a substantial amount of digital labor to other countries (Horton et al. 2018). This development is partly due to the widespread prevalence of gig work in the country. Based on tax filings, reliance on part-time self-employment is growing across the United States (Abraham et al. 2018), and in 2018, nearly 25% of US adults had experience working in the gig economy (Edison Research 2018). In general, online job platforms have thrown Western gig workers into direct competition with those in developing countries, a relatively new dynamic that requires further study (Beerepoot and Lambregts 2015).

In the Global South, gig work has grown rapidly and attracted controversy for its local economic consequences. Workers across Sub-Saharan Africa and other developing regions are increasingly reliant on informal work, including gig work (Anwar and Graham 2020). In one survey, 68% of gig workers from low- and middle-income countries reported that online gig work is a primary source of income for their households (Graham et al. 2017a). Digital freelancing is particularly prevalent in countries like the Philippines and India, though this work increasingly substitutes for traditional jobs in many regions struggling with high unemployment or weak local job markets (Dicken 2015).

9.2.2 *Experiences of Precarity*

Although the global expansion of gig work has contributed to the rise of a middle class in nations like India, Kenya, and Ghana, concerns remain about the highly unequal distribution of resources within the digital economy (Fish and Srinivasan 2012). Despite optimistic narratives about the "win-win" nature of fulfilling labor demand in the Global North with informal labor in the Global South, these ties have proven to be "an engine of vulnerable employment," accelerating workforce casualization and eroding basic worker protections in many low- and middle-income countries (Meagher 2016). Even as multinational corporations have implemented responsibility codes or other means of monitoring labor conditions, bowing to pressure from activists and NGOs, workers in the informal economy reap few if any benefits from these protections (Barrientos 2008). Although some workers experience success in the gig economy, many others continue to face long hours, insecure employment, and meager compensation (Graham et al. 2017b).

Cross-national research reveals some broad similarities in how work precarity is experienced, specifically as a personal problem that "can only be overcome by individual effort and investment" (Mrozowicki and Trappmann 2021). Like their counterparts in the Global South, digital freelancers in the US, UK, Australia, and the EU are attracted to gig work by the promise of flexibility and freedom, and they too struggle with lack of basic benefits like healthcare, unemployment insurance, and family leave (Thompson 2018). Inconsistent contracts and lack of recourse against client demands often mean gig workers cannot manage their hours (Anwar and Graham 2020). Since gig workers also find themselves "normatively disembedded from social protections" intended to safeguard their rights, they are forced to rely on interpersonal networks for economic or other support (Wood et al. 2019b). Further, wages are subject to downward pressure by fierce competition and perceptions of powerlessness, which lead many workers to underbid when negotiating contracts (Graham et al. 2017b:145).

9.2.3 Subjective Perception and Status in Markets

Despite the similar challenges with precarious employment facing gig workers around the world, workers based in the Global North and especially the US benefit from a symbolic association with high-status professional competence. Culture influences valuation of goods and services in part through "social performances of value," or the "symbolic and relational work, and impression management strategies" that signify prestige, quality, or aesthetic desirability to potential buyers (Bandelj and Wherry 2011). Successful production of this "symbolic value" allows economic actors to command higher prices, as in the paradigmatic case of countries known for specific exports like wine (Ponte and Daviron 2011). On the world stage, strategic management of national image is an indispensable means to gain competitive economic advantage, particularly for countries with reputations "spoiled" by war, social unrest, or perceived lack of development (Rivera 2008). Therefore, global markets traverse not only national territories, but also their corresponding "cultural territories" (Jijon 2019) of unevenly distributed symbolic capital that affects the prices of exports, labor, and services like tourism.

Since market competition is mediated by attributions of value (Boltanski and Thévenot 2006), how actors understand and carry out these evaluations is fundamental to economic activity. In particular, the maintenance of symbolic boundaries that demarcate superior and inferior groups plays a major role in sustaining wage and status differentials (Lamont and Fournier 1992). These distinctions are the cultural foundation for social and professional closure, supplying justifications for inequalities between genders (Bolton and Muzio 2007) and by national origin (Jenkins and Reddy 2016). Even so, scholars have called for more research elaborating how processes of interpretation and status differentiation occur within markets (Beckert 2013). Jijon (2019) further cites the need to examine how globalization is experienced, arguing that meaning must be contextualized within global cultural diffusion and circulating images, narratives, and beliefs about the "foreign Other."

9.2.4 Status Differentiation in Gig Work

Perceptions of status also play a crucial role in economic transactions in online gig work markets, both across and within platforms. Digital platforms operating in competitive markets often seek to differentiate themselves from competitors by developing "platform identities" based on distinctive app or website design and consumer segments (Cennamo 2021). Within digital platforms, status is largely based on reputation scoring, which is the major factor affecting hiring decisions (Xu 2015). Further, the impact of these ratings on wages and likelihood of receiving contracts is strongest for highly skilled work like web development (Beerepoot and Lambregts 2015).

Status differentiation based on work reputation has profound consequences for workers' experiences on digital platforms. Many report facing significant pressure to avoid negative evaluations by putting in long hours and quickly answering customer requests, even to the point of exhaustion and sleep deprivation (Wood et al. 2019a). In many cases, high-reputation middlemen leverage their high status by subcontracting work, further diminishing pay rates for newer workers, and preventing them from building up their own scores (Graham et al. 2017a). As a result, the quantification of status hierarchies through reputation scoring is among the most successful "algorithmic management" techniques designed to increase productivity and intensify competition between gig workers (Wood et al. 2019a).

In addition to individual reputation, collective reputation based on country of origin plays a major role in hiring decisions within global labor markets. Xu (2015) finds that employers on digital labor platforms generalize first experiences with workers from a country to all workers from that country, substantially influencing future hiring decisions based on very limited information. Many gig workers in low- and middle-income countries report facing discrimination based on country of origin, and analysis of transaction data confirms "clients on average assume that workers from low- and middle-income countries provide less valuable work than workers from high-income countries" (Graham et al. 2017a: 7–8). Some digital workers have encountered ads specifically prohibiting South Asians from bidding, while many African workers describe overcoming pervasive stereotypes about their language abilities and internet access (Graham et al. 2017b). On the previously named-oDesk, an online job platform where Western workers compete directly with Indian and Filipino contractors, Beerepoot and Lambregts found some evidence of wage convergence due to competitive pressures, but ultimately "discriminatory recruitment practices" based on national origin ensured that Western workers received the highest absolute wages (Beerepoot and Lambregts 2015: 252–253).

9.2.5 Exclusive Platforms and Defining Elite Gig Workers

Perceptions of status are particularly salient within high status gig platforms that promise clients access to the most skilled or credentialed workers. Unlike lower status gig platforms, such as TaskRabbit, the personal assistant site, or Instacart, the grocery shopping platform, high status gig platforms market themselves as exclusive online talent enclaves. Catalant (previously called HourlyNerd), Graphite (formerly known as SpareHire), and TopTal (a portmanteau of Top and Talent), market themselves as offering "on-demand business expertise" and "top business talent." For instance, Graphite markets itself as providing "Access [to] the world's best independent talent, trained by Big-3 firms and vetted by our internal talent team" (Graphite n.d.). Graphite's 6,000 experts include more than 800 with big-three consulting experience

(Bain, McKinsey, and BCG), and more than 1,600 with an MBA from a top ten business school (Our Experts n.d.). Significantly more than half of their workers (65%) hold advanced degrees (Our Experts n.d.). Likewise, Catalant offers access to more than "70,000 independent experts" used by "more than 30% of the Fortune 100... to power strategic plan execution, enterprise portfolio management, centers of excellence, organizational redesigns... and post-merger integrations"(What We Do n.d.).

Unlike traditional gig platforms with few barriers to access beyond a background check and attendance at an orientation session, these elite platforms emphasize their exclusivity. Graphite notes that only 1 in 20 experts are accepted, and that the workers are "extremely talented individuals [who] are choosing to work independently" (Our Experts n.d.). TopTal requires applicants to complete a multi-stage three- to five-week application process, ultimately only accepting 3% of applicants. These platforms combine symbolic boundaries with restrictive access to achieve a partial form of professional closure that raises the perceived value of the labor offered to clients (Boussard 2018).

Although their education levels, incomes, and cultural capital would easily qualify these workers as members of the elite (Khan 2012), we focus our categorization on the work itself. We define elite gig work as offering a $100 an hour "minimum wage" or $1,000 a day, and offering work that is considered to be highly lucrative and prestigious such as management consulting, strategic planning, and financial services. Traditionally seen as desirable jobs for top business school graduates, this work can now be found via gig platform, without any of the traditional workplace benefits, protections, or opportunities for advancement.

Our data shows that US-based elite gig workers respond to international competitive pressures by invoking a global status hierarchy based on merit, positioning themselves as more competent and prestigious than workers based in the Global South. US-based workers also associate the exclusive platform identities cultivated by high status gig platforms with this meritocratic status hierarchy, further cementing their distinctiveness in the global market for professional gig work. These interpretive boundaries bolster the symbolic value of US-based workers' labor in a manner that reflects and reinforces prevailing wage differentials between white collar gig workers based in the US and the rest of the world, particularly the Global South.

9.3 Research Methodology

The 35 respondents for this study were recruited from the TopTal, Graphite, and Catalant platforms. Only workers who had at least one completed, and reviewed, project on their respective platform at the time of recruitment were eligible. Each participant completed a short demographic survey before being interviewed in a participant-directed semi-structured interview (Weiss 1994). Interviews were conducted between June 2019 and early March 2020, with most interviews conducted in person. Interview questions were open-ended and included such topics as how the worker became involved with the high status gig economy and experience with other platforms, the challenges they encountered, memorable experiences, and their views on the future of work.

All interviews were audio-recorded, transcribed, and index-coded before being coded inductively and analyzed for patterns (Deterding and Waters 2018). To preserve confidentiality, all respondents were assigned pseudonyms based on the most popular names from their birth year. To encourage participation in the study, workers were given a $50 gift card incentive and, for in-person interviews, offered lunch or a small meal.

Respondents included 25 males and 10 females. Twenty-five participants identified as white, six as Asian, one as Hispanic, one as racially mixed, and one as American.[1] Their ages ranged from 25 to 66 with an average age of 39.4. Their household incomes were high: 28 had an income of more than $100,000, including 16 with household incomes north of $200,000 and only two had incomes below $99,999. Twenty-four identified themselves as having a graduate degree, including two participants with doctorates and one with a law degree. All participants had at least a four-year college degree.

9.4 Findings

9.4.1 Brand Cachet: "I Have a Good Brand Name"

Cognizant of the power of prestigious American brands in finance, management consulting and media, respondents spoke of highlighting their experience with such companies in their online profiles. Much like Airbnb hosts and TaskRabbit workers rely on crafting desirable online listings to garner an income (Ravenelle 2016), elite gig workers also engage in digital impression management to market themselves to potential clients, often listing brand names or highlighting trendy concepts in order to secure interest and income. As noted by Bröckling (2016) and Gershon (2016, 2017), crafting an "entrepreneurial self" through self-branding is increasingly crucial to seeking and landing a job, particularly as workers are made to shoulder greater personal responsibility for their economic outcomes (Beck 1992).

As Josh, 39, a self-described "corporate development executive" explained, "I would say 'I'm an ex-McKinsey consultant who's worked across the innovation landscape on the operational side and investor side.'" Workers also linked their past affiliations at well-known successful companies with their own brand as independent contractors and consultants. Seth, 36, a freelance consultant who gets most of his work from his business network, noted: "I am lucky in that I have a good brand name, so I do very well on Catalant ... I do very well on Catalant and SpareHire relative to others. I say 'I was a former market lead at McKinsey.'" Seth believes that the McKinsey brand name helps potential clients to believe that he's "clearly very smart." Additionally, Seth, who had previously worked in Hollywood, further increased his success rate by referencing his work with a well-known film production company when aiming for media clients. As he explained, "The hit rate that I get back from that when I'm applying to projects is very high."

Current affiliations were also seen as an asset, as Chris, 35, a consultant and adjunct instructor at an elite university explained. "It's a ladder .... the program I teach in is a really selective program. And then it's [Ivy League business school] and, and for better or worse, like people say, 'Oh, you know, [Ivy League business school] allows him to teach. You must be really smart,'" he said. "I don't think there's any truth to that ... but like I think people, that's how people see it. For better or worse. So I think that really helped." Chris's reference to a "ladder" further illustrates his awareness of status stratification. With high levels of education, and previous and current experiences working for well-known and successful companies, these workers were often aware that they were valuable commodities. As Seth, the former Hollywood employee, notes:

> I can be quite a valuable name to put on a proposal. There are some people that I work with who I said, 'Listen, if you ever need to check with me if I can ever be valuable just as a name on a proposal, go for it. You don't have to commit to bring me into the project if and when it closes. We can have that conversation then, and if it makes sense on both sides, great. If it helps you win business, put my name there because it's, for me, it does me no harm to put my name there.'

As part of this brand cachet, and their perceptions of themselves as "valuable commodities," workers often found themselves in demand with international clients in Saudi Arabia, Dubai, Singapore, Switzerland, Canada, and Germany. This international demand meant the opportunity to travel extensively, and respondents routinely spoke of managing consulting engagements across multiple continents and time zones. Occasionally, demand from wealthy overseas clients created conflicts with workers' self-perceptions of themselves as competent, but also ethical workers, as Mark, 49, a "generalist consultant" soon realized after engaging in consulting to the royal family in Saudi Arabia:

> I just wasn't ramped up on – I mean, I should have been – I'm actually a news person. I wasn't ramped up on the horrible things that the royal family had done, especially Khashoggi, etc. Only after we took the project and people had reactions to me telling them what I was working on, I was like, "Oh, yeah, okay. Probably shouldn't do that. That's, yeah, that's bad." Other than that, I wouldn't take projects that I feel bad about.

Although workers like Mark sought to add symbolic value to their labor by adopting prestigious self-branding, they often cast work that seemed ethically dubious as mercenary, and therefore incompatible with their high-status sophistication. Even so, respondents generally appreciated the platforms' ability to facilitate access to international clients, though such sites increased the porosity of international borders and increased competition for elite consulting work, as the next section will show.

9.4.2 Competing with Cheaper Labor: "I Was Competing against People ... Working for $2 an Hour"

Respondents expressed concern that the proliferation of lower-status platforms also made it harder for them to gain lucrative work commensurate with their position at the top of the international status hierarchy. Bret, 37, a marketing and design consultant, noted that the platforms enabled "much more of a free-for-all" with "50 proposals from different small businesses," with not just a large number of applicants, but also "randoms from all over the world." With so many applicants competing, especially those from lower-cost countries, hourly rates are a concern. As George, 59, a management consultant, explained:

> They serve to tamp down billable rates because you have a lot of competition between people pitching for particular jobs. I find that myself, I'm maybe competing against some guy in Bangalore who is willing to work for $10 an hour.

Although these high-status workers typically join online platforms to enjoy flexibility, pivot into a new field, or to recover after a bad corporate experience (Ravenelle et al. 2021), the consultants saw themselves and their output as especially skillful, deserving a wage premium over labor from developing countries. Cari, 42, a director of marketing for a book publisher who used online work to supplement her salary, explained:

> I found UpWork, which was good except I was competing against people who were in Thailand or the Philippines, working for $2 an hour, and it wasn't somebody who could charge the prices I wanted to charge and people at my level. I felt like Graphite was a better version and a more high-level version.

Even when workers didn't view themselves as working in direct competition with international workers, they still viewed the projects and rates on less elite platforms as undesirable. "I've done one project through Upwork but it was just, like, 'help us make this PowerPoint look pretty' and frankly Upwork is like, 'we recommend charging $18 an hour,' and I was, like, 'no that sounds terrible,'" said Sophia, 32, a management consultant. "You can't have the same kind of work through them and it's very much kind of, like, the lower-tier stuff." Anastasia, 41, CEO of her own consulting company, was blunt in her assessment about the appropriateness of certain platforms.

> Even though I figured out how to make Upwork work, I realized that it's not ... it doesn't cater to the market I really want to target. It's almost ... for a long time I was and still am, a little bit – embarrassed is not the right word – but I do recognize that [Upwork] is misaligned with my branding.

Seth, 36, had an even more visceral reaction to Fiverr, another online gig platform. "I set up a profile to see what it is, and then immediately said, 'No, that's not the work I do.'"

Respondents' brands were not just their personal affiliations, intimately connected with their sense of self (Gershon 2016), but also how they viewed themselves in the global marketplace. Particularly upon entry into white collar gig work, these workers struggled with direct competition from low-wage contractors in other countries. They eventually rationalized their decision to join elite platforms not simply in terms of material benefits but also symbolic alignment with their high-status position in the international hierarchy. Additionally, in an effort to counteract the race to the bottom on rates, respondents tried to emphasize their "brand" as members of an elite, with prestigious qualifications and high-level skill sets justifying wage premiums, as illustrated in the final section.

9.4.3 Skilled and Deserving a Wage Premium: "The Value That You Can Bring Versus Other People"

Workers sought to justify higher wages through appeals to the meritocratic nature of the international status hierarchy, not only by highlighting their personal skill but also by asserting the inferior quality of labor performed by workers based outside the US. In order to secure higher wages and to further build their brand identity as elite workers, respondents spoke of finding work that was "appropriate" in terms of the expected skills and income. For instance, Nathan, 46, who focused on growth strategy consulting, described his attraction to an elite platform as a good fit on multiple levels.

> I like that they had the appropriate kinds of projects that interested me. A lot of them are within my field, I'm a generalist but I do like IT consulting, there's a lot of variety there but there's a lot of stuff that I feel qualified to do ... And the rates are appropriate. Some are not at all – but I choose not to apply to them – but some are within the sweet spot.

To counteract the race to the bottom in terms of wages, and to further support their own identity as an elite worker with prestigious qualifications, respondents cast themselves as uniquely capable of performing skilled labor, as opposed to the menial and interchangeable tasks carried out by contractors based in other countries. George, 59, noted that it was, "A big challenge in terms of conveying your value and making sure that everyone understands the value that you can bring versus other people," clarifying that he differed in:

> my experience in decision-making responsibility. Also, frankly, the level of work I do, right? So, somebody can call themselves a management consultant, but they're really just doing clerical work for a particular client. And that's a different type of job than I do.

The idea that workers in other countries, in addition to charging less, were offering inferior work that was missing a critical component was further echoed by other respondents. In this way, the elite workers further compare and contrast their skills and experience with those of workers who they viewed as less capable or innovative. "I've done a lot of independent consulting since graduating with my MBA. I looked at other platforms, some of them, the rates are just ridiculously low," said Nathan, 46, the growth strategy consultant.

> I did a lot of writing of business plans back in the day and a lot of Indian companies or Indian individuals, people in India, they would just undercut you. When we would put out for our business plans, we were charging between $10,000 and $15,000. They'd be doing it for $200.

In addition to charging less, Nathan viewed the work as insufficient, further adding:

> They're just doing cookie-cutter. I mean fill in your name and it's kind of like doing your taxes on Turbo Tax kind of thing. It just doesn't work. We would get a lot of clients actually that went with these kinds of organizations and we'd have to clean up their mess. 'Cause not all businesses are the same. Even if you have two restaurants, they're vastly different models.

Through their reasoning about foreign competition, US gig workers construct an international status hierarchy to defend and benefit from the "symbolic value" (Ponte and Daviron 2011) attributed to professional labor from the United States. While some respondents described workers from the Global South as incapable of completing jobs to an acceptable standard, other elite workers saw them as performing menial tasks very different from their own highly skilled labor. For instance, Bret, 37, who spoke of "a free-for-all" with the online consulting platforms, limited his experience on non-elite platforms to hiring workers, describing the experience as "transactional." He further clarified,

> And I would only use them for no-brainer transactional stuff. Last week I hired someone on Fiverr, and it was because I needed something overnight. So I hired someone in Southeast Asia, and they did Photoshop work. It required no direction. It was just, "Here's three bullet points of what you need to do, here's the files." As long as they're ready in the morning, we're good. And we're done, and from what I understand it was a good rate for them, and it was incredibly cheap for me.

Juxtapositioning their work as analytical and skillful, versus transactional and clerical, further enables respondents to see themselves and their output as especially skillful, deserving a wage premium over labor from developing countries. This perception further justifies migration to platforms that cater to higher-end clients, including foreign governments and multinational firms, which enables them to better capitalize on their American credentials. Lastly, the respondents' reasoning legitimates their belief in an international status hierarchy by grounding it in meritocratic principles.

9.5 Conclusion

Although digital work platforms seek to suppress collective consciousness and labor organizing by cultivating a sense of "placelessness" among workers (Lehdonvirta 2016), our data shows that in the context of international competition, US gig workers understand the digital labor market in terms of national identity within an international status hierarchy. By asserting the relevance of national origin for the quality of services, US freelancers devalue the labor of workers in the Global South while upholding the prestige (and higher wages) associated with skilled digital labor from the United States. This interpretive process supports and reflects nation-based professional closure in the international gig economy, contributing to US gig workers' ability to garner consistently high ratings (Xu 2015) and to command the highest absolute wages for many skilled digital services (Beerepoot and Lambregts 2015). We find that US gig workers draw symbolic boundaries (Lamont and Fournier 1992) that attribute competence and expertise to US workers, while casting workers from the Global South as offering substandard or potentially counterfeit services. By positioning themselves at the top of an international status hierarchy justified on the basis of merit, US gig workers draw upon and reinforce the "symbolic value" (Ponte and Daviron 2011) of perceived quality and prestige associated with skilled digital labor from the United States.

Beginning with decades-old concern about the consequences of "twenty-first century offshoring" (Levy 2005:687), the institutionalization of a global labor market for professional services has increasingly prompted white collar workers in the US to rethink their labor in the context of international competition. US-based elite gig workers occupy a unique position within this market. Even as the conditions of digital freelance labor expose them to many of the same problems as white collar gig workers in other countries, including long hours and lack of social benefits (Mrozowicki and Trappmann 2021), US-based gig workers enjoy the highest absolute wages (Beerepoot and Lambregts 2015) and preferential hiring over workers from other countries (Xu 2015). This combination of economic precarity and relative privilege leads US-based workers to territorialize their understanding of the global labor market in terms of a meritocratic status hierarchy corresponding to national origin. They deploy symbolic boundaries to distinguish their own skilled and competent work from the menial and substandard labor of workers from the Global South, contributing to the professional closure of consulting work that occurs on exclusive gig work platforms (Boussard 2018).

One question that naturally arises is how successful and elite these respondents are if they agree to an interview for $50. Given the frequency with which they discuss the need for high pay commensurate with their elite qualifications, giving an hour or more of one's time for $50 raises questions about the potential precarity of the respondents. While few respondents declined the incentive, for most, the opportunity to contribute to research – and the promise that they would be able to access the published findings – was appealing. Most research on gig work has focused on lower status platforms, resulting in a dearth of information on the experiences of elite workers. Additionally, for those workers who regularly interview others as part of their consulting work, there may have been a sense of "research karma" and a desire to experience the interview process as an interviewee.

Our research answers the call for further attention to the role of interpretation and social status in markets (Beckert 2013; Podolny and Signals 2008; Wherry 2014), while detailing how globalization shapes perceptions of the "foreign Other" (Jijon 2019) within an international economic context. This study also elaborates questions about how Western workers understand and subjectively contend with the "new international division of labor" (Cho 1985), particularly how they perceive foreign competition within labor markets – an area needing further study (Beerepoot and Lambregts 2015). Future research can extend examination of these perceptions to workers in other countries and regions, as well as to those gig workers in the Global South who are involved in novel forms of transnational labor organizing (Lehdonvirta 2016; Lindell 2013).

Acknowledgements

This research was funded by the Ewing Marion Kauffman Foundation. The contents of this publication are solely the responsibility of Grantee. The authors wish to thank the Ewing Marion Kauffman Foundation for their support and Erica Janko and Savannah Newton for their assistance in index coding.

Note

1 It is relatively common for individuals in the United States to reject racial categorization and opt instead for national identification (DiAngelo 2018).

References

Abraham, Katharine G., John C. Haltiwanger, Kristin Sandusky, and James R. Spletzer. "Measuring the Gig Economy: Current Knowledge and Open Issues." *National Bureau of Economic Research* (2018). doi: 10.3386/w24950.

Agrawal, Ajay, John Horton, Nicola Lacetera, and Elizabeth Lyons. "Digitization and the Contract Labor Market: A Research Agenda." In *Economic Analysis of the Digital Economy*, edited by A. Goldfarb, S. M. Greenstein, and C. E. Tucker, 219–256. Illinois: University of Chicago Press, 2015.

Anwar, Mohammad Amir, and Mark Graham. "Between a Rock and a Hard Place: Freedom, Flexibility, Precarity and Vulnerability in the Gig Economy in Africa." *Competition & Change* (2020). doi: 10.1177/1024529420914473.

Bandelji, Nina, and Frederick F. Wherry. "Introduction: An Inquiry into the Cultural Wealth of Nations." In *The Cultural Wealth of Nations*, edited by N. Bandelj, and F. F. Wherry, 1–20. Stanford: Stanford University Press, 2011.

Barrientos, Stephanie. "Contract Labour: The 'Achilles Heel' of Corporate Codes in Commercial Value Chains." *Development and Change* 39, no.6 (2008):977–990. doi: 10.1111/j.1467-7660.2008.00524.x.

Beck, Ulrich. *Risk Society: Towards a New Modernity*. London: Sage Publications, 1992.

Beckert, Jens. "Capitalism as a System of Expectations: Toward a Sociological Microfoundation of Political Economy." *Politics & Society* 4, no.3 (2013):323–350. doi: 10.1177/0032329213493750.

Beerepoot, Niels, and Bart Lambregts. "Competition in Online Job Marketplaces: Towards a Global Labour Market for Outsourcing Services?" *Global Networks* 15, no.2 (2015):236–255. doi: doi: 10.1111/glob.12051.

Blinder, Alan S., and Alan B. Krueger. "Alternative Measures of Offshorability: A Survey Approach." *Journal of Labor Economics* 31, no. S1 (2013):S97–S128. doi: 10.1086/669061.

Boltanski, Luc, and Laurent Thévenot. *On Justification: Economies of Worth*. New Jersey: Princeton University Press, 2006.

Bolton, Sharon C., and Daniel Muzio. "Can't Live with 'Em; Can't Live without 'Em: Gendered Segmentation in the Legal Profession." *Sociology* 41, no. 1 (2007):47–64. doi: 10.1177/0038038507072283.

Boussard, V. "Professional Closure Regimes in the Global Age: The Boundary Work of Professional Services Specializing in Mergers and Acquisitions." *Journal of Professions and Organization* 5, no. 3 (2018):279–296.

Bröckling, Ulrich. *The Entrepreneurial Self: Fabricating a New Type of Subject*. London: SAGE Publications Ltd, 2016.

Cennamo, Carmelo. "Competing in Digital Markets: A Platform-Based Perspective." *Academy of Management Perspectives* 35. No.2 (2021): 265–291. doi: 10.5465/amp.2016.0048.

Cho, Soon Kyoung. "The Labor Process and Capital Mobility: The Limits of the New International Division of Labor." *Politics & Society* 14, no.2 (1985):185–222. doi: 10.1177/003232928501400203.

De Stefano, Valerio. "The Rise of the Just-in-Time Workforce: On-Demand Work, Crowdwork, and Labor Protection in the Gig-Economy." *Comparative Labor Law & Policy Journal* 37, no.3 (2015):471–504.

Deterding, Nicole M., and Mary C. Waters. "Flexible Coding of In-depth Interviews: A Twenty-first-century Approach." *Sociological Methods & Research* 50, no.2 (2018):708–739. doi: 10.1177/0049124118799377.

DiAngelo, Robin. *White Fragility: Why It's so Hard for White People to Talk about Racism*. Massachutes: Beacon Press, 2018.

Dicken, Peter. *Global Shift: Mapping the Changing Contours of the World Economy*, 7th edition. New York: Guilford Press, 2015.

Edison Research. *The Gig Economy: From the Marketplace-Edison Research Poll December 2018*. Somerville, NJ: Edison Research, 2018.

Egger, Hartmut, and Peter Egger. "Labor Market Effects of Outsourcing under Industrial Interdependence." *International Review of Economics & Finance* 14, no. 3 (2005):349–363. doi: 10.1016/j.iref.2004.12.006.

Fish, Adam, and Ramesh Srinivasan. "Digital Labor Is the New Killer App." *New Media & Society* 14, no. 1 (2012):137–152. doi: 10.1177/1461444811412159.

Friedman, Thomas L. *The World Is Flat: A Brief History of the Twenty-First Century*, 1st edition. New York: Farrar, Straus and Giroux, 2005.

Gershon, Ilana. "'I'm Not a Businessman, I'm a Business, Man': Typing the Neoliberal Self into a Branded Existence." *HAU: Journal of Ethnographic Theory* 6, no. 3 (2016):223–246. doi: 10.14318/hau6.3.017.

Gershon, Ilana. *Down and Out in the New Economy: How People Find (or Don't Find) Work Today*. Illinois: University of Chicago Press, 2017.

Graham, M., V. Lehdonvirta, A. Wood, H. Barnard, I. Hjorth, and D. Peter Simon. *The Risks and Rewards of Online Gig Work at the Global Margins*. Oxford: Oxford Internet Institute, 2017b.

Graham, Mark, Isis Hjorth, and Vili Lehdonvirta. "Digital Labour and Development: Impacts of Global Digital Labour Platforms and the Gig Economy on Worker Livelihoods." *Transfer: European Review of Labour and Research* 23, no.2 (2017a):135–162. doi: 10.1177/1024258916687250.

"Graphite". Graphite. n.d. Accessed June 17, 2021. www.graphite.com/.

Horton, John, William R. Kerr, and Christopher Stanton. "Digital Labor Markets and Global Talent Flows." In *High-Skilled Migration to the United States and Its Economic Consequences*, edited by G. H. Hanson, W. R. Kerr, and S. Turner, 71–108. Illinois: University of Chicago Press, 2018.

Jenkins, Tania M., and Shalini Reddy. "Revisiting the Rationing of Medical Degrees in the United States." *Contexts* 15, no.4 (2016):36–41. doi:10.1177/1536504216684820.

Jijon, Isabel. "Toward a Hermeneutic Model of Cultural Globalization: Four Lessons from Translation Studies." *Sociological Theory* 37, no.2 (2019):142–161. doi: 10.1177/0735275119850862.

Katta, Srujana, Adam Badger, Mark Graham, Kelle Howson, Funda Ustek-Spilda, and Alessio Bertolini. "(Dis)Embeddedness and (de)Commodification: COVID-19, Uber, and the Unravelling Logics of the Gig Economy." *Dialogues in Human Geography* 10, no. 2 (2020):203–207. doi: 10.1177/2043820620934942.

Khan, Shamus. "The Sociology of Elites." *Annual Review of Sociology*, 38, no.1 (2012):361–377. doi: 10.1146/annurev-soc-071811-145542.

Lamont, Michèle, and Marcel Fournier, eds. *Cultivating Differences: Symbolic Boundaries and the Making of Inequality*. Illinois: University of Chicago Press, 1992.

Lehdonvirta, Vili. "Algorithms That Divide and Unite: Delocalisation, Identity and Collective Action in 'Microwork'." In *Space, Place and Global Digital Work*, edited by J. Flecker, 53–80. London: Palgrave Macmillan UK, 2016.

Levy, David L. "Offshoring in the New Global Political Economy." *Journal of Management Studies* 42, no. 3 (2005):685–693. doi: 10.1111/j.1467-6486.2005.00514.x.

Lindell, Ilda. *Africa's Informal Workers: Collective Agency, Alliances and Transnational Organizing in Urban Africa*. New York: Zed Books, 2013.

Meagher, Kate. "The Scramble for Africans: Demography, Globalisation and Africa's Informal Labour Markets." *The Journal of Development Studies* 52, no.4 (2016):483–497. doi: 10.1080/00220388.2015.1126253.

Mrozowicki, Adam, and Vera Trappmann. "Precarity as a Biographical Problem? Young Workers Living with Precarity in Germany and Poland." *Work, Employment and Society* 35, no.2 (2021):221–238. doi: 10.1177/0950017020936898.

"Our Experts." Graphite. n.d. Accessed June 17, 2021. www.graphite.com/info/ourexperts.

Podolny, Joel M. *Status Signals: A Sociological Study of Market Competition*. New Jersey: Princeton University Press, 2008.

Ponte, Stefano, and Benoit Daviron. "Creating and Controlling Symbolic Value: The Case of South African Wine." In *The Cultural Wealth of Nations*, edited by N. Bandelj, and F. F. Wherry, 197–221. Stanford: Stanford University Press, 2011.

Ravenelle, Alexandrea J. "A Return to Gemeinschaft: Digital Impression Management and the Sharing Economy." In *Digital Sociologies*, edited by J. Daniels, K. Gregory, 25–43. Bristol: Policy Press, 2016.

Ravenelle, Alexandrea J. *Hustle and Gig: Struggling and Surviving in the Sharing Economy*. Berkeley, CA: University of California Press, 2019.

Ravenelle, Alexandrea J., Erica C. Janko, and Ken C. Kowalski (2021). "Gigging with an MBA: Elite Entrepreneurship in the Gig Economy." In *Digital Entrepreneurship and the Sharing Economy*, edited by E. Vinogradov, B. Leick, and D. Assadi, 145–159. New York: Routledge.

Rivera, Lauren A. "Managing 'Spoiled' National Identity: War, Tourism, and Memory in Croatia." *American Sociological Review* 73, no.4 (2008):613–634. doi: 10.1177/000312240807300405.

Thompson, Beverly Yuen. "Digital Nomads: Employment in the Online Gig Economy." *Glocalism: Journal of Culture, Politics and Innovation* 1 (2018):1–26. doi: 10.12893/gjcpi.2018.1.11.

Weiss, Robert S. *Learning From Strangers: The Art and Method of Qualitative Interview Studies*. New York: The Free Press. 1994.

"What We Do". Catalant. n.d. Accessed July 23, 2020. https://gocatalant.com/about-2/

Wherry, Frederick F. "Analyzing the Culture of Markets." *Theory and Society* 43, no. 3 (2014):421–436. doi: 10.1007/s11186-014-9218-3.

Wood, Alex J., Mark Graham, Vili Lehdonvirta, and Isis Hjorth. "Good Gig, Bad Gig: Autonomy and Algorithmic Control in the Global Gig Economy." *Work, Employment and Society* 33, no. 1 (2019a):56–75. doi: 10.1177/0950017018785616.

Wood, Alex J., Mark Graham, Vili Lehdonvirta, and Isis Hjorth. "Networked but Commodified: The (Dis)Embeddedness of Digital Labour in the Gig Economy." *Sociology* 53, no.5 (2019b):931–950. doi: 10.1177/0038038519828906.

Xu, Guo. "How Does Collective Reputation Affect Hiring? Selection and Sorting in an Online Labour Market." *SSRN Scholarly Paper*, ID 2609898. Rochester, NY: Social Science Research Network, 2015.

10 Understanding Residential Sorting through Property Listings

A Case Study of Neighborhood Change in Charlotte, NC 1993–2018

Isabelle Nilsson and Elizabeth C. Delmelle

10.1 Introduction

Neighborhood dynamics are largely driven by the in and out migration of residents in conjunction with changes to the built environment. Residential location decisions are therefore a core determinant of the neighborhood change process. From an economic perspective, residents select a home subject to their budgetary constraints that maximizes their satisfaction in terms of the home's characteristics and the bundle of local amenities associated with that property (Brasington, 2014; Tiebout, 1956). There is a body of research that has sought to understand how residents, differentiated by income, age, race, or lifecycle status, value certain urban amenities when making these decisions. This research helps to explain spatial restructuring within urban areas by that may give rise to social inequalities and displacement by examining urban sorting patterns according to characteristics of residents, including recent shifts in the importance of urban amenities in attracting wealth back to urban centers (Glaeser & Gottlieb, 2006; Baum-Snow & Hartley, 2020).

Most of our current understanding of the role of various amenities in shaping neighborhood change has relied on traditional data sources such as the census, combined with spatial variables on measurable amenities hypothesized to influence residential sorting (e.g. walkability, distance to natural amenities, breweries, coffee shops, etc.). However, the use of more novel, user-generated datasets are opening up exciting avenues for understanding changes in more real-time than retrospective census surveys and they may encompass hidden or more difficult to measure attributes of a location.

In this chapter, we explore the use of the text, also referred to as public remarks, of property advertisements as one such novel dataset that holds the potential to provide insight into changes occurring in a neighborhood in near real-time, and to bring to light amenities that help explain sorting patterns by race and income. To achieve this, we blend property listing text with a classification of neighborhoods according to their annual trends in mortgage applicants by race and income using data from the Home Mortgage Disclosure Act. We build upon our prior research that used a *k*-means approach to classify neighborhoods at static time points (Delmelle & Nilsson, 2021) by providing an

DOI: 10.4324/9781003191049-13

alternative way of classifying neighborhoods that considers their longer-term trajectories of mortgage applicants by race and income in an attempt to better capture neighborhood change. We do this by introducing the use of latent growth curve modeling as a means of explicitly categorizing neighborhoods according to their longer-term trajectories. We then apply the same text analysis procedure as in Delmelle and Nilsson (2021) to understand the relative importance of various words used in the advertisements by type of neighborhood and explore how well the text can predict the type of neighborhood it belongs to. We find this new classification technique yields more interpretable and less noisy results in the corresponding text analysis and leads to better predictions in some cases. Finally, we discuss the potential of this classification technique and data sources for advancing neighborhood change prediction efforts.

10.2 Background

We begin our background section with a discussion on what is known about residential location choice decisions by different socioeconomic, demographic, and racial groups and the role of amenities in shaping these decisions. This informs our hypotheses about which amenities we might expect to be advertised more prominently in certain types of neighborhoods. We then turn to the existing literature on real estate listings and discuss the role that realtors may play in swaying location decisions, potentially serving as gatekeepers to particular neighborhoods. Finally, we discuss the current state of the art for predicting neighborhood changes and discuss how text analysis of real estate listings may advance these efforts.

10.2.1 Residential Location Choice and Amenities

There are several ways to think about the residential location decisions that give rise to neighborhood changes. Economic theories suggest that households seek to maximize their housing satisfaction, given their own budgetary constraints, and will subsequently sort themselves based on the bundle of available local amenities associated with a property that meets these criteria (Brasington, 2014; Tiebout, 1956). Traditionally, households were thought to make tradeoffs between commute lengths and housing and lot sizes, resulting in an ordered sorting pattern observed throughout many cities across the United States: one that featured a largely minority and poor urban core with residents living in smaller homes, unable or unwilling to afford higher commuting costs and larger more expensive suburban homes, and a wealthy and largely White suburb (Alonso, 1964; Muth, 1969). Since those foundational theories, however, the urban landscape has become more complex with younger and wealthier residents re-populating previous poor and minority neighborhoods, and pockets of prosperity and wealth more fragmented throughout a city, shaking up the neat, overly-simplified monocentric urban models that described cities of the early 1900s (Baum-Snow & Hartley, 2020; Delmelle, 2019; Florida & Adler, 2018; Glaeser & Gottlieb, 2006; Lee et al., 2019).

Explanations behind this increasingly fragmented urban socioeconomic landscape have homed in on shifting amenity preferences associated with changes in the nation's demographic and racial makeup. While the post-war baby boom generation settled in new homes constructed during the suburban housing growth of the 1950s and 1960s, their children, the premillennial and millennial generations, have exhibited significantly different family structures and housing patterns than their parents. On average, they are less likely to be married, have fewer children and often have children later in life, are more highly educated, and have lived with their parents longer than previous generations (Clark, 2019). The increasing share of highly educated, child-less residents has led to an amplified demand for center city locations with more creative-cultural amenities over the past several decades (Glaeser & Gottlieb, 2006; Baum-Snow & Hartley, 2020; Lee et al., 2019; Couture & Handbury, 2020). Millennials in particular have demonstrated an increased preference for urban amenities such as walkability, transit, and mixed-used developments as compared to their parent's generation (Moos, 2014). The presence of historic and architecturally interesting housing to be redeveloped is a key attribute alongside these amenities in anticipating where reinvestment and population shifts may occur (Danielsen & Lang, 2010). New urban developments have catered towards these demographic trends by largely constructing housing suitable for child-less households leaving center city family options relatively limited (Ehlenz et al., 2020).

While demand for urban living has been on the rise, growth continues to be strong among wealthy Whites for suburban locations (Myers, 2016). Lifecycle trends continue to significantly impact housing and location choices in traditional ways, even if the millennial generation has spent a longer duration of their life in a child-less phase. Research has shown that once members of this generation do settle down and start a family, their location decisions are really not that different from those of their predecessor generation (Lee, 2020). There is some speculation that suburban locations that feature some 'urban' amenities including walkability and mixed land-uses will be in high demand (Ehlenz et al., 2020). So-called new urbanist or new sub-urbanist developments have increased in popularity, especially in rapidly growing sunbelt cities where developers are willing to experiment with new forms of development, and where a lack of historic downtown locations to revitalize has established the need to create urbanity anew (Delmelle et al., 2014; Ehlenz et al., 2020).

The desire for urban amenities is not the full story of recent location choices. A rise in Homeowner's Associations nation-wide and their associated 'exclusive' amenities such as golf courses have increasingly attracted a very affluent and racially homogenous population (Clarke and Freedman, 2019). These types of suburban amenities have been charged with perpetuating patterns of racial and income segregation in cities – they often include large membership fees that serve to exclude some who may be able to afford housing costs, and they typically feature amenities desired by a wealthier and Whiter population (Strahilevitz, 2006). High-amenity, minority neighborhoods are a limited option across most of the United States. As noted by Bayer and McMillan

(2005), to live in a high amenity neighborhood, Black homebuyers must opt to live in predominantly White neighborhoods. Older-suburban neighborhoods that lack the character of historic center-city homes and are devoid of many of today's desirable amenities have increasingly become the more affordable housing options for poorer and minority residents (Hanlon, 2008).

The current state of knowledge on the association of various urban amenities by various racial and socioeconomic groups is limited by a lack of comprehensive data on all types of available amenities that may serve to sort residents. Most studies have selected a limited sample of measurable amenities (Lee et al., 2019) or have grouped neighborhoods into bundles of unobserved characteristics to study their changing value through time (Lee et al., 2021). In this work, we explore the use of real estate listing text (i.e., public remarks) to extract key amenities used to advertise properties in different types of neighborhoods. We hypothesize that the amenities of listings will vary according to the racial and income profile of homebuyers, according to the residential sorting literature discussed above.

10.2.2 Realtors and Housing Markets

When residents select a home to purchase, they often rely on the expertise of real estate agents to help guide them through the search process. Real estate agents or brokers generally work on commission, earning a share of the purchase price of the home and it is therefore in their best interest when selling a property to seek out a homebuyer who will pay the highest amount for a home. Thus, the actions of realtors in showing properties to prospective homebuyers and in how they market a property play a role in shaping broader patterns of residential sorting. When listing a property, brokers have several tools at their disposal for attracting potential buyers including the words used to describe the property and accompanying photographs. Some research on these advertising tactics has demonstrated that the words used in public remarks have a larger impact on selling price and time on the market than pictures (Luchtenberg et al., 2019). Moreover, research has also shown that higher-end homes with more specific characteristics tailored towards a particular homeowner tend to contain less information in their property listings than more generic, lower-end homes (Bian et al., 2021). The motivation for this is to maximize the possibility of a person viewing a higher-end property that may have very taste-specific attributes.

Text, or the multiple listing service (MLS) public remarks, has been incorporated into hedonic modeling efforts to explain or predict home prices and listing time on the market. Research has shown that using the public remarks can improve prediction, but that not all terms are equally important, and their importance may vary across space (Nowak et al., 2021; Nowak & Smith, 2017; Shen & Ross, 2021). Seo et al. (2020) showed that terms related to the quality of a structure in particular was an important determinant of prices.

As the research in this chapter is concerned with linking the words used in MLS public remarks with neighborhood changes according to race and income, it is pertinent to also discuss the role of realtors in contributing to patterns of racial and income segregation in US cities. Much has been written about the discriminatory history of real estate practices prior to the passage of the 1968 Fair Housing Act where denying access to housing and neighborhoods based on race was both legal and widely practiced by real estate agents and mortgage lenders (Massey, 2005). Historical practices of rating neighborhoods based on their housing quality, characteristics of residents – including the presence of immigrants and non-whites – and location with respect to industry, established segregated maps dictating where financial investment occurred. White and suburban neighborhoods were deemed less risky to financial institutions and were therefore beneficiaries of affordable loans whereas minority, city center neighborhoods were locations of disinvestment. These 'redlining' maps created in the 1930s have had a lasting impact on patterns of racial and income segregation to this day (Aaronson et al., 2021; White et al., 2021).

The passage of the Fair Housing Act of 1968 formally outlawed discrimination by race in the real estate market. However, practices that circumvented the law enabled patterns of segregation to remain entrenched in US cities. Residential steering, or the act of showing different properties in different neighborhoods according to the race of the home seeker is one such practice that has been regularly uncovered by real estate audits (Galster & Godfrey, 2005; Ondrich et al., 2003). Research has shown that minority home-seekers are disproportionately shown properties in lower-income and lower-amenity neighborhoods with worse school quality (Galster & Godfrey, 2005).

While more research has examined the direct actions of real estate agents in showing or not showing residents certain properties according to their race, less attention has been paid to the words used to advertise properties, who is purchasing those properties, and the connection to broader patterns of neighborhood change. As mentioned, real estate agents seek to maximize their profits when selling a property, both to increase their own commission, and also on behalf of the homeowner, their client. To do so, they use marketing techniques such as text and pictures to attract a certain homebuyer. Some prior research has suggested that the words used to market properties differs across an urban language – that there exists a local dialect in real estate listings (Pryce & Oates, 2008). Kennedy et al. (2021) studied rental advertisement text in Seattle according to the racial composition of neighborhoods and found more incentives listed in non-White neighborhoods and had a greater focus on transportation while listings in predominantly White neighborhoods emphasized trust and connections to neighborhood culture and history. In a previous analysis of residential property listings using the same dataset featured in this chapter, of a sample of properties obtained from Zillow for one-month in Charlotte, North Carolina, we linked text with neighborhoods classified according to the racial and income profile of mortgage applicants (Delmelle & Nilsson, 2021). In that analysis, we found more unique characteristics of historical homes (e.g. parquet flooring) in neighborhoods with indicators of gentrification – or those with an increase in White homebuyers in minority neighborhoods. Listings in neighborhoods with a majority of Black homebuyers were less likely to mention schools and were less likely to feature the name of the neighborhood compared to Whiter and wealthier neighborhoods.

10.3 Case Study

Our case study on the connection between property advertisement text and neighborhood characteristics is performed on a one-month sample of property listings in Charlotte, North Carolina, USA, in September of 2019. Charlotte is a fast-growing sunbelt city with a competitive housing market – the demand for housing has outpaced supply, making the last decade a sellers' market (Childress Klein Center for Real Estate, 2020). Charlotte's population grew 21 percent during the decade from 2010 to 2020 from 731,400 to 885,700 residents (Chemtob & Off, 2020). Demand has been strong in the urban core – locally referred to as 'uptown' – as gentrification pressures have accompanied the development of a new light rail line and accelerated in older, walkable environments where the presence of breweries and new restaurants have increased (Delmelle et al., 2021). Simultaneously, demand has remained strong in outlying suburbs and towns (Chemtob & Off, 2019). Consistent with other fast-growing southern cities, the types of developments in these suburbs has varied from traditional, single-family dwellings, to denser, more walkable, 'new urbanist' designs (Delmelle et al., 2014). As a result of these simultaneous dynamics, poverty has increased in older, first and second-ring suburban neighborhoods (Delmelle et al., 2021).

10.3.1 Methods and Data

In this case study, we classify neighborhoods based on their trends (or trajectories) in home mortgage applicant income and share of minority (Black and Hispanic/Latino) applicants between 1993 and 2018 using latent class growth analysis and cross-classification between class trajectories along these two variables. Once neighborhoods have been assigned to a cluster based on their trends in these two variables, we analyze the words used in the public remarks of real estate listings in the different types to generate insights into what home features and neighborhood amenities are highlighted across neighborhood types. Public remarks are the description of the property such as:

> Beautifully renovated full brick ranch home with a new wood privacy fence. Stainless steel appliances. White shaker style kitchen cabinets. Beautiful hardwood floors. New HVAC system and insulated windows. Great location! Less than 1 mile to NODA, Charlotte Historic Arts & Entertainment District. Approx. 5 miles from Lynx Light Rail. The HOA covers roof, crawl space & exterior. HOA Dues $119/ monthly to Shamrock Green HOA.

We compare the findings in this case study with those from our previous study (Delmelle & Nilsson, 2021) where we used the same real estate dataset and text-analysis approach but classified neighborhoods on a different set of variables including both mortgage applicant characteristics and census data. In our previous study, we also only considered more recent changes in these variables (difference between 2013 and 2018) and did not consider the longer-term trajectories of changes in neighborhood characteristics. Since neighborhoods tend to be slow to change (Nilsson & Delmelle, 2018; Wei & Knox, 2014) and may follow different trajectories of change, we want to capture this using a dataset that includes a longer time period and a classification methodology that considers the temporal trend in neighborhood homebuyer characteristics rather than changes between two arbitrary selected points in time.

10.3.1.1 Neighborhood Classification

Neighborhoods are classified based on the trends in annual mean applicant income and share of minority applicants over the time period 1993–2018. Data comes from the Home Mortgage Disclosure Act (HMDA) dataset and includes (among other variables) the income, race, and ethnicity of applicants. While an even longer time period would have been desired since a complete change in neighborhood characteristics can take decades to occur, HMDA data is only available from 1993 onward (FFIEC, 2021). However, it is one of the few datasets on neighborhood characteristics that is published annually at small geographies over such an extended period of time. The geographic identifier for home mortgage applications is the census tract, which will be serving as our proxy for neighborhood. One of the advantages of the HMDA data is that it is released annually and it also gives us an indication of potential change in current characteristics by showing the characteristics of those applying to buy homes in the neighborhood. Hence, we limit the analysis to home purchase loans and exclude applications considering for example refinance or home improvement as it is more of an indicator of who currently lives in the neighborhood and may skew towards neighborhoods with more capital. We also exclude mortgage application records with edit failures.

The classification is performed using a latent class growth model or latent class growth analysis (LCGA). This type of modeling framework focuses on the distribution of outcomes conditional on time, i.e., the distribution of outcome trajectories $P(Time_i)$, where Y_i is individual i's longitudinal sequence of outcomes and $Time_i$ represents the time at which individual i's response is recorded. The latent class trajectory model assumes that individuals belong to different subpopulations (latent classes) where each class has a unique growth trajectory of unknown order J (Jones & Nagin, 2013; Zwiers et al., 2017).

Here we estimate a linear latent class growth model with a fixed intercept and slope where the dependent variable is a function of time. To estimate the model, we use the FLXMRglmfix function in the FlexMix package in R (Gruen et al., 2020). This function allows us to specify general linear models for latent class regression and classification. Due to no prior assumption about the functional form of different classes, we specify it as a linear model based on a visual inspection of trends by neighborhood, especially for the mean applicant income variable. We set the residual term to be equal across classes which is the default in popular latent class growth analysis software such as Mplus (Mplus, 2021). We apply a stepwise procedure using the stepFlexmix function (Gruen et al., 2020) which allows us to fit models with an increasing number of latent classes and compare them using common model selection criteria including Akaike information criterion (AIC) or Bayesian information criterion (BIC) (Leisch, 2004). Here we limit the maximum number of classes to five in order to make the cross-classification and interpretability of our results feasible.[1] Finally, finite mixture models with a fixed number of classes are usually estimated with the expectation–maximization (EM) algorithm with a maximum likelihood framework. Since the EM algorithm converges only to the next local maximum of the likelihood, it should be run repeatedly using different starting values (Leisch, 2004). Here we estimate models with 1 to 5 classes up to 50 iterations each to reach convergence and to prevent local maxima.

Once we have a model for each number of classes, we determined the appropriate number of classes for each dependent variable based on model fit statistics, visualization of resulting trends, and distribution of observations by class. We then use cross-classification to further group neighborhoods based on their combination of income and ethnic/racial trajectories.

10.3.1.2 Text Analysis and Prediction

To analyze the public remarks of real estate listings in the different clusters to generate insights into what home features and neighborhood amenities are being highlighted across neighborhood types, we apply the same methodological approach as in Delmelle and Nilsson (2021). This enables us to compare results and what insights can be gained from using two different approaches to classifying neighborhoods. We also use the same sample of property advertisements collected from the website Zillow (Zillow, 2021). It contains all properties listed for sale in Charlotte, NC, in September 2019. The address of the listings was geocoded to assign them a neighborhood type.

The public remarks in the listings were cleaned from stop words (e.g., "a", "the", "is", "and", etc.), generic words associated with properties (e.g., "bedroom", "bathroom"), and several words were combined to improve the interpretation of the results (e.g., "stainless steel" became "stainlesssteel"). We also removed words that showed up in less than ten property listings in order to remove, for example, typos and outliers. Finally, we replaced all neighborhood and street names as well as names of shops, restaurants, and entertainment establishments unique to Charlotte with generic placeholders such as "neighborhoodname", "streetname", "shop", "restaurant", and "entertainment". All cleaning and management of the data was performed using the TidyText package in R (Silge & Robinson, 2017).

To predict what neighborhood cluster a property belongs to we use a binomial logistic regression model combined with LASSO regularization to select variables (i.e., words) to include in the model (Hastie et al., 2021). Binomial logistic models allow us to better understand the influence of specific words in predicting neighborhood cluster type as well as the relative importance of different words used in the public remarks for that neighborhood type compared to all other neighborhood types instead of an arbitrary base case. We split the data into a training (75%) and testing (25%) set and estimate the models on the training dataset using the glmnet function in R (Hastie et al., 2021). Using the parameters from the models estimated using the training dataset, we make predictions with regards to which neighborhood type a property in the testing dataset belongs to. Of particular interest for this study are the words that show to be most discriminant for each neighborhood type (i.e., the largest positive and negative predictors for each neighborhood type) and we pay particular attention to these later in the analysis.

10.4 Results

10.4.1 Neighborhood Classification

All models for both variables converged before 50 iterations were reached. After comparing the models based on BIC, distribution of observations by classes, rootograms of posterior class probabilities,[2] and making sure that classes showed qualitatively different trajectories, we decided on a three-class model for each variable. While the BIC was slightly lower for a greater number of classes (indicating a better fit), the rootograms showed better separation of classes under fewer classes. We also had to consider the number of observations within each class for the cross-classification and, more importantly, for the prediction portion of the analysis. The three classes for each variable are shown in Figures 10.1 and 10.2.

As expected, as incomes rise over time, all three trajectories have a positive slope. However, the intercept and rate of change varies across classes. It shows trajectories where neighborhoods with higher incomes in the beginning of the study period are becoming increasingly wealthy at a sfaster pace than other neighborhoods while neighborhoods with initially lower incomes experienced smaller increases in incomes over time. Some more distinct changes in trajectories are shown in terms of racial composition of neighborhoods. We identify three classes, or trajectories: neighborhoods with high shares of minority applicants in 1993 that continue to experience an influx of minority applicants over time and those that had high shares of minority applicants in the beginning of the study period but that experienced a sharp decrease in minority applicants over time. Finally, we identified one set of neighborhoods that initially had a relatively small share of minority applicants but that experienced an increasing share of minority applicants over time. This is not surprising as recent census data suggests that a vast majority of Charlotte neighborhoods increased in diversity with only a small share, a certain group of neighborhoods, experienced a reduction in racial diversity (Off & Wright, 2021).

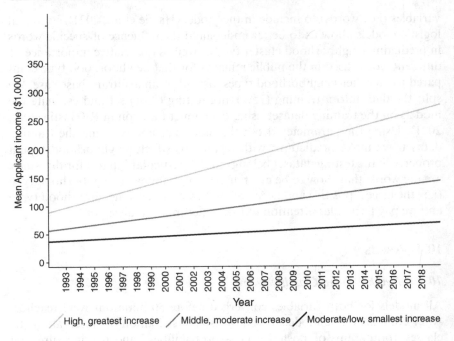

Figure 10.1 Mean applicant income classes.

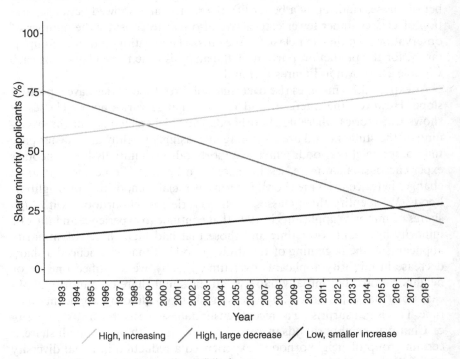

Figure 10.2 Share minority applicant (%) classes.

Using cross-classification, neighborhoods were grouped into clusters based on their applicant income and minority applicant trajectories. This originally resulted in six classes (Table 10.1), which was later combined into the five classes that are mapped in Figure 10.3 due to the small number of (only two) neighborhoods in one of the original cross-classified clusters.

While the "true" number of latent classes/trajectories is unknown, the final geographical distribution of our cross-classified classes corresponds well with the authors' local knowledge of the city as well as the five classes generated in our previous study (Delmelle & Nilsson, 2021).

The final cross-classified clusters are named and described accordingly:

1 *White and wealthy (income class 1, minority class 1)*
These neighborhoods have the highest applicant incomes initially and the largest increase in income over time. They also had the smallest initial share of minority applicants and have experienced the smallest increase in minority applicants over time. This cluster include neighborhoods such as Myers Park and Eastover in the southern parts of the city often referred to among locals as the "Wedge of Wealth" as well as parts of Davidson and Cornelius in the Lake Norman area in the north (CharlotteFive, 2018; Lacour, 2019; Leading on Opportunity, 2017). These are the least dense and most white (90%) neighborhoods according to recent census data (see Table 10.2). They also have a median household income that is well above the other clusters and more than double the median household income for the city which in 2019 was $62,817 (Census, 2021).

2 *White higher income (income class 2, minority class 1)*
These neighborhoods are in the second income class with upper middle to higher incomes but not as high as income class 1 and have experienced a moderate increase in income over time. This cluster include neighborhoods in the towns of Davidson, Cornelius, and Huntersville in the north, the wards of uptown, and Plaza Midwood just southeast of uptown as well as the outer ring of the "Wedge of Wealth" and the city's most southern neighborhoods including Piper Glen Estates and Providence Plantation. Neighborhoods in this cluster have the second highest median income and second highest share of white, non-Hispanic residents of the

Table 10.1 Cross-classification

Applicant income class	Share minority applicants class		
	1. Low, smaller increase	2. High, large decrease	3. High, increasing
1. High, greatest increase	9	0	0
2. Middle, moderate increase	32	2	0
3. Moderate/low, smallest increase	15	13	37

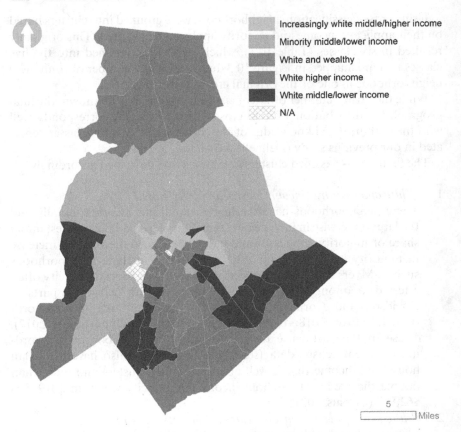

Increasingly white middle/higher income
Minority middle/lower income
White and wealthy
White higher income
White middle/lower income
N/A

5 Miles

Figure 10.3 Mapped cluster by neighborhood across Mecklenburg County, North Carolina, where the City of Charlotte is the county seat.

Note: Two neighborhoods are missing data for one or more of the time periods. One of these, the larger hashed neighborhood in the map, is the location of Charlotte Douglas International Airport.

clusters according to recent census estimates (see Table 10.2). They also have the largest share of family households closely followed by cluster 1.

3   *Increasingly white middle/higher income (income classes 2 and 3, minority class 2)*
This cluster includes former minority neighborhoods in and around the center city, many which have been undergoing gentrification in the past couple of decades including neighborhoods like NoDa, Belmont, Villa Heights, Wesley Heights, and Wilmore. These neighborhoods have experienced an influx of White, higher-earning millennials but some are still very diverse (Dunn, 2017; Logan, 2018). These are neighborhoods in transition from being lower-income, majority-minority neighborhoods to becoming gentrified hence the lower income reported in Table 10.2. Here we combined census tracts that were in the higher income classes 2 and 3 but that both used to have a high share of minority applicants which has dramatically decreased since 1993. The reason for including the two

Table 10.2 2018 American Community Survey five-year estimates for the five clusters

	White and wealthy	White higher income	Increasingly White middle/ higher income	White middle/ lower income	Minority middle/lower income
Population density	2,743	2,879	4,058	3,374	2,861
Median household income ($)	132,438	95,325	41,009	59,547	51,372
White (not Hispanic or Latino) population (%)	90%	73%	27%	52%	24%
Family households (%)	63%	64%	47%	58%	61%
1-person non-family households (%)	31%	28%	40%	32%	30%

higher income neighborhoods in this category is due to many shared built environments, demographic and socioeconomic characteristics and the low number in the combined income class 2/minority class 3. Furthermore, while the other neighborhoods are classified in income class 3, it is important to note that many of the in-movers to these neighborhoods are young professionals that are in the beginning of their careers and usually childless. Hence, while their incomes may not be as high, they are of higher socioeconomic status (e.g., education) and more often single households meaning that applicant income is based on only one income and not two as in many of the more family oriented, suburban neighborhoods as reflected in the percent one-person households in Table 10.2.

4 *White middle/lower income (income class 3, minority class 1)*
   This cluster contains suburban and increasingly suburban former rural areas including the towns of Pineville and Mint Hill. It also includes some neighborhoods closer to the center-city. They are majority white, and contain a mix of working- and middle-class neighborhoods with a large share of family households. They have historically had a low share of minority home mortgage applicants and have only experienced a modest increase in the number of minority applicants since 1993.

5 *Minority middle/lower income (income class 3, minority class 3)*
   This cluster consists of mainly majority-minority neighborhoods and middle- and lower-income neighborhoods including much of the struggling areas referred to among locals as the "Crescent" (Lacour, 2019;

Leading on Opportunity, 2017). However, this cluster also contains more suburban, middle income neighborhoods with higher shares of minorities. This is the only cluster that had neighborhoods classified as having a high initial share of minority applicants and also had the largest increase in minority applicants over time (Figure 10.2). It has the second lowest median household income of all clusters according to recent census estimates (Table 10.2), one that is below the city median household income. It also has the lowest share of White, non-Hispanic residents.

Overall, these clusters are very similar to those found in Delmelle and Nilsson (2021). It is important to note here that in our previous analysis we were able to use more disaggregated census tracts based on the 2010 delineation due to our use of 2013 and 2018 HMDA and census data in the classification. In the current study, we have to use tract delineations from 1990 to accommodate the longer time-series dating back to 1993 and avoid the issues associated with the arbitrary disaggregation of HMDA data to newer tract boundaries. The *White and Wealthy* cluster shows very similar characteristics and mapped patterns to the *White-Increasingly High Income* cluster in Delmelle and Nilsson (2021). Similarly, the *White Higher Income* cluster in this study shows a very similar pattern to the *White Higher Income* cluster in the previous study. While there is some correspondence between the *White Homebuyers in Minority Neighborhoods* cluster from Delmelle and Nilsson (2021) and our *Increasingly White Higher/Middle Income* cluster in this study, the previous classification shows a greater spread and fragmentation out from the center-city and in towards some first ring suburbs including some neighborhoods in the "Crescent". This is likely due to the reliance on more recent data in the previous study as these neighborhoods have more recently started to feel pressures of gentrification. In our current classification, many of these are still considered majority-minority neighborhoods for a large share of the longer time period under consideration. The two clusters of *Increasingly Black* and *Hispanic Homebuyers in Minority Neighborhoods* in Delmelle and Nilsson (2021) show a very similar pattern (an extended "Crescent") to our current *Minority Middle/Lower Income* cluster which includes both of these groups. Finally, the majority of the neighborhoods in our current *White Middle/Lower Income* cluster are classified as *White Higher Income* in our previous classification. Again, likely to the use of more recent data in the previous analysis (ignoring the historical income pattern) as well as differences in methodologies and variables included.

In short, despite differences in spatial and temporal disaggregation, number of variables and longitudinal extent of the data used in the classification used in this study and the one used in Delmelle and Nilsson (2021), the number and characteristics of the resulting clusters generated by the two different approaches to classification are very similar with some nuances.

10.4.2 Text Analysis

We now turn to the analysis of the text used in the public remarks of real estate listings in the different clusters generated in the previous section. First, we show the most commonly used words in public remarks by neighborhood clusters. As in Delmelle and Nilsson (2021), the neighborhood name placeholder ranks highest in all clusters with the minority homebuyer cluster showing less of an importance of the neighborhood name compared to the second most commonly used word. The common words for this cluster also reflect the suburban nature of its neighborhoods with spacious, garage, family, and car among the top words. While higher-end features such as granite and hardwoods are being mentioned, this is the only cluster that has carpet mentioned.

The White, higher income clusters have, besides "neighborhood name", "hardwood", and "granite" ranked highly, and other words that do not show up in the lower income clusters such as "custom" and "private". The low density of the wealthiest cluster can also be observed in the word "acre" making the top 15-word list. The *White higher income* cluster has the word "walk" ranked highly suggesting a denser, more walkable, 'new urbanist' design to these suburban neighborhoods (Delmelle et al., 2014). The urban environment of the *Increasingly white higher/middle income* neighborhoods are apparent with "downtown" being the second most commonly used word as well as "walk", again, likely indicating the walkability in these near-center-city neighborhoods. It also includes higher end home features like "custom". The changing nature of these neighborhoods and the properties in them are reflected by "renovated" being the fourth most common word. These findings by neighborhood cluster show close similarity with the different neighborhood clusters in Delmelle and Nilsson (2021) (Figure 10.4).

Figure 10.5 shows the most discriminant words by cluster. i.e., the words with the largest (positive and negative) coefficients by neighborhood cluster. The strongest (positive) predictors for the *White and wealthy* cluster includes high-end "subzero" appliances, the word "constructed" which likely indicates newly constructed, "professional", "European" style, and attribute words such as "tastefully". While more affordable options such as "laminate" are negatively associated. The negative association with the word "transitional" likely points to the stable nature of these wealthier neighborhoods where little change has occurred over time in terms of socioeconomic composition. The strongest positive predictor of the *White Higher Income* cluster is "location" and "school" also shows up as an important predictor likely in reference to the quality of the schools in these areas. "Craftsmanship", "cherry", and "colonial" points to the higher-end features of the properties and the negative association with words "busy" is indicative of the more suburban nature. Also, the negative association with "tenant" and "investor" could suggest a low share of renter occupied and delipidated homes available for investors to flip.

The more urban nature of the *Increasingly White Middle/Higher Income* cluster shows in the words positively associated with properties in these

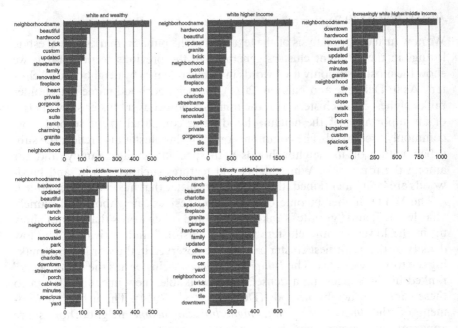

Figure 10.4 The 20 most frequently used words by cluster (sorted based on frequency in descending order).

neighborhoods including "downtown", the "blueline" light rail and the "sky-line". It also shows the negative association with more suburban and fami-ly-oriented amenities such as "cul-de-sac", "school", "mall" and "shopping". Densification through the construction of "duplexes" in these "hottest" and "rapidly" changing neighborhoods are also indictive of their transitional nature. The high, positive association with the word "rehab" also points to this trend.

  ⸱ On the other hand, the most highly predictive words of the *Minority mid-dle/lower Income* cluster points to their more suburban nature with "subdivi-sion" among the most positively associated words as well as proximity to the "airport" and "transit", where the latter may try to attract transit-dependent households. These findings are in line with those by Kennedy et al. (2021) who found that rental advertisement descriptions in non-White neighbor-hoods had a greater focus on transportation. While the airport may not sig-nal means of daily transportation, it is a major employer where many lower-wage workers are employed (Delmelle et al., 2021). As noted in Delmelle and Nilsson (2021), the mentioning of the specific neighborhood name appears to hold more prestige in the Whiter and higher-income neigh-borhoods while it is negatively associated with minority neighborhoods as shown in these results. The disparities in school quality between White and minority neighborhoods is indicated by the results with "school" being negatively associated with the minority neighborhood cluster. The results also contain words hinting at perhaps fewer capital investments in these

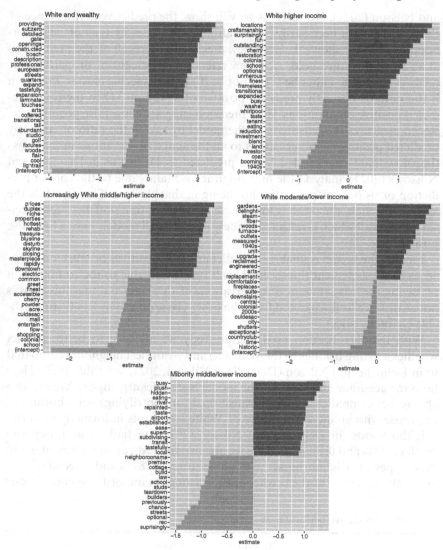

Figure 10.5 The 15 most positively and negatively associated words, respectively, by cluster.

neighborhoods with "build", "builders", and "teardown" (to build something new on the property) being negatively associated. These findings support the statements made by Hanlon (2008) who find that older-suburban neighborhoods that lack the character of historic center-city homes and are devoid of many of today's desirable amenities have increasingly become the more affordable housing options for lower-income and minority residents.

Generally, these results paint a more distinct picture of the changes (or lack thereof) going on across neighborhoods in Charlotte, NC, and fits better with the local knowledge of the authors compared to the findings in Delmelle and Nilsson (2021). However, as we will show, the predictive performance of the models is not necessarily better, although it varies by type of neighborhood. Table 10.3 shows the accuracy, precision and recall of the predictions at an 80 percent threshold. Accuracy refers to the percent of properties correctly predicted to either be (1) or not be (0) in that specific cluster. Precision is the ratio of correctly predicted positive observations (1) to the total predicted positive observations. High precision suggests a low false positive rate. Recall measures the sensitivity of the model by taking the ratio of correctly predicted positive (1) observations to all the actual observations in that class (i.e., how many cases coded 1 were predicted as 1). Just as in Delmelle and Nilsson's (2021) analysis, the high accuracy is mainly due to the correct prediction of 0s (i.e., of properties not being in that cluster). That is, while the accuracy is high, the recall is low. There is usually a trade-off between precision and recall which is reflected in these results as all models have a low false positive rate.

Finally, the model is unable to predict any true positives of cluster 4 properties at an 80 percent threshold. This is most likely due to the combination of having the least amount of observations of this class and that this class is not as distinct/well separated from the other classes (i.e., has less distinguishable characteristics).

With the classification used in this study we are able to better predict whether a listing belongs to the White(st) and wealthiest neighbors compared to in Delmelle and Nilsson (Delmelle & Nilsson, 2021, see Table 10.3). Here both the accuracy and recall of the model is significantly higher. We are also able to better predict the more transitional (or gentrifying) neighborhoods, i.e., those with an increasing share of White applicants in formerly minority neighborhoods. In Delmelle and Nilsson (2021) we had an accuracy and recall of 78% and 0.1% while here we are able to predict listings belonging to these types of neighborhoods with an accuracy of 87% and a recall of 7%. While the accuracy of the predictions for listings in minority neighborhoods

Table 10.3 Predictive performance

	Accuracy	*Precision*	*Recall*
1. White and wealthy	94%	71%	4%
2. White higher income	74%	91%	4%
3. Increasingly white middle/higher income	87%	89%	7%
4. White middle/lower income	N/A	N/A	N/A
5. Minority middle/lower income	68%	87%	24%

is worse in this study, the recall is much higher which indicates that we are better able to predict true positives.

10.5 Concluding Remarks

In this chapter we presented an approach to studying the role of housing and neighborhood amenities in residential sorting and neighborhood change. We extended this approach, originally presented in Delmelle and Nilsson (2021), by incorporating a longer temporal perspective (dating back to 1993) and a neighborhood classification approach which is more aimed at capturing change over time, namely latent class growth modeling. Classifying neighborhoods based on their longitudinal trajectories is rare in the neighborhood change literature. In the case study presented in this chapter, we showed that this classification technique yields more interpretable results compared to the corresponding text analysis in Delmelle and Nilsson (2021) and leads to better predictions in some cases, but not all.

The text analysis shows patterns of the spatial restructuring that are occurring in many North American cities that include the suburbanization of minority and lower-income neighborhoods with suburban characteristics including 'subdivision' being among the most positively associated words with property listings in these neighborhoods. Similar to our previous analysis in Delmelle and Nilsson (2021) we found that listings in neighborhoods with a majority of minority home buyers were less likely to mention schools and were less likely to feature the name of the neighborhood compared to Whiter and wealthier neighborhoods. Unlike in our previous analysis, we found that minority neighborhoods were less likely to feature words hinting at capital investments such as "build", "builders", and "teardown" (to build something new on the property) – thus illuminating the often difficult to capture process of disinvestment in these neighborhoods.

We were also better able to predict neighborhoods in transition or undergoing gentrification using the classification proposed in this study. The urban and rapidly changing nature of these neighborhoods were also better captured in the text analysis based on this new classification scheme including positively discriminant words such as "downtown", "blueline" (Charlotte's light rail line), "skyline", "hottest", "rapidly", and "rehab".

With respect to urban amenities and their connection to sorting patterns, in this analysis, we found that the wealthiest and Whitest neighborhoods were negative associated with amenities such as light rail and the 'arts', and were perhaps surprisingly negatively associated with 'golf', despite the literature's suggestion that golf courses perpetuate segregation by race and income. Rather, the wealthiest neighborhoods in Charlotte appear to be more isolated with the word 'gate' significantly associated with them. These are larger, more exclusive estates, as signaled by their advertisements. The notion that residential sorting by lifecycle remains strong can be extracted by these advertisements. Whiter and wealthier suburban locations frequently mention schools in their property advertisements while neighborhoods exhibiting signs of gentrification – an increasing share of White residents in minority neighborhoods close to the city center have a negative association with schools. This potentially reaffirms the idea that cities are increasingly becoming segregated by age and that center-city properties are not marketed towards families seeking out high school quality. Rather they are tailored towards those valuing proximity to new transit (the Blue Line) and views of the skyline.

Overall, while this analysis has used a relatively small sample of property advertisements as an illustrative approach for matching amenities with neighborhood trajectories, we find the improved ease in interpreting results holds the potential for advancing several avenues of future research. First, longitudinal data on property listings could provide important empirical insights onto the changing importance of amenities alongside neighborhood trajectories. Next, we envision that this framework can advance efforts to predict neighborhood changes in more real time. If property listings begin to be advertised to a demographic that differs from its current one, based on the predictions explored here, we can anticipate where changes may occur even before the movement of residents has taken place. Understanding these changes in near real time, rather than waiting for retrospective surveys as is typically done in neighborhood change studies will provide key knowledge to urban planners and decision makers in allocating resources or anticipating shifting demands (to bus services, school assignments, etc.). Doing so in a timely manner is crucial in order to implement cost-effective interventions to prevent increased social inequalities and potential displacement before changes become entrenched (Chapple & Zuk, 2016).

Notes

1 We also attempted a higher number of classes but ended up with overlapping classes and small classes in each cluster together with stagnating AIC and BIC values.
2 Rootograms are a type of histogram of the posterior class probabilities that allows modelers to visually assess the cluster structure. They are similar to histograms with the difference being that the height of the bars in rootograms correspond to square roots of counts rather than the raw counts. A peak at probability 1 indicates that a mixture component is well separated from the other components, while no peak at 1 and/or a concentration of observations in the middle of the probability interval indicates overlap with other components (Leisch, 2004).

References

Aaronson, D., Faber, J., Hartley, D., Mazumder, B., & Sharkey, P. (2021). The long-run effects of the 1930s HOLC "redlining" maps on place-based measures of economic opportunity and socioeconomic success. *Regional Science and Urban Economics, 86*, 103622.

Alonso, W (1964). *Location and Land Use: Toward a General Theory of Land Rent*. Cambridge, MA: Harvard University Press.

Baum-Snow, N., & Hartley, D. (2020). Accounting for central neighborhood change, 1980–2010. *Journal of Urban Economics, 117*, 103228.

Bayer, P., & McMillan, R. (2005). Racial sorting and neighborhood quality (No. w11813). National Bureau of Economic Research.

Bian, X., Contat, J. C., Waller, B. D., & Wentland, S. A. (2021). Why disclose less information? Toward resolving a disclosure puzzle in the housing market. *The Journal of Real Estate Finance and Economics.* https://doi.org/10.1007/s11146-021-09824-6

Brasington, D. M. (2014). Housing choice, residential mobility, and hedonic approaches. In *Handbook of Regional Science*, ed. M. M. Fischer and P. Nijkamp, 147–165. Berlin Heidelberg: Springer-Verlag.

Census (2021). QuickFacts: Charlotte city, North Carolina. www.census.gov/quickfacts/charlottecitynorthcarolina

Chapple, K., & Zuk, M. (2016). Forewarned: The use of neighborhood early warning systems for gentrification and displacement. *Cityscape, 18*(3), 109–130.

CharlotteFive (2018). What's the wealthiest zip code in Charlotte? *CharlotteFive*, June 18, www.charlotteobserver.com/charlottefive/c5-around-town/article236144008.html

Chemtob, D., & Off, G. (2019). Charlotte jumps to 16th biggest city in U.S. But neighboring towns saw bigger surge. In WBTV. www.wbtv.com/2019/05/24/charlotte-jumps-th-biggest-city-us-neighboring-towns-saw-bigger-surge/

Chemtob, D., & Off, G. (2020). Charlotte passes San Fran, now 15th largest US city. In Courier-Tribune. www.courier-tribune.com/story/news/coronavirus/2020/05/22/charlotte-passes-san-fran-now-15th-largest-us-city/112314852/

Childress Klein Center for Real Estate. (2020). 2020 the state of housing in Charlotte report. University of North Carolina at Charlotte. https://realestate.uncc.edu/research/state-housing-charlotte-report

Clark, W. (2019). Millennials in the housing market: The transition to ownership in challenging contexts. *Housing, Theory and Society, 36*(2), 206–227.

Clarke, W. & Freedman, M. (2019). The rise and effects of homeowners associations. *Journal of Urban Economics, 112*, 1–15.

Couture, V., & Handbury, J. (2020). Urban revival in America. *Journal of Urban Economics, 119*, 103267.

Danielsen, K. A., & Lang, R. E. (2010). Live work play: new perspectives on the downtown rebound. *disP-The Planning Review, 46*(180), 91–105.

Delmelle, E. C. (2019). The increasing sociospatial fragmentation of urban America. *Urban Science, 3*(1), 9.

Delmelle, E. C., & Nilsson, I. (2021). The language of neighborhoods: A predictive-analytical framework based on property advertisement text and mortgage lending data. *Computers, Environment and Urban Systems, 88*, 101658.

Delmelle, E. C., Nilsson, I., & Schuch, J. C. (2021). Who's moving in? A longitudinal analysis of home purchase loan borrowers in new transit neighborhoods. *Geographical Analysis, 53*, 237–258.

Delmelle, E. C., Zhou, Y., & Thill, J. C. (2014). Densification without growth management? Evidence from local land development and housing trends in Charlotte, North Carolina, USA. *Sustainability, 6*(6), 3975–3990.

Dunn, A. (2017). In Charlotte's trendy neighborhoods, a culture clash of Black and White, rich and poor. *Charlotte Agenda*, July 20, www.charlotteagenda.com/97973/charlottes-trendy-neighborhoods-culture-clash-Black-White-rich-poor/

Ehlenz, M.M., Pfeiffer, D., & Pearthree, G. (2020). Downtown revitalization in the era of millennials: How developer perceptions of millennial market demands are shaping urban landscapes. *Urban Geography, 41*(1), 79–102, DOI: 10.1080/02723638.2019.1647062.

FFIEC – Federal Financial Institutions Examination Council (2021). History of HMDA. www.ffiec.gov/hmda/history2.htm

Florida, R., & Adler, P. (2018). The patchwork metropolis: The morphology of the divided postindustrial city. *Journal of Urban Affairs, 40*(5), 609–624.

Galster, G., & Godfrey, E. (2005). By words and deeds: Racial steering by real estate agents in the US in 2000. *Journal of the American Planning Association, 71*(3), 251–268.

Glaeser, E. L., & Gottlieb, J. D. (2006). Urban resurgence and the consumer city. *Urban Studies, 43*(8), 1275–1299.

Gruen, B., Leisch, F., Sarkar, D., Mortier, F., & Picard, N. (2020). Package 'flexmix'. https://cran.r-project.org/web/packages/flexmix/flexmix.pdf

Hanlon, B. (2008). The decline of older, inner suburbs in metropolitan America. *Housing Policy Debate, 19*(3), 423–456.

Hastie, T., Qian, J., & Tay, K. (2021). An introduction to glmnet. https://cran.r-project.org/web/packages/glmnet/vignettes/glmnet.pdf

Jones, B. L., & Nagin, D. S. (2013). A note on a Stata plugin for estimating group-based trajectory models. *Sociological Methods & Research, 42*, 608–613.

Kennedy, I., Hess, C., Paullada, A., & Chasins, S. (2021). Racialized discourse in Seattle rental ad texts. *Social Forces, 99*(4), 1432–1456.

Lacour, G. (2019). Breaking down the wedge in Charlotte. *Charlotte Magazine*, March 19, www.charlottemagazine.com/breaking-down-the-wedge-in-charlotte/

Leading on Opportunity (2017). The Charlotte-Mecklenburg opportunity task force report. March 2017, www.fftc.org/sites/default/files/2018-05/LeadingOnOpportunity_Report.pdf

Lee, H. (2020). Are millennials coming to town? Residential location choice of young adults. *Urban Affairs Review, 56*(2), 565–604.

Lee, J., Irwin, N., Irwin, E., & Miller, H. J. (2021). The role of distance-dependent versus localized amenities in polarizing urban spatial structure: A spatio-temporal analysis of residential location value in Columbus, Ohio, 2000–2015. *Geographical Analysis, 53*(2), 283–306.

Lee, Y., Lee, B., & Shubho, M. T. H. (2019). Urban revival by Millennials? Intraurban net migration patterns of young adults, 1980–2010. *Journal of Regional Science, 59*(3), 538–566.

Leisch, F. (2004). FlexMix: A general framework for finite mixture models and latent class regression in R. *Journal of Statistical Software, 11*(8), 1–18.

Logan, L. (2018). Like an ex, Villa Heights isn't the neighborhood we used to know. *Charlotte Observer*, July 9, www.charlotteobserver.com/charlottefive/c5-people/article236117218.html

Luchtenberg, K. F., Seiler, M. J., & Sun, H. (2019). Listing agent signals: Does a picture paint a thousand words? *The Journal of Real Estate Finance and Economics, 59*(4), 617–648.

Massey, D. S. (2005). Racial discrimination in housing: A moving target. *Social Problems, 52*(2), 148–151.

Moos, M. (2014). "Generationed" space: Societal restructuring and young adults' changing residential location patterns. *The Canadian Geographer/Le Géographe canadien, 58*(1), 11–33.

Mplus (2021). Mplus. www.statmodel.com/

Muth, R.F. (1969). *Cities and Housing: The Spatial Pattern of Urban Residential Land Use*. Chicago, IL: University of Chicago Press.

Myers, D. (2016). Peak millennials: Three reinforcing cycles that amplify the rise and fall of urban concentration by millennials. *Housing Policy Debate, 26*(6), 928–947.

Nilsson, I., & Delmelle, E. C. (2018). Transit investments and neighborhood change: On the likelihood of change. *Journal of Transport Geography, 66,* 167–179.

Nowak, A., & Smith, P. (2017). Textual analysis in real estate. *Journal of Applied Econometrics, 32*(4), 896–918.

Nowak, A. D., Price, B. S., & Smith, P. S. (2021). Real estate dictionaries across space and time. *The Journal of Real Estate Finance and Economics, 62,* 139–163.

Off, G., & Wright, W. (2021). Charlotte is more diverse than ever, census data shows. But not in every neighborhood. *Charlotte Observer,* August 30, www.charlotteobserver.com/news/local/article253787038.html

Ondrich, J., Ross, S., & Yinger, J. (2003). Now you see it, now you don't: why do real estate agents withhold available houses from black customers? *Review of Economics and Statistics, 85*(4), 854–873.

Pryce, G., & Oates, S. (2008). Rhetoric in the language of real estate marketing. *Housing Studies, 23*(2), 319–348.

Seo, Y., Im, J., & Mikelbank, B. (2020). Does the written word matter? The role of uncovering and utilizing information from written comments in housing ads. *Journal of Housing Research, 29*(2), 133–155.

Shen, L., & Ross, S. (2021). Information value of property description: A machine learning approach. *Journal of Urban Economics, 121,* 103299.

Silge, J., & Robinson, D. (2017). *Text Mining in R: A Tidy Approach.* Sebastopol, CA: O'Reilly Media, Inc.

Strahilevitz, L. J. (2006). Exclusionary amenities in residential communities. *Virginia Law Review, 92,* 437.

Tiebout, C. (1956). A pure theory of local expenditures. *Journal of Political Economy, 64*(5), 416–424.

Wei, F., & Knox, P. L. (2014). Neighborhood change in metropolitan America, 1990 to 2010. *Urban Affairs Review, 50,* 459–489.

White, A. G., Guikema, S. D., & Logan, T. M. (2021). Urban population characteristics and their correlation with historic discriminatory housing practices. *Applied Geography, 132,* 102445.

Zillow (2021). www.zillow.com/

Zwiers, M., van Ham, M., & Manley, D. (2017). Trajectories of ethnic neighbourhood change: Spatial patterns of increasing ethnic diversity. *Population, Space and Place, 24,* e2094.

11 Making the Right Move

How Effective Matching on the Frontlines Maintains the Market for Bribes[1]

Diana Dakhlallah

Bribery impacts almost half the world's inhabitants (Transparency International, 2019, 2016). Differences in the prevalence of bribery across countries have often been interpreted as differences in how culturally accept-able it is (Barr and Serra, 2010; Fisman and Miguel, 2007). While people in these places might say they can't get things done without bribing, they don't mean they find it acceptable. Ethnographies of places where bribery is "banalized" describe it as predatory (Hoang, 2018), violent (Gupta, 2012), and perverting the fabric of society (Smith, 2008). If the prevailing sentiment is that bribery is bad, and the reality is that it is widespread, then *how* does bribery happen?

We have an answer for bribes that flow through thick social ties. For exam-ple, Vietnamese entrepreneurs go to great lengths to come up with "creative gift giving strategies, which are highly stressful and intensely personal, to obfuscate the exchanges in their relationships with state officials" (Hoang, 2018: 667). Customizing strategies—such as gift exchange, bundling with innocuous transfers, and brokerage—requires repeat interactions (Hoang, 2018; Lomnitz, 1988; Rossman, 2014). The chosen strategy, coupled with the forged relationship, neutralizes the disagreeableness and risk of bribing.

However, the majority of bribes flow through thin social ties. Service-seekers (clients) obtain a driver's permit (Smith, 2008), a land deed (Gupta, 2012), or healthcare (Vian, 2008). In these markets, characterized by short-term and discretionary face-to-face interactions, bribes are spot transactions. Individuals typically don't know each other and payments are primarily about sticking to or improving on existing rules (e.g. Transparency International, 2016). Bribes are predominantly unsanctioned monetary transfers, rather than personalized gifts, from clients to service-providers (providers) to access services or improve their quality. The information thin-ness of such encounters makes it difficult for individuals to identify an accommodating exchange partner (Schelling, 2011), especially since the modal bribe-giver is not violating the rules. As such, more can go wrong than right. Yet, individuals exchange bribes and rarely challenge these exchanges overtly, even though bribe-givers have good reason to do so.

How individuals manage to seamlessly complete these fraught transactions has been partially answered. To address this question, bribery is theorized in

DOI: 10.4324/9781003191049-14

terms of expected value, where the potential reward is worth the potential cost (Becker and Stigler, 1974; Olken and Pande, 2012). This leads to a focus on selective targeting—individuals use social information to figure out who is receptive. In theory, this makes sense, but the little ethnographic work that exists suggests that people are quite bad at selective targeting (de Sardan, 1999; Gupta, 1995; Miller et al., 2001; Smith, 2008). At best, they can identify a type of person who might be receptive, but variance in types is still big enough (Irlenbusch and Villeval, 2015), that in expected value terms, bribery would happen significantly less frequently than it does, if at all. The question of *how* people choose to engage in spot bribe transactions—and what their choices imply for individuals and for the market for bribes—remains unanswered.

While identifying the right target is critical, figuring out the right way to engage with that target is perhaps even more so, to reduce undesirable consequences. In this chapter, I argue and demonstrate that how well-matched first moves are to the socioeconomic status of the transacting parties is a decisive micro-mechanism for neutralizing conflict in spot bribe exchange. To this end, I theoretically develop "first moves": how invitations to exchange bribes are extended and by whom. I then examine how status dynamics between providers and clients condition the choice of first moves. Socioeconomic status, as with status markers in general, captures people's social image and the way they expect to be treated by others; violating these expectations causes conflict (Bourdieu, 2013; Granovetter, 2007). The "match" between first moves and status determines the expected value of moving in on the bribe action. For completeness, I look at instances of "ineffective matches," where neutralization fails and conflict ensues.

This study uses survey data on bribe exchanges in the Moroccan healthcare sector. For context, out of 36 African countries covered by Afrobarometer surveys in 2016, Morocco had one of the highest bribery rate at 47% (Figure 11.1a). About 10% of Moroccans who pay bribes report these payments, a reporting rate that is comparable to other countries (Figure 11.1b). Morocco is characterized by weak enforcement; the probability of taking action in response to reports is one of the lowest (Figure 11.1c). It is easy in such a context to dismiss effective matching as a concern for exchange partners. This chapter demonstrates such dismissals are misguided and highlights the importance of paying attention to how bribery happens. I analyze how bribes flow between healthcare-providers (providers) and healthcare-seekers (clients) during the provision of care. These data are unique for three reasons:

(1) unlike existing survey data on bribery, they contain information on who solicits a bribe first and how that solicitation is conducted; (2) they capture a sizeable number of bribe transactions, which gives us enough variation on solicitation strategy (first moves)—our main outcome of interest—to unpack and understand; (3) they have information about both providers and clients, which enables us to examine the demand and supply sides, respectively, of these transactions in the same context and establish a more complete account of the relational dynamics of these exchanges.

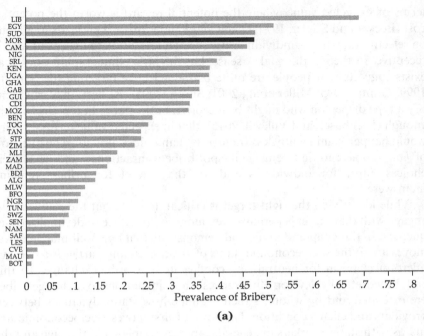

(a)

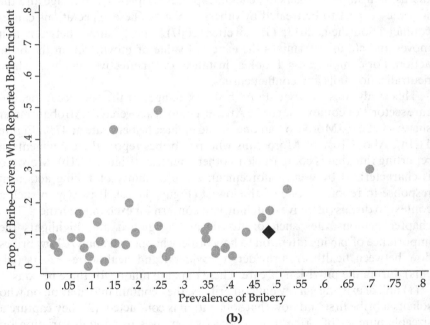

(b)

Figure 11.1 (a) Proportion of respondents surveyed who paid bribes in at least one of six public sectors across 36 African countries, conditional on being in contact with the sector(s) within the past year of the interview; (b) Proportion of bribe-payers who reported their bribe payments to an authority over the country-level prevalence rate.

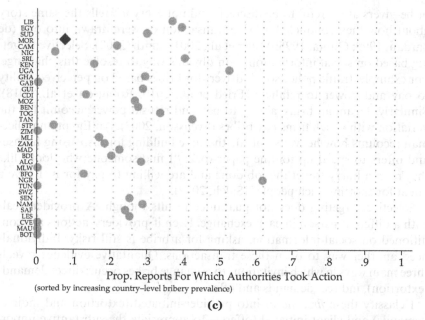

Prop. Reports For Which Authorities Took Action
(sorted by increasing country–level bribery prevalence)

(c)

Figure 11.1 (c) Proportion of reports for which authorities took action, sorted by increasing country-level prevalence. Morocco is in black; see B.1 for more information.

Data Source: Afrobarometer Round 6, 2016.

This study is novel in the way it theoretically develops and tests the logic of first-mover behavior in bribe exchange. It makes several contributions to our understanding of bribery. For individuals, status-sensitive first moves comprise a micro-mechanism that creates the accommodating exchange partners sought out by providers. This explains how bribery can be simultaneously distasteful, risky, and scalable. For the market, status-sensitive first moves explain the low reporting rates by clients. This muting keeps bribery under the radar of principals and makes bribery sticky. Furthermore, attention to how bribery happens reveals the use-value of a new outcome: first moves. This outcome calls into question standard interpretations of prevalence rates by giving us a more textured account of bribery in a given context. It also furnishes us with analytical tools to (re) consider the effectiveness of existing policy approaches, and suggests new and more targeted ones for addressing bribery.

11.1 Selective Targeting, First Moves, and Status

Even where bribery is considered common, figuring out who to ask for or who to offer a bribe is challenging. Qualitative accounts of bribery indicate that in a given context not everyone makes a bribe payment, bribe-takers and bribe-givers are careful to be discreet, and not everyone tells the same story about how they conducted a bribe transaction or were drawn into one (de Sardan, 1999; Gupta, 1995; Miller et al., 2001; Smith, 2008). Selective targeting based on social information is an obvious way to resolve this challenge. For example, traffic police solicit drivers for bribes based on perceived ability to pay and power to retaliate (Fried et al., 2010; Robinson et al., 2018). Similarly, Ugandan firms' ability to pay and refusal power account for the variation with which firms pay bribes (Svensson, 2003). An Egyptian policeman recounts how he waits outside the state building for dispensing licenses and offers to speed up routine paperwork: "I find someone who looks like they're in a hurry to get the job done and are willing to pay for it. I have a talent for spotting such people" (Saleh, 2012).

Selective targeting does not guarantee results though. A provider deals with a different person in each exchange. Even if providers' actions are conditioned on social information, asking for a bribe is still risky. Individuals need another way to open these transactions. Qualitative evidence reveals three main ways in which invitations to exchange bribes occur: direct demands (extortion), indirect demands and offers.

I classify these *first moves* into provider-initiated (extortion and indirect demands) and client-initiated (offers). To appreciate the substantive import of these different solicitation strategies for exchange outcomes, recognize that the dilemma for providers is that, once they reveal their intent, they have self-identified as bribe-takers. Individuals care about the receptivity of the target, as well as how they look to themselves and to others (Bénabou and Tirole, 2011; Gino et al., 2011; Irlenbusch and Villeval, 2015; Rossman, 2014). For example, an Indian land-records keeper willingly comes out empty-handed from an indirect bribe negotiation with two villagers because they did not know how to give the bribe (Gupta, 1995). Thus, the irreversibility of a first move is contingent on how explicit self-identification of interests is on the one hand (Schelling, 2011) and, if explicit, how manageable the consequences of the revelation are on the other.

One approach is to ensure plausible deniability. Providers resort to indirect language (Pinker et al., 2008) or suggestive behavior. For example, Nigerian checkpoint police ask for money by saying "We are loyal, sir" or "I salute you" followed by an expecting look, and if the police officer finds a driver's papers are in order, he might ask to look under the hood to check the engine and chassis number (Smith, 2008). The exit option afforded by the ambiguity of indirect asks allows providers to hedge their bets. They make money from clients who understand and oblige them, and no money from those who do not, while simultaneously reducing the possibility of a scandalous confrontation from explicitly propositioning resistors. The confronted would incur reputational costs and may experience punitive consequences. I also consider the use of an intermediary as an indirect demand. While the very existence of the intermediary is an explicit articulation of bribe intent, the social distance created between bribe-givers and bribe-takers retains the plausible deniability transacting parties might care for.

Theoretically, an indirect approach is an optimal strategy. However, unconcealed bribe requests occur with a high enough frequency to warrant attention. Since a fundamental feature of spot bribe interactions is their information thinness, making a first move that is irreversible regarding intent and consequences is often a bad idea. If bribe-takers are more likely to suffer from first-mover disadvantage when they extort, it is important to identify under what circumstances they might choose to make such high-risk invitations. By extension, when clients are first-movers, providers are in their best-case scenario. Because offers unambiguously articulate a willingness to bribe, they eliminate the uncertainty of guesswork and absorb the risk of a sour encounter. Note that even when clients offer, they must also do the relevant analysis to identify and correctly engage with a receptive provider, since not all providers accept bribes (Lomnitz, 1988; Pinker et al., 2008; Rasul and Rogger, 2018).

To succinctly capture the analytical discussion of first-mover disadvantage from the perspective of the bribe-taker, let us rank the non-material utility uN of the different moves as: $uN_{extortion} < uN_{indirect} \leq uN_{offers}$. An example of uN is saving face—the ability of a provider to save face is smallest with extortion and greatest with offers. On the other hand, directly asking for something, rather than suggesting or waiting to be offered, is a surer strategy to make more money, so the material utility uM of the different moves is ranked as: $uM_{extortion} > uM_{indirect} > uM_{offers}$. Accounting for uM and uN at once, indirect demands should be the go-to strategy for providers who want bribes.

Notice that if selective targeting were enough to solve the identification problem, then extortion would dominate other strategies. Explicit asks will deliver the material rewards desired from $uM_{extortion}$ while netting out the social costs associated with $uN_{extortion}$. If selective targeting is not enough, then indirect asks dominate. Yet qualitative accounts indicate that indirect demands do not. As such, it is important to understand what determines when one solicitation strategy is used over another in bribe encounters, i.e., what determines the choice of the first move in these high-risk, short-lived transactions.

To do so, I consider how status dynamics between providers and clients condition the choice of first moves. Socioeconomic status signals the way individuals expect to be treated in a given context; the violation (or observance) of these expectations is tied to consequential penalties (or rewards) (Bourdieu, 2013; Granovetter, 2007; Lomnitz, 1988). This understanding of positional social significance and its broader ramifications can be strategically leveraged in bribe exchanges. Being relationally discerning expands the repertoire of first moves available to providers. It allows them to choose plausible deniability over the irreversible self-identification of extortion with some bribe-givers but not others, or forego a first move, leaving it up to the chance of interfacing with a "charitable" client. In this way uN of a given solicitation strategy is more adaptably and profitably defined on a case-by-case basis.

We expect to observe the following patterns in the data. The probability of extortionate advances decreases with client status and the probability of offering increases with client status. Because plausible deniability maintains an exit option for providers, it affords them a broader slice of the client pool to move in on for a bribe opportunity. As such, the probability of resorting to indirect demands, while decreasing with status, will be significantly less sensitive to client status than extortion or offers. For a solicitation strategy to effectively manage the consequences of these fraught transactions, the choice of the first move should be sensitive to social information; this is consistent with the idea that different kinds of people are associated with different kinds of risks (Gambetta, 2011; Goffman, 1982; Schelling, 2011).

11.2 Empirical Strategy

11.2.1 Data

The survey data used in this study capture self-reported experiences with bribery during the provision of healthcare services. These data are the result of a 2010/2011 study commissioned by the Moroccan anti-corruption agency L'Instance centrale de prévention de la corruption (ICPC). At the time, about three-quarters of clients who visited healthcare establishments on an annual basis were not repeat healthcare-seekers (Ministère de la Santé, 2011).

A total of 1157 clients from five major cities were surveyed. Surveyors placed themselves in the vicinity of healthcare establishments and randomly selected passers-by for survey administration; see A.1 for text used to solicit survey participation. Surveyors did not keep track of the non-response rate of passers-by. This would have provided a rough approximation of the willingness of Moroccans to discuss corruption. Initiatives such as Transparency Maroc's Observatoire de la corruption indicate that bribery is a topic that is discussed openly, which should mitigate response bias concerns. Furthermore, the bribe rate documented in these data is consistent with prevalence calculations using Afrobarometer surveys; see B.2 for more information.

Conditional on having sought care in a healthcare establishment within the year prior to the date of the interview, respondents were asked questions about their experience accessing healthcare services; see A.2 for the services covered. Three kinds of establishments deliver healthcare in Morocco: hospitals, community healthcare centers, and clinics. Surveys were administered in French and Darija, a colloquial variety of Arabic specific to Morocco.

11.2.2 Outcome Variables

Bribe Exchange Respondents are asked about how they obtained healthcare services. For example, [To access a health establishment] during your last visit, did you: (1) access normally without corruption or the help of personnel; (2) engage in a corrupt exchange; (3) proceed with the assistance of an influential person: personnel of the establishment; (4) proceed with the assistance of an influential person: an authority; (5) exchange benefits-in-kind; (6) other, please specify; (7) no response. The text in brackets changes depending on the service sought. Respondents are allowed one response. I group responses into four main categories: normal access (items 1 and 6), bribe exchange (items 2 and 5), assistance from influential party (items 3 and 4), and no response (item 7). For the analyses, I recode this variable into a binary outcome, whether respondents provision healthcare services by bribing (Yes, No).

Solicitation Strategy (First Moves) Independent of whether they exchange a bribe during healthcare provision, respondents are asked: In general, how is corruption performed? Respondents are allowed up to two response choices: payment is (1) demanded by personnel directly; (2) offered by you; (3) demanded through a third party; (4) demanded in an implicit/indirect manner. I focus on the first response choice; only 9.5% (110/1157) of respondents specify a second answer. As a robustness check, I replace the first move of clients who report two solicitation strategies with their second choice. I also limit my analyses to respondents who report at least one bribe exchange during healthcare provision. The distribution of solicitation patterns reported by respondents who paid bribes for healthcare services and those who did not is similar. In light of the way the question is phrased, even if a client reports more than one bribe payment she is assigned one solicitation strategy. The estimated models account for this, as I describe below.

11.2.3 *Explanatory Variables*

Status of Client I focus on the effects of key socioeconomic characteristics that capture client status: monthly household income, level of education achieved, whether they use influential contacts to obtain care, and whether the client is a poverty certificate carrier.

- Monthly household income is composed of five income brackets (in Moroccan dirhams, where 10 dirhams ≈ 1 USD) : $x < 5K$, $5K \leq x < 10K$, $10K \leq x < 15K$, $15K \leq x < 20K$, $x \geq 20K$, and no response (8.38% of respondent sample). For a more interpretable income effect, I transform the categorical specification into a continuous one by calculating the midpoint for the close-ended income intervals and performing a Pareto transformation to calculate the midpoint of the open-ended interval (Hout, 2004). The average monthly household income for this sample is 5,737 (SD 6,297) dirhams. I standardize this variable for the regression analyses (Gelman, 2008). For a more accurate effect of income in the absence of size of dependents, I use years of age and marital status as controls in the multivariate models.
- Education is composed of seven levels: none, primary, secondary, baccalauréat or equivalent, baccalauréat +2, baccalauréat +3 (bachelor degree), baccalauréat +4/5 or more (graduate degree), and no response (2.59% of respondent sample). I recode the categorical specification in terms of the number of years of education. The average years of education is 10.8 (SD 5.8) years. I also standardize this variable for the regression analyses (Gelman, 2008).
- The income and education levels of respondents in this survey sample are comparable to national figures at the time (Haut-Commissariat Au Plan, 2009).

- A beneficiary may have accessed one service normally but used connections for another, or may have bribed in one service and used connections in another. Respondents who indicate assistance from an influential authority outside the establishment or from personnel inside the healthcare establishment are coded as "Yes," otherwise they are assigned a "No." This variable is derived from the same question the bribe exchange outcome variable is derived from (described above). About 7% (83/1157) of clients resort to connections to obtain care.
- The poverty certificate targets households falling below a certain poverty threshold and helps them access public healthcare services at reduced fees. At the time these data were collected, poverty certificate issuance required that beneficiaries visit officials in their municipality and the Ministry of Interior. About 28% (314/1157) of clients are poverty certificate carriers. About 36% (112/314) of carriers in the data paid bribes to obtain their certificate; the analyses account for this.

Table C1 presents descriptive statistics for the respondent sample. Table C2 presents the pairwise correlations between the status variables.

Status of Provider Unique to this dataset is the ability to examine to whom bribes flowed. The occupational category of healthcare workers not only organizes the delivery of care but also structures the nature of social interactions between healthcare-providers and healthcare-seekers in status-sensitive ways (Abbott 2014; King and Nembhard 2016). Clients are asked to indicate with whom they exchanged a bribe: doctor, nurse, support agent, security agent, other, and no response. This list expresses a decreasing status gradient, moving from doctors (highest) to security agents (lowest).

11.2.4 Estimation Models

For the analysis of selective targeting (Section 11.3.2), I estimate logit regression models that predict the probability of exchanging a bribe during healthcare provision. Since clients can exchange more than one bribe throughout their care path, this model captures the probability of making at least one bribe payment, controlling for the number of services sought by the client.

For the analysis of how status conditions first moves (Section 11.3.3), due to the discrete and unordered nature of first moves, I estimate a multinomial logit model, which predicts the log-odds of a given solicitation strategy as a function of client status. I recode solicitation strategy into three categories because of the few reports of third-party use by respondents. I collapse indirect demand and third-party use into a single category, which results in the outcome categories extortion, indirect demand and offer. An analysis excluding clients who resorted to intermediaries gives similar results. About half of clients who pay bribes make more than one bribe payment, therefore I cluster the standard errors by client. To account for the fact that clients who make more than one bribe payment are assigned one solicitation strategy, I incorporate two controls for robustness. One control variable captures the total number of bribe payments made by a client throughout her care path. The second control variable captures the variability in the occupational status of providers bribed by a client. For this, I construct a status variability score in which the higher the score the greater the status difference in providers bribed. Approximately 70% of clients have no variability in the occupational status of providers they bribed, either because they made only one bribe payment or bribed the same provider types. The full suite of models estimated for this analysis is documented in Table D1.

11.3 Results

11.3.1 How Bribes Flow: Variety in Practice

About 46% of clients make at least one bribe payment during services provision. The average number of services sought per client is 4.69 (SD 1.71). Clients make an average of 1.94 (SD 1.13) bribe payments. They pay on average 412 (SD 750) Moroccan dirhams in bribes, which is equivalent to about 12% of clients' monthly household income (unadjusted for household size). Table A1 summarizes reasons for bribing.

Table 11.1 gives the distribution of first moves, conditional on paying bribes. About two-thirds of first moves are provider-initiated. The demand-side (providers) for bribes is a significantly stronger push factor compared to the supply-side (clients). A little less than a third of clients are extorted, another third are indirectly asked, and about a quarter offered bribes.[2] Examples of extortion are "scratch your pocket," "don't you have anything to give?" or "put something in my pocket." Indirect demands include "I need someone to help me too" in response to a client's request for help, "your eye is your balance," "the issue is in your hands," "coffee," or "sweets." Examples of offers include "I told her to approach and slipped the money discreetly in her pocket" or "I told her to look under the pillow." Intermediaries are rarely used. Even in a country where enforcement is considered weak and where the discourse about corruption is focused on its banalization (Aboudrar 2014; ICPC 2011), a large portion of providers hedge their bets in these interactions by maintaining plausible deniability and relying on offers. While punishment may not be a credible threat, the disapproval of others and avoiding

Table 11.1 Bribe exchange: prevalence and distribution of first moves

Outcome	Clients
Paid at least one bribe	45.89% (531/1157)
Solicitation strategy	
1. Provider-initiated	
a. Direct demand (extortion)	28.63% (152/531)
b. Indirect demand	
Dyadic	33.15% (176/531)
Third-party use	1.69% (9/531)
2. Client-initiated (offer)	26.18% (139/531)
No response	10.36% (55/531)

Data source: ICPC Survey, Morocco.

defining oneself as corrupt—concerns that correspond to social and self-reputation, respectively—are barriers to complete impunity. These are barriers that have been documented in other distasteful transactions (Anteby 2010; Bénabou and Tirole 2011; Gino et al. 2011; Rossman 2014).

It is noteworthy that hedging bets not only reduces the chances of an unfavorable outcome but is more profitable per exchange. The average bribe size is 164 (SD 401), 225 (SD 507) and 269 (SD 502) dirhams for extortion, indirect demand and offering, respectively.[3]

11.3.2 Selective Targeting

To examine who pays bribes, let us turn to Model 3 in Table 11.2. First, income captures two countervailing effects: clients' ability to pay and their power to effectively exact retribution through institutional and non-institutional channels (Fried et al., 2010; Robinson et al., 2018). The statistically significant negative coefficient on the squared income term captures these countervailing effects. This indicates the presence of an inflection point, below which the returns from targeting wealthier clients are positive as their ability to pay increases. Above this point, the returns from targeting wealthier

Table 11.2 Logit models predicting who pays bribes during healthcare services provision

	(1)	(2)	(3)
Education (SD units)	0.232	0.275	0.202
	(0.937)	(1.108)	(0.771)
Education squared	−0.072	−0.070	−0.017
	(−0.212)	(−0.207)	(−0.047)
Income (SD units)	0.542	0.568	0.599
	(1.505)	(1.573)	(1.591)
Income squared	−0.407**	−0.411**	−0.430**
	(−2.436)	(−2.456)	(−2.494)
Used connections	−1.367***	−1.337***	−1.412***
	(−4.522)	(−4.423)	(−4.477)
Poverty certificate (PC)		0.379**	
		(2.179)	
Bribed for PC			1.636***
			(5.475)
Did not bribe for PC			−0.553*
			(−1.956)
Controls	Suppressed	Suppressed	Suppressed
Constant	−0.099	−0.192	−0.686
	(−0.229)	(−0.440)	(−1.433)
Observations	971	971	903
χ^2(df)	156.189 (16)	160.977 (17)	196.085 (18)

t statistics in parentheses. P-values: * $p < 0.10$, ** $p < 0.05$, *** $p < 0.01$.
Coefficients in log-odds form. Controls: number of services sought, marital status, years of age, medical coverage, city, urbanicity, and establishment type. The difference in number of observations across models is due to non-responses on whether the poverty certificate was obtained by bribing. Reference categories: unmarried, hospital, not poverty certificate carrier, no medical coverage, Rabat and rural. Data source: ICPC Survey, Morocco.

clients are negative as their power and ability to exact retribution becomes a greater concern for providers. This tradeoff between what counts as a valuable versus risky target has been replicated in other studies (Fried et al., 2010; Robinson et al., 2018; Svensson, 2003). Education, on the other hand, is not statistically significant.

Second, if clients can avoid paying bribes they do. For example, they rely on influential connections. Those who resort to connections during service provision are significantly less likely to pay bribes. This translates to a 0.23 (CI: 0.14, 0.32) probability of paying bribes for clients with contacts versus 0.48 (CI: 0.45, 0.51) for those without connections. Third, the poverty certificate captures the tail end of the socioeconomic status spectrum. Since over a third of poverty certificate carriers paid bribes to obtain their certificate, it also gives us a window into how the quality of prior experience with state institutions shapes future interactions. Clients who obtained their poverty certificates by bribing exhibit a 0.76 (CI: 0.67, 0.85) probability of bribing when seeking healthcare. Contrast this to 0.33 (CI: 0.23, 0.43) for those who did not bribe to get their certificate and 0.44 (CI: 0.41, 0.48) for non-carriers. This finding suggests that strategic choices are learned choices from experience.[4]

11.3.3 First Moves and Status Dynamics

How does client socioeconomic status determine the choice of a first move? Figure 11.2, which is based off of the multinomial logit estimates of Model 4 in Table D1, gives the predicted probability of a given solicitation strategy as a function of client income, conditional on paying bribes during healthcare provision. The patterns are consistent with predictions.

The probability of extorting drops precipitously from 0.66 (CI: 0.32, 0.99) at an income level of one standard deviation below the mean to 0.10 (CI: 0.03, 0.18) at one standard deviation above the mean, stabilizing thereafter. Indirect demand is weakly sensitive to client status, consistent with the ability to exploit plausible deniability. The probability of resorting to indirect asks modestly drops from 0.43 (0.18, 0.68) at an income level of one standard deviation below the mean to 0.37 (0.25, 0.48) at one standard deviation above the mean, stabilizing thereafter. The likelihood of offers by clients increases substantially from 0.06 (0.00, 0.12) at one standard deviation below the mean to 0.48 (0.35, 0.60) at an income level of one standard deviation above the mean, stabilizing thereafter.

Effective matches are a two-sided affair. Out of the 1028 instances of bribe exchange (recall it is possible for a client to make more than one bribe payment), 15.86% is attributable to doctors, 63.04% to nurses, 7.10% to support agents, 9.63% to security agents, and 4.37% to other/no response. The lion's share of bribes flows to physicians and nurses, the workers with the specialized skills and qualifications to offer healthcare services. Clients are most willing to exchange bribes with doctors, presumably the healthcare workers they believe deliver the most value: doctors are significantly more likely to be offered bribes compared to nurses, while support and security agents are

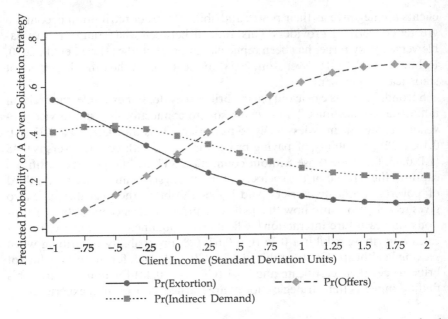

Figure 11.2 Predicted probabilities of a given solicitation strategy over income level of client, conditional on paying bribes during healthcare provision. Probabilities are derived from the multinomial logit estimates of Model 4 in Table D1.

Data source: ICPC Survey, Morocco.

significantly less likely to be offered bribes than nurses (see Table D2). This sensitivity to provider status is consistent with the price discrimination observed regarding bribe size, which increases with provider status, starting at an average bribe of 27 dirhams for security agents and ending at an average 415 dirhams for doctors (see Figure D1).

11.3.4 Ineffective Matches

I find that status conditions the choice of first moves. This strategic work reinforces the risk-reward tradeoff logic of selective targeting. For completeness, I now examine "ineffective matches" by looking at instances of bribery that were reported to the authorities by clients. When a mismatch between the choice of the first move and client status occurs, some form of confrontation is expected to ensue.

Out of 1,157 respondents, a mere 1.12% filed a formal complaint, including clients who paid bribes and those who did not. By considering first moves, it is possible to make sense of the low reporting rates.[5] If about a quarter of solicitations are client offers, then the probability of confrontation is reduced by a quarter. If roughly a third of bribes are indirect demands, then plausible deniability leads to a commensurate reduction in conflict. Finally, if extortion constitutes almost a third of bribe asks, then almost a third of all bribe interactions carry serious potential for conflict. Once factored into the impact of effective matches, the proportion of bribe interactions that results in reporting (and confrontation more generally) is reduced further. I find clients who reported bribe attempts were predominantly extorted (69%), highly educated (69% with at least a bachelor's degree) and well-off (85% with a monthly household income of at least 5,000 drihams); see Table D3.

11.4 Discussion

Individuals who want to engage in spot bribe exchange need to identify and engage with each other effectively. They face a heightened identification problem because of the impersonal and short-term features of their encounters. These individuals do not have the luxury of thick social ties that furnish them with better information and absorb the risk of mishaps (Hoang, 2018; Lomnitz, 1988; Rossman, 2014). The status-sensitive deployment of first moves is a critical micro-mechanism that creates the accommodating exchange partner they need.

Consistent with prior research, I find that providers selectively target clients by weighing the tradeoff between their riskiness and ability-to-pay. The novelty of this study's results lies in the theorization and analysis of first moves and how the status of individuals conditions first-mover behavior. I find the choice of a first move depends on client status: extortion is negatively correlated with client status and offers are positively correlated with client status, while indirect demands are significantly less status sensitive. The status sensitivity of first moves absorbs the residual risk from selective targeting to effectively shape and manage exchange outcomes. I find that reported bribery attempts are associated with ineffective matches in which there occurs a mismatch between the choice of the first move and client status.

Morocco suffers from weak institutions (World Bank Governance Indicators, 2010/2016) and human capital shortages in its healthcare sector (Ministère de la Santé, 2011; Ministère de la Santé, 2012), a combination that is often theorized as encouraging impunity in behavior. Yet exchange partners are careful despite the low credibility in the threat of punishment. At a bribe exchange probability of roughly 1-in-2, the institutional context should absorb the risk of misbehavior since institutions establish the rules and expectations of recurrent behavior (North, 1991; Viterna and Robertson, 2015). However, exercising strategic choice in first moves indicates that providers do not have carte blanche to receive bribes. This is important for two reasons.

First, culture and institutions reinforce each other (Alesina and Giuliano, 2015; Fisman and Miguel, 2007; Viterna and Robertson, 2015); as such, in equilibrium, their effects should move in the same direction. Yet the strategic logic of status-sensitive first moves in a high prevalence-low enforcement setting indicates otherwise. This suggests that focusing on the micro-mechanisms that undergird bribe exchanges can be a fruitful way to detect the (de) coupling or tension between cultural and institutional factors that shape them. For example, the prevalence of bribery may be treated as a proxy for the rules and expectations of recurrent behavior (i.e., the extent of institutional rootedness) and the pronouncedness of strategic behavior as an expression of the degree to which these transactions have graduated outside the realm of disreputability (i.e., the extent of cultural rootedness).

Second, the salience of strategic behavior in a low enforcement setting indicates that individuals, and in particular providers, care about the social consequences should their actions be exposed; "[I]f punishment was merely the price tag attached to crime, nobody would feel ashamed when caught (Elster, 1989: 105)." The fact that actors care about the social consequences of their actions is good news because it means that policymakers have more levers at their disposal to address bribery. Recent work leverages reputational concerns among healthcare-providers inside hospitals to reduce bribery (Dakhlallah, 2021). Interestingly, it also means that anti-corruption awareness campaigns overestimate the extent to which people find bribery acceptable, and are often preaching to the choir.

The status-sensitive deployment of first moves gives us a compelling explanation for why bribe markets are sticky. Since bribery is typically not accessible to third-party observers, reporting by clients plays an important role in mitigating the opacity around this practice. Their voices are an important source of information on which authorities take action (Becker and Stigler, 1974; Bergemann, 2017; Olken and Pande, 2012). For example, policy initiatives introduce devices that facilitate reporting such as corruption hotlines and provider name tags (ICPC, 2011). The strategic logic of effective matches neutralizes conflict and explains the low take-up rate of reporting by clients. A hospital director recounts that when he first began his position at his hospital, he knew that bribes were exchanged during service provision and it bothered him, but during his time as director not one patient came forward to complain (field conversations). The fact that news of a bribe payment does not get past the point of exchange reinforces the stickiness of bribery; as such, the status sensitivity of first moves maintains the market for bribes. Furthermore, locating the problem of sustaining bribery at the level of individual interactions reveals that, while bribe extraction networks may exist in some settings (Shleifer and Vishny, 1993; Wade 1982), explicit collusion among providers is not necessary to maintain bribe markets. The decentralized mechanism of mitigating risk one interaction at a time by being mindful of how bribe-givers are engaged with is sufficient.

This chapter shows how strategic interpersonal dynamics mute reporting in a low enforcement setting. These effects should replicate, and perhaps in a stronger fashion in higher enforcement environments, where confronted individuals would suffer both the reputational and punitive costs of their actions. Furthermore, if muting happens across prevalence environments, sense can now be made of what at first blush is a puzzling observation, that reporting is uncorrelated with prevalence levels (Figure 11.1b, where $r = 0.14$, *p-value* = 0.19). Effective matches generate the independent relationship between prevalence and reporting rates.

Policy levers currently deployed to curtail bribery are blunt at best (Fisman and Golden, 2017; Mungiu-Pippidi, 2015; Olken and Pande, 2012). Salvaging reform efforts to improve organizational capabilities requires prioritizing an understanding of the context-specificity of what needs to be reformed to create customized local solutions (Andrews et al., 2017; Hirschman, 2014; Schrank, 2015; Tendler et al., 1997; Viterna and Robertson, 2015). This cautions us against the tendency to locate enabling and sustaining factors of bribery in "container" concepts such as culture, institutions and governance that defy precise corrective policies. Container concepts ignore heterogeneity. As such, existing strategies for making sense of bribery do not focus attention on micro-level variation in these transactions. First moves give us a concrete and analytically tractable engagement with context to generate more precise policy interventions. Consider the following two examples.

First, consider the claim that bribe offers by clients capture their willingness to exchange bribes. This suggests that at an aggregate level, the proportion of first moves that are client-initiated is one way to capture social approval of bribery. Theoretically speaking, internalized compliance with a social practice should cut across segments of society. It should not be sensitive to enabling factors such as income, yet it is (Figure 11.2). To make an argument about a socially accepted practice, offers should be independent of socioeconomic status. The flip-side of this argument is that bribes should not be forced out of clients either, and especially not in a status-sensitive way (Figure 11.2). Paying attention to how bribery happens is a way to establish whether bribery is culturally-driven behavior or not.

Now, consider the following thought experiment. You are a policy analyst whose task is to recommend possible interventions to reduce bribery. You are presented with two countries, A and B, in which half of clients report bribe exchange during service provision. In A, 70% of exchanges are initiated by providers compared to 20% in B which are initiated by clients. Should you approach bribery in the same way in countries A and B? You are most likely to say no. In reflecting on why you think they are different, you might say that bribery in A appears to be more coercive on average than in B. You might then argue that clients in B appear to disapprove less of bribery than clients in A since they overwhelmingly offer bribes. As such, you might recommend a targeted intervention with providers in A and with clients in B. Suppose you are given more information: three-quarters of provider-initiated exchanges reported in A are based on extortion and one-quarter is due to indirect demands. Contrast this to another state of the world for A, call it A′, where the reverse pattern is observed. If you agree that extortion is a solicitation strategy that is more conducive to conflict, you conclude that providers in A enjoy significantly more impunity compared to those in A′. In an attempt to come up with the appropriate intervention in targeting providers, you might want to resolve the first-order issue of trying to understand what it is about the state apparatus or state-society relations that is causing this impunity in A compared to A′ which is, in effect, making it costless for providers to extract bribes in such an emboldened manner.

Notice that the simple conceptual exercise of varying the distribution of solicitation strategies gears us toward not only different interpretations of but also different loci of investigation of and intervention in the bribery equilibrium presented. In the absence of the thick descriptive detail that direct observation and ethnographies give us, the distributions and relational dynamics of first moves can be leveraged towards more meaningful cross-context comparisons in lieu of only relying on prevalence rates. From the outset, they allow us to make comparative statements without assuming cultural or institutional underpinnings for a prevalence rate; rather these "containers" become analysis endpoints.

11.5 Limitations

This study has several limitations, all of which suggest directions for future research. First, these data give us a unique opportunity to systematically explore the promise of expanding the set of bribery outcomes to first moves and examine what conditions first-mover behavior. A better version of these data would have collected solicitation patterns at the level of the transacting pair, rather than at the respondent-level, since clients make multiple bribe payments. I attempted to mitigate this data limitation through robustness analyses and controls where appropriate. Another limitation of these data—and this is a more general problem in data collection efforts on bribery—is the absence of attempted but unrealized exchanges. This would generate a more complete analysis of relational dynamics in these transactions. Thus, if strategic deployment of first moves is critical to alleviate unpalatability, the observed patterns should hold for both realized and unrealized advances. The worst possible outcome is to be refused and confronted. Third, I do not make causal arguments that link certain distributions of first moves to certain prevalence rates, nor about how certain distributions may make bribery more or less sticky. These are exciting new areas for research that carry implications for how policy efforts may be directed and designed. They reinforce the importance of making first moves and data at the level of transacting pairs routine features of data collected on bribery across contexts. The Afrobarometer and Transparency International's Global Corruption Barometer are existing platforms that could initially support experimenting with the integration of these data features.

Notes

1 Thanks to John-Paul Ferguson, Sarah Bond, Laura Doering, Warner Henson II, Paolo Parigi, Brian Rubineau, and Roman Galperin for their careful reading. Previous versions of this manuscript received invaluable feedback from Mark Granovetter, Andrew Walder, Ariela Schachter, Kate Weisshaar, and Patrick Bergemann. The support of Stanford University's Abbasi program for Islamic Studies and the Stanford Interdisciplinary Graduate Fellowship program are gratefully acknowledged. This study would not have been possible without the data provided by the L' Instance centrale de prévention de la corruption (ICPC) in Morocco.

2 Replacing the first move of clients who report two solicitation strategies with their second choice delivers a slightly different but comparable distribution to what is presented in Table 11.1: extortion 19.39%, dyadic indirect 39.55%, third party indirect 1.69%, and offers 28.25%. The substantive conclusions drawn remain the same.

3 Income and geographic controls do not change these estimates much. While indirect demands and offers are more profitable per exchange, establishing that indirect demands and offers are more profitable overall would require an estimate of unrealized attempts associated with each kind of solicitation strategy. Solicitation data for unrealized attempts are not available.

4 This is also consistent with the finding that poverty certificate carriers who bribed to obtain their certificate are more likely to *offer* bribes during healthcare provision compared to non-carriers; see Table D1. However, the possibility cannot be ruled out that the subset of poverty certificate holders who bribed to get their certificate have a higher propensity to bribe compared to those who did not.

5 Reporting is one form of retribution. The existence of alternative informal forms of retribution, e.g. shaming, is also sure to provide a partial explanation for the low reporting rates observed.

Appendix

A Materials

A.1 Solicitation Text Used to Recruit Respondents for Survey

The original text in French reads as:

Bonjour Monsieur/Madame,

Je suis [nom de l'enquêteur]. Nous sommes mandaté par l'ICPC pour lancer une étude sur le phénomène de la corruption dans le secteur de la santé. Les résultats de cette étude permettront d'émettre les recommandations pour éradiquer ce fléau dans les établissements de santé et par conséquent garantir aux citoyens la disponibilité et la qualité des services de santé dans un cadre d'équité et de transparence.

Compte tenu de l'importance du sujet, permettez-moi de solliciter un peu de votre temps et votre précieuse coopération qui sont plus que nécessaires pour le succès de l'enquête.

The translated text reads as:

Good day Sir/Madam,

My name is [surveyor name]. We are mandated by the ICPC to launch a study on the phenomenon of corruption in the healthcare sector. The results of this study will allow us to make recommendations to eradicate this scourge in healthcare establishments and as a consequence guarantee citizens the availability and quality of healthcare services in a framework of equity and transparency.

Given the importance of the subject, permit me to ask for some of your time and your valuable cooperation which are more than necessary for the success of this study.

A.2 Kinds of Services Covered in Survey

The survey asks about clients' (i.e., healthcare-seekers') experiences in 11 kinds of services: access, information and orientation, emergency services, non-urgent specialized consultations, laboratory analysis and radiology, medications, hospitalization, operations planning, operations equipment, blood transfusion, and payment. Different clients have different combinations of services in their care path.

A.3 Reasons for Resorting to Bribe Exchange

See Table A1.

Table A1 Reasons cited by respondents for bribing during healthcare services provision. Frequencies correspond to number of respondents. Percentages sum to 100 for each service

Reasons	Frequency	%
Information and orientation		
To have information	33	42.31
To be better oriented and accompanied	41	52.56
To avoid the queue	4	5.13
To have supplementary information	0	0.00
Other	0	0.00
Emergency services		
To avoid the queue	98	48.51
To be seen by a physician	59	29.21
To have an accelerated appointment	33	16.34
To be privileged in the choice of physician	7	3.47
To bypass being transferred to a closer establishment	0	0.00
Other	5	2.48
Non-urgent consultations		
To obtain a prescription for supplementary medication for reimbursement	2	2.82
To have a certificate	45	63.38
To have a certificate to use falsely	24	33.80
To be privileged in the choice of physician	0	0.00
Radiology and analysis		
To have the results in time	34	52.31
To avoid the queue	21	32.31
To not do it elsewhere and more expensively	9	13.85
To have supplementary analyses conducted	0	0.00
Other	1	1.54
Medications		
To obtain free medication	40	86.96
To obtain more medication than prescribed	5	10.87
To avoid the wait for medication that need to be prepared	0	0.00
Other	1	2.17
Hospitalization		
To have a bed	38	24.52
To have a supplementary bed to accompany a patient	28	18.06

(Continued)

Table A1 (Continued)

Reasons	Frequency	%
To have a higher quality bed, covers etc.	34	21.94
To have an individually occupied room	7	4.52
To have food	2	1.29
To have supplementary food for a person accompanying a patient	6	3.87
To have better quality and/or more food	6	3.87
Other	34	21.94
Operations planning		
To have a date	12	14.63
To have a closer date	48	58.54
To have a physician of your choice	5	6.10
Other	17	20.73

Data source: ICPC Survey, Morocco.

Notes: Surveyors did not ask for reasons for every service, which is why the total of this table is less than the total of bribe exchanges reported.

B Afrobarometer Data

The Afrobarometer is a pan-African research institution, established in 1999, that conducts public attitude surveys on social, political and economic issues across African countries on a regular basis. More information can be found here: www.afrobarometer.org/about. National probability samples are used to ensure that survey respondents are representative of all citizens of voting age in a country.

B.1 Calculations for Figure 11.1

Figure 11.1 is derived from Round 6 of the Afrobarometer, which was published in 2016.

Figure 11.1a is derived from the following questions : "And how often, if ever, did you have to pay a bribe, give a gift, or do a favor for a:

- teacher or school official in order to get the services you needed from the schools? (*Question* Q55B)
- health worker or clinic or hospital staff in order to get the medical care you needed? (*Question* Q55D)
- government official in order to get the document you needed [for document or permit]? (*Question* Q55F)
- government official in order to get the document you needed [for water, sanitation or electric services]? (*Question* Q55H)

- police officer in order to get the assistance you needed, or to avoid a problem like passing a checkpoint or avoiding a fine or arrest? (*Question* Q55J)
- judge or court official in order to get the assistance you needed from the courts?" (*Question* Q55L)

To calculate the country-level prevalence rates, I divide the number of respondents who reported paying bribes in any of the six public sectors—education, healthcare, permits, utilities (i.e., water, sanitation and electricity), police, and courts—by the number of respondents who reported having contact with the sectors.

Figure 11.1b is derived from the question: "If you ever paid a bribe for any of the services discussed above [Q55B, Q55D, Q55F, Q55H, Q55J or Q55L], did you report any of the incidents you mentioned to a government official or someone in authority? (*Question* Q56)." To calculate the proportion who reported bribe payments, I divide the number of respondents who reported the incidents by the number of respondents who paid bribes.

Figure 11.1c is derived from the question: "Which of the following happened the most recent time that you reported a bribery incident? Authorities took action against the government officials involved (*Question* Q57A)." To calculate the proportion of reports for which authorities took action, I divide the number of respondents who indicated that authorities took action by the number of respondents who reported the incidents.

B.2 Comparing Bribe Rates during Healthcare Provision between the ICPC and Afrobarometer Data

I compare bribe rates between the ICPC survey data and two rounds of the Afrobarometer surveys, Round 6 (R6, Moroccans were surveyed in 2015) and Round 5 (R5, Moroccans were surveyed in 2013).

The bribery question in R5 is worded slightly differently from R6 (see Q55D above). In R5, respondents are asked "In the past year, how often, if ever, have you had to pay a bribe, give a gift, or do a favor to government officials in order to: Get treatment at a local health clinic or hospital?" (*Question* Q61C).

To calculate the bribe rates for healthcare from Q55D-R6 and Q61C-R5, I divide the number of respondents who report making at least one bribe payment during healthcare provision by the number of respondents who made contact with the healthcare sector. In 2013, 57% of Moroccans who sought healthcare reported making at least one bribe payment, compared to 38% of Moroccans in 2016. In the ICPC survey data, I find that about 46% of clients who sought healthcare services in 2010/2011 made at least one bribe payment (see Table 11.1 in main text)—a prevalence rate that falls within the range revealed by the Afrobarometer surveys for Morocco.

C Descriptive Statistics of Sample

Table C1 Descriptive statistics for explanatory and control variables

Variables	% or Mean (SD)	Frequency
Explanatory variables		
Monthly household income (in Moroccan dirhams)	5,737 (6,297)	1060
Education (in years)	10.8 (5.8)	1127
Poverty certificate carriers (% respondents)	27.79	314
Poverty certificate obtained by bribing (% poverty certificate carriers)		
Yes	35.67	112
No response	26.75	84
Used connections (% respondents)	7.17	83
Control variables		
No. of services sought per respondent	4.68 (1.73)	1157
Age (in years)	35.7 (12)	1150
Female (% respondents)	42.35	490
Married (% respondents)	64.86	716
Medical coverage (% respondents)	41.83	484
Establishment type (% respondents)		
Hospital (public)	71.05	822
Community Health center (public)	10.80	125
Clinic (private or semi-private)	18.15	210
Urban residence (% respondents)	93.09	1077
City of residence (% respondents)		
Rabat	27.05	313
Casablanca	29.73	344
Tangier	16.77	194
Marrakech	15.99	185
Oujda	10.46	121
No. of respondents		1157

Data source: ICPC Survey, Morocco.

Table C2 Pairwise correlations between client status variables

	Income	Education	Poverty certificate	Used connections
Income	1			
Education	0.352***	1		
Poverty Certificate	−0.205***	−0.289***		
Used Connections	0.0522	0.0562	−0.0125	1

*$p < 0.05$, **$p < 0.01$, ***$p < 0.001$.
Note: Standardized income and education variables used for the pairwise correlations.

D Analyses

D.1 First Moves and Client Status

See Table D1 below, which is the basis for results depicted in Figure 11.2 of the main text.

Table D1 Multinomial logit model predicting solicitation strategy by client status

Base outcome: Indirect demand	(1)		(2)		(3)		(4)	
	Extortion	Offers	Extortion	Offers	Extortion	Offers	Extortion	Offers
Education (SD units)	0.141	−0.453	0.144	−0.541	0.171	−0.527	0.206	−0.517
	(0.416)	(−1.204)	(0.395)	(−1.347)	(0.467)	(−1.312)	(0.556)	(−1.283)
Income (SD units)	−0.598	1.806***	−0.556	1.909***	−0.555	1.928***	−0.559	1.931***
	(−0.941)	(2.809)	(−0.865)	(2.924)	(−0.856)	(2.942)	(−0.860)	(2.936)
Income squared	0.125	−0.542*	0.144	−0.525*	0.142	−0.533*	0.149	−0.534*
	(0.382)	(−1.778)	(0.440)	(−1.720)	(0.428)	(−1.743)	(0.448)	(−1.741)
Poverty certificate (PC)	−0.282	0.754**						
	(−0.944)	(2.380)						
Bribed for PC (Ref = not carrier)			−0.743*	0.720*	−0.554	0.746**	−0.550	0.745**
			(−1.850)	(1.950)	(−1.342)	(1.977)	(−1.322)	(1.975)
Did not bribe for PC			−1.047*	−0.815	−1.079*	−0.825	−1.015*	−0.810
			(−1.735)	(−1.092)	(−1.777)	(−1.106)	(−1.671)	(−1.085)
Total No. of bribe payments made					−0.288**	−0.014	−0.418***	−0.066
					(−2.225)	(−0.106)	(−2.746)	(−0.441)
Status variability score across providers bribed							0.544*	0.258
							(1.795)	(0.796)
Controls	Suppressed		Suppressed		Suppressed		Suppressed	
Constant	−2.341***	−1.369	−2.296***	−1.287	−1.852**	−1.234	−1.772**	−1.202
	(−3.194)	(−1.615)	(−2.714)	(−1.359)	(−2.128)	(−1.284)	(−2.021)	(−1.249)
Observations	413		381		381		381	
χ^2(df)	123.878		124.026		129.819		133.110	
	(28)		(30)		(32)		(34)	

t statistics in parentheses.

* $p < 0.10$, ** $p < 0.05$, *** $p < 0.01$.

Analysis sample limited to clients who pay bribes during healthcare services provision. The functional form of the model is $\log(P(Y = m)/P(Y = n)) = \beta m0 + \beta m\mathbf{X}$ where m and n are mutually exclusive categories of the dependent variable, $\beta m0$ is the intercept of category m, $\mathbf{B}m$ is the vector of coefficients relevant to the vector of explanatory and control variables $\mathbf{X}$. Coefficient in log-odds form. A Wald test for combining alternatives was conducted to test the collapsibility assumption. Test statistics were significant for all pairwise combinations, therefore the criterion is satisfied. Controls: marital status, years of age, medical coverage, city, urbanicity, and establishment type. Reference categories: unmarried, hospital, not poverty certificate carrier, no medical coverage, Rabat, and rural.

Data Source: ICPC Survey, Morocco.

D.2 First Moves and Provider Status

In the main text, I write that effective matches are a two-sided affair.

Table D2 reports estimates of multinomial logit models that predict the log-odds of a reported solicitation strategy (extortion, indirect demand, and offers) as a function of provider status. Standard errors are clustered by client since a given healthcare-seeker can pay bribes to more than one provider. About half of clients who paid bribes made more than one bribe payment. Of this group, about one-third of them report payments to different types of healthcare-providers. As such, Model 2 includes controls for the number of bribes paid per client and a status variability score that captures variability in the occupational status of healthcare-providers that clients bribed. The substantive conclusions are the same across models.

Table D2 Multinomial logit predicting solicitation strategy by provider status

Base outcome: indirect demand	(1)		(2)	
	Extortion	*Offers*	*Extortion*	*Offers*
Doctor	−0.012	0.613**	−0.157	0.517**
	(−0.048)	(2.556)	(−0.624)	(2.255)
Nurse (= Ref)	—	—	—	—
Support agent	−0.334	−0.778*	−0.334	−0.848**
	(−0.922)	(−1.813)	(−0.921)	(−2.038)
Security agent	0.345	−0.620*	−0.030	−0.874***
	(1.287)	(−1.654)	(−0.127)	(−2.936)
Total No. of bribe payments made			−0.421***	−0.077
			(−3.244)	(−0.515)
Status variability score across providers bribed			0.513*	0.364
			(1.665)	(1.073)
Controls	Suppressed		Suppressed	
Constant	−1.949***	−3.646***	−0.982	−3.605***
	(−3.164)	(−4.850)	(−1.501)	(−3.945)
Observations (clusters)	796 (409)		796 (409)	
χ^2 (df)	88.383 (20)		98.807 (24)	

t statistics in parentheses.
* $p < 0.10$, ** $p < 0.05$, *** $p < 0.01$.
Coefficients in log-odds form. Sample restricted to bribe transactions during healthcare services provision. Standard errors are clustered by client. Respondents who reported "Other/No Response" for the status of the provider were excluded from the analysis since the interpretation of this category is not clear. This amounts to the exclusion of 45 bribe exchanges (40 interactions with "No Response" and 5 interactions with "Other"). The Wald test for non-collapsibility was statistically significant. Controls: city, urbanicity, and establishment type. Reference categories: Rabat, rural, and hospital.

D.3 Bribe Size and Provider Status

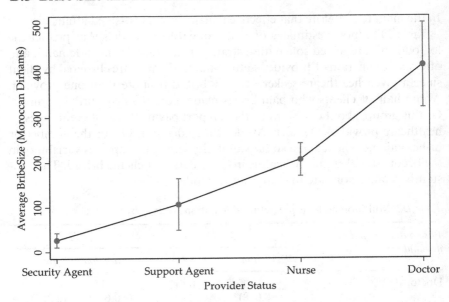

Figure D1 Plotted is the average bribe size (in Moroccan dirhams) by provider status. 10 dirhams 1 USD. These estimates come from an OLS model that predicts bribe size as a function of increasing provider status, controlling for establishment type, city and urbanicity. Standard errors are clustered by client since a client can make more than one bribe payment. Data Source: ICPC Survey, Morocco.

D.4 Ineffective Matches

Table D3 Characteristics of reported interactions. 64 respondents preferred to not provide an answer to whether they reported a corrupt provider. Recall 531 out of 1157 respondents bribed during healthcare provision. Therefore 1.13% (6/531) and 1.12% (7/626) correspond to the reporting rates of those who bribed and those who did not, respectively.

	Frequency	*%*
Whether bribed during healthcare provision		
Yes	6	46
No	7	54
Education level		
None	1	7.69
Secondary	1	7.69
Baccalauréat (or equiv)	1	7.69
Baccalauréat +2	1	7.69
Baccalauréat +3	3	23.08
Baccalauréat +4/5 or plus	6	46.15
Monthly household income (dirhams)		
$x < 5K$	1	7.69

(Continued)

Table D3 (Continued)

	Frequency	%
5K ≤ x < 10K	9	69.23
10K ≤ x < 15K	1	7.69
15K ≤ x < 20K	−1	−7.69
x ≥ 20K	1	7.69
No Response		
Solicitation Strategy		
Extortion	9	69.23
Indirect demand	2	15.38
Offers	1	7.69
Not specified	1	7.69
Provider status	2	16.67
Doctor		
Nurse	6	50
Support agent	—	—
Security agent	1	8.33
Other	3	25
Total	13	100

Data Source: ICPC Survey, Morocco.

References

Abbott, A. (2014). *The system of professions: An essay on the division of expert labor.* University of Chicago Press.

Aboudrar, M. A. (2014). *Prévention et lutte contre la corruption: où en sommes nous?* Instance Centrale de Prévention de la Corruption.

Alesina, A. and P. Giuliano (2015). Culture and institutions. *Journal of Economic Literature 53* (4), 898–944.

Andrews, M., L. Pritchett, and M. Woolcock (2017). *Building state capability: Evidence, analysis, action.* Oxford University Press.

Anteby, M. (2010). Markets, morals, and practices of trade: Jurisdictional disputes in the us commerce in cadavers. *Administrative Science Quarterly 55* (4), 606–638.

Barr, A. and D. Serra (2010). Corruption and culture: An experimental analysis. *Journal of Public Economics 94* (11), 862–869.

Becker, G. S. and G. J. Stigler (1974). Law enforcement, malfeasance, and compensation of enforcers. *The Journal of Legal Studies 3* (1), 1–18.

Bénabou, R. and J. Tirole (2011). Identity, morals, and taboos: Beliefs as assets. *The Quarterly Journal of Economics 126* (2), 805–855.

Bergemann, P. (2017). Denunciation and social control. *American Sociological Review 82* (2), 384–406.

Bourdieu, P. (2013). *Distinction: A social critique of the judgement of taste.* Routledge.

Dakhlallah, D. (2021). *Collective reputation and bribe exchange: A field experiment.* R & R.

de Sardan, J. P. O. (1999). A moral economy of corruption in Africa? *The Journal of Modern African Studies 37* (1), 25–52.

Elster, J. (1989). Social norms and economic theory. *Journal of Economic Perspectives 3* (4), 99–117.

Fisman, R. and M. Golden (2017). *Corruption: What everyone needs to know.* Oxford University Press.

Fisman, R. and E. Miguel (2007). Corruption, norms, and legal enforcement: Evidence from diplomatic parking tickets. *Journal of Political economy 115* (6), 1020–1048.

Fried, B. J., P. Lagunes, and A. Venkataramani (2010). Corruption and inequality at the crossroad: A multimethod study of bribery and discrimination in Latin America. *Latin American Research Review 45* 76–97.

Gambetta, D. (2011). *Codes of the underworld: How criminals communicate*. Princeton University Press.

Gelman, A. (2008). Scaling regression inputs by dividing by two standard deviations. *Statistics in Medicine 27* (15), 2865–2873.

Gino, F., M. E. Schweitzer, N. L. Mead, and D. Ariely (2011). Unable to resist temptation: How self-control depletion promotes unethical behavior. *Organizational Behavior and Human Decision Processes 115* (2), 191–203.

Goffman, E. (1982). *Interaction ritual-essays on face-to-face behavior*. 1st Pantheon Books.

Granovetter, M. (2007). The Social Construction of Corruption. In *On capitalism*, ed. Victor Nee and Richard Swedberg, 152–174. Stanford University Press.

Gupta, A. (1995). Blurred boundaries: The discourse of corruption, the culture of politics, and the imagined state. *American Ethnologist 22* (2), 375–402.

Gupta, A. (2012). *Red tape: Bureaucracy, structural violence, and poverty in India*. Duke University Press.

Haut-Commissariat Au Plan, R. A. M. (2009). Étude sur les classes moyennes au Maroc: Les classes moyennes Marocaines: caractéristiques, évolution et facteurs d'élargissement. Technical report.

Hirschman, A. O. (2014). *Development projects observed*. Brookings Institution Press.

Hoang, K. K. (2018). Risky investments: How local and foreign investors finesse corruption-rife emerging markets. *American Sociological Review 83* (4), 657–685.

Hout, M. (2004). *Getting the most out of the GSS income measures*. National Opinion Research Center.

Icpc, L. (2011). étude sur le phenomene de la corruption dans le secteur de la sante: evaluation et diagnostic. Technical report, Royaume du Maroc, Le Premier Ministre.

Irlenbusch, B. and M. C. Villeval (2015). Behavioral ethics: How psychology influenced economics and how economics might inform psychology? *Current Opinion in Psychology 6*, 87–92.

King, M. and I. M. Nembhard (2016). Role change and interaction dynamics in hierarchical groups: A field experiment in healthcare. In *Academy of Management Proceedings*, Volume 2016, pp. 14974. Academy of Management Briarcliff Manor, NY 10510.

Lomnitz, L. A. (1988). Informal exchange networks in formal systems: A theoretical model. *American Anthropologist 90* (1), 42–55.

Miller, W. L., Å. B. Grødeland, and T. Y. Koshechkina (2001). *A culture of corruption?: Coping with government in post-communist Europe*. Central European University Press.

Ministère de la Santé, R. D. M. (2011). Santé en chiffre 2010, edition 2011. Technical report.

Ministère de la Santé, R. D. M. (2012). Stratégie sectorielle de la santé 2012-2016. Technical report.

Mungiu-Pippidi, A. (2015). *The quest for good governance: How societies develop control of corruption*. Cambridge University Press.

North, D. C. (1991). American economic association. *The Journal of Economic Perspectives 5* (1), 97–112.

Olken, B. A. and R. Pande (2012). Corruption in developing countries. *Annual Review of Economics 4* (1), 479–509.

Pinker, S., M. A. Nowak, and J. J. Lee (2008). The logic of indirect speech. *Proceedings of the National Academy of sciences 105* (3), 833–838.

Rasul, I. and D. Rogger (2018). Management of bureaucrats and public service delivery: Evidence from the nigerian civil service. *The Economic Journal 128* (608), 413–446.

Robinson, A. L., B. Seim, et al. (2018). Who is targeted in corruption? disentangling the effects of wealth and power on exposure to bribery. *Quarterly Journal of Political Science 13* (3), 313–331.

Rossman, G. (2014). Obfuscatory relational work and disreputable exchange. *Sociological Theory 32*, 43–63.

Saleh, Y. (2012). Hopes for a new Egypt marred by pervasive corruption. *Reuters*.

Schelling, T. (2011). *The strategy of conflict*. Literary Licensing.

Schrank, A. (2015). Toward a new economic sociology of development. *Sociology of Development 1* (2), 233–258.

Shleifer, A. and R. W. Vishny (1993). Corruption. *The Quarterly Journal of Economics 108* (3), 599–617.

Smith, D. J. (2008). *A culture of corruption*. Princeton University Press.

Svensson, J. (2003). Who must pay bribes and how much? Evidence from a cross section of firms. *The Quarterly Journal of Economics 118* (1), 207–230.

Tendler, J. et al. (1997). *Good government in the tropics*. Johns Hopkins University Press.

Transparency International (2016). People and corruption: Middle east and North Africa survey 2016. Technical report, Global Corruption Barometer, Transparency International.

Transparency International (2019). Global corruption barometer data.

Vian, T. (2008). Review of corruption in the health sector: Theory, methods and interventions. *Health Policy and Planning 23* (2), 83–94.

Viterna, J. and C. Robertson (2015). New directions for the sociology of development. *Annual Review of Sociology 41*, 243–269.

Wade, R. (1982). The system of administrative and political corruption: Canal irrigation in south India. *The Journal of Development Studies 18* (3), 287–328.

World Bank Governance Indicators (2010/2016). https://databank.worldbank.org/source/worldwide-governance-indicators

Part IV
The Challenge of Peripherality

12 Measuring the Interaction between the Interregional Accessibility and the Geography of Institutions

The Case of Greece

Dimitrios Tsiotas and Vassilis Tselios

12.1 Introduction

The regional economy is a composite concept defined within a multilevel conceptual context. It incorporates the geographical (spatial) background and the geomorphological features of the regional system, the functionality of the spatial economy, the characteristics and attributes of the underlying living structures and society, and the temporal dimension related to the evolution of the regions and other spatial units over time (O'Sullivan, 2007; Rodrigue et al., 2013). The regional economic systems are very complex and enjoy a multidisciplinary conceptualization. Accordingly, regional economics is a field that continuously transforms into a separate discipline called regional science (Fischer and Nijkamp, 2014). The composite nature of regional science contributes to a broader understanding of the spatial-economic systems, leading to a conceptual integration in the study of regional economies by incorporating the disciplinary approaches emerging by the dynamics of the continuously changing world. Within this dynamically complex context, the economies, institutions, and territories, which are pillars of the structure and functionality of the regional economic systems, evolve and change toward the directions, trends, and norms of the changing world (MacKinnon et al., 2009; Rodriguez-Pose, 2013, 2020). Today, in the era of connectivity, the structural and functional attributes of the spatial-socioeconomic systems are revisited within the conceptual context of inter-connectedness (Tsiotas and Tselios, 2021). Worldwide, this condition emerged due to the technological evolution of transportation and communication that led to transportation costs' reduction and a consequent shrunk of spatial distances (Rodrigue et al., 2013).

Network science (Barabasi, 2016; Ducruet and Beauguitte, 2014; Newman, 2010) is a modern discipline conceptualizing connectivity using the network paradigm. The development of this discipline facilitated the effective modeling of socioeconomic and spatial interaction systems into graphs, consisting of pair-sets of nodes and edges (links) (Rodrigue et al., 2013). Under the perspective of network science, the pillars of the economy, institutions, and territories can enjoy an interpretation within the framework of interconnectedness. From one side, economic processes have been challenged by the

DOI: 10.4324/9781003191049-16

transition towards post-industrial economies, growing globalization, and financial instability (Rodriguez-Pose, 2013, 2020), thus driving into global connectivity. From another aspect, institutions face challenges due to several externalities that affect routines, interaction patterns, and cognitive orientations (MacKinnon et al., 2009; Rodriguez-Pose, 2013). Such challenges facilitate the development of connectivity. From a third perspective, territories experience increasing self-definition pressures due to globalization (Rodriguez-Pose, 2020) and need to operate within a globally interconnected environment. All these aspects illustrate the interrelation between the spheres of economy, institutions, and territories. This relationship is comprehensively examined from the economic perspective (Amin and Thrift, 1995; MacKinnon et al., 2009; Rodriguez-Pose, 2013) but not within the context of interconnectedness through the network paradigm. This situation raises an attractive paradox describing that these three spheres are highly and mutually interconnected components in the structure and functionality of the globalized economy. However, in terms of scientific knowledge, these spheres appear "disconnected" because their linkages have not yet been studied within a solid context integrating the network and connectivity paradigm (which drives the conceptualization of modern society).

Moreover, within the context of peripherality (Danson and De Souza, 2012; Spiekermann and Neubauer, 2002), the relationship between economy, institutions, and territories can become more comprehended. In conceptual terms, peripherality is a property of a place to be peripheral, relative to others considered central. Due to its relativity, it is a complex property enjoying (Danson and De Souza, 2012): (a) geographical, related to the distance to connect with the core; (b) structural, produced through structures of dependency from the center; (c) functional, combining economic with socio-cultural and discursive aspects; and (d) miscellaneous, where each periphery combines different aspects and fields of peripherality, making though classification difficult; interpretations. Peripherality has a double definition (Danson and De Souza, 2012; Spiekermann and Neubauer, 2002). First, it is static because it is related to geographical and structural features. Second, it is dynamic because it is related to functional, such as social, economic, political, or similar, features. This double configuration is fully compatible with the two-dimensional definition of a network model, consisting of a structural (related to the topology and geometry) and a functional (related to the type of flows or information exchange) component. Therefore, the network paradigm can provide a framework for deeper comprehending the relationship between institutions and peripherality and the broader interrelation between economy, institutions, and territories. In particular, it facilitates an integrated structural and functional modeling of connectivity systems, allowing an interpretation of the network elements in terms of central and peripheral configuration.

Aiming to contribute to a better understanding of this framework, this chapter constructs (at the interregional scale – NUTS III) an aggregate network model of multimodal transportation in Greece and considers aspects of the country's institutional configuration; to detect whether these pillars are interconnected, and in what level this is applicable. The study builds on the background of multimodal interregional transportation; due to its ability to illustrate an integrated complex system of spatial and socioeconomic functionality (Lavrinenko et al., 2019). In this context, transportation is seen as a derived demand between regional markets producing flows of socioeconomic interest (labor, trade, and recreation). The Greek multimodal transportation network consists of four transportation layers; the road network of interregional connections, the national railway network, the maritime network of ferry connections, and the domestic air transport network. On this multimodal transportation system, we consider different proxies of both formal and informal institutional factors in Greece; representing family size (Basco, 2015); religion (Haynes, 2007); prison capacity (Glasmeier and Farrigan, 2007); media infrastructures (Picard, 2009); education (Bille and Schulze, 2006; Cohen, 2017); army sites (Artavani et al., 2015); and industrial capital (Cooke et al., 2005). The rationale of these institutional proxies is to approximate as spherically as possible the institutional configuration in Greece. In particular, although a universal definition of institutions is not yet concerted (Hodgson, 2006, 2015), it is commonly accepted that they are defined based on the rules governing and configuring the structures of social interactions in the integrated systems. Provided that institutions are by default social configurations but not vice versa, it is necessary to consider diverse dimensions (Goodin, 1996; Hodgson, 2006, 2015) of a society's institutional structure. Such dimensions can be the disciplinary or penitentiary (Skowronski and Talik, 2021; Spierenburg, 2007); demographic (Basco, 2015; Bille and Schulze, 2006; Cohen, 2017); doctrine (Haynes, 2007); citizenship (Janowitz, 1976); political (Napoli, 2003; Schudson, 2002); cultural (Haynes, 2007); and productive (Xiaoming, 2006) institutional aspect; to approximate thereby the regulated social structures' multifaceted configuration. Moreover, these social configuration dimensions are simultaneously determinants of economic and regional development (Rodrigue et al., 2013; Rodriguez-Pose, 2013, 2020), namely of the utmost goal of transport infrastructures' development (Polyzos and Tsiotas, 2020). Therefore, the conceptual link between the institutions' configuration and transportation network infrastructures can be found based on economic and regional development, which is their common utmost goal.

Within this context, this chapter studies the degree to which the spatial configuration of the available institutional proxy-variables is related to the network structure of the multimodal transportation network in Greece; and examines possible interaction patterns between the interregional accessibility and the geography of institutions. Based on the triplet relationship "institutions → economic and regional development ← transport infrastructures", the purpose of this study is to uncover aspects of the indirect linkage between institutions and transport infrastructures and thus to provide insights into the coexistence of transport accessibility and the institutional configuration, for the country and its peripheral areas, on their common basis of economic and regional development. The overall approach proposes a framework integrating the concepts of economy, institutions, and space. The remainder of this chapter is structured as follows; Section 12.2 describes the methodology of the study, the available data, and the variables included in the analysis. Section 12.3 shows the results and provides a discussion within the milieu of regional economics and development. Finally, Section 12.4 provides the conclusions.

12.2 Methodology and Data

The analysis aims to uncover aspects of the indirect linkage between institutions and transport infrastructures, based on the triplet relationship "institutions $\rightarrow$ economic and regional development $\leftarrow$ transport infrastructures". To do so, we configure one group of institutional and another one of transportation network variables; and we apply empirical (correlation and t-testing) analysis to detect possible relations between these two groups of variables. Network variables are constructed based on a multilayer graph model, the interregional multimodal Greek transportation network, whereas the other variables are composed of secondary data. The following paragraphs describe, in more detail, the network modeling, the configuration of the available variables, and the methods used in the empirical analysis.

12.2.1 Multilayer Graph Modeling

The interregional multimodal transportation network in Greece (GMTN$\equiv$$G$) is modeled to a multilayer graph (Boccaletti et al., 2014; Kivela et al., 2014), which is composed of four (4) layers, as is shown in Figure 12.1. The first layer $G_1(1'115;2'289)$ represents the interregional road accessibility network of the Greek cities (Greek Road Network – GRN); and is modeled to a geo-referenced primal graph, as is shown in Figure 12.1a. In the first layer, the $n_1 = 1'115$ in number nodes express the Greek cities, settlements, and urban units with a population greater than 500 citizens, while the $m_1 = 2'289$ links (edges) express the potential of direct road accessibility between urban units, in Greece. This network is non-directed, and its weights represent spatial distances. This layer was constructed on data from the Google Mapping Services website (Google Maps, 2021), where the nodes are geo-referenced (according to the WGS Web Mercator projection) at the coordinates (longitude, latitude) of the city centers. The first layer is not a connective graph but an aggregate network consisting of 156 components. This structure is a result of the hybrid mainland and insular morphology of the country. Further, the biggest component is the road network of mainland Greece, whereas the other 155 represent local road networks of the Greek islands that lack road transport connectivity with the mainland.

The second $G_2(107,107)$ layer represents the interregional railway network in Greece (Greek RAil Network – GRAN). It is an undirected graph constructed in the L-space representation (Barthelemy, 2011; Marshall et al., 2018). In this layer, the $n_2 = 107$ nodes represent *railway intersections*, and the $m_2 = 107$ links represent railway routes running between successive nodes, as

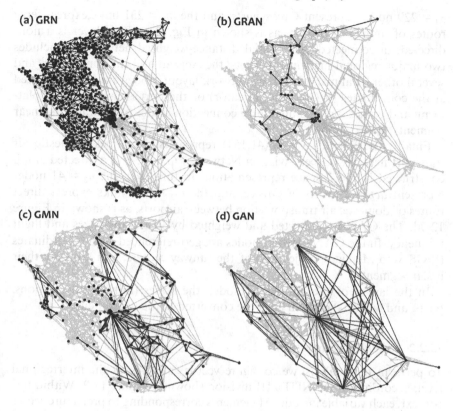

Figure 12.1 The multilayer network-model of the interregional multimodal transport
accessibility in Greece (GMTN), composed of (a) the road network
GRN(1′115;2′289), (b) the rail network GRAN(107;107), (c) the maritime network GMN(229;231), and (d) the air transport network
GAN(41;154) layer. The components (nodes, edges) included in each
layer are dark highlighted, comparatively to their inactive background
GMTN shown in light grey.

is shown in Figure 12.1b. The GRAN is non-directed and disconnected, and
its weights express spatial distances. The data for the construction of this
layer was extracted from the geodata.gov.gr (2010) database, which is a web
platform that provides open geospatial data and services for Greece and
serves as a national open-data catalog. In the network layout (Bastian et al.,
2009), nodes are geo-referenced and correspond to the exact coordinates of
the physical rail intersections (defined by the WGS Web Mercator projection),
whereas links express linear segments between nodes. The GRAN is a *disconnected* graph including (a) two major interregional rail network components
of the north and south mainland Greece, and (b) the local urban components
of the *electric*, *suburban*, and *metro* railways in the Athens region.

The third $G_3(229,231)$ layer represents the national network of maritime
ferry connections in Greece (Greek Maritime Network – GMN). It is an
undirected graph constructed in the L-space representation. In the GMN, the

n_3 = 229 nodes represent Greek *ports*; and the m_3 = 231 links express ferry routes of annual operation, as is shown in Fig.12.1c. This layer is a non-directed, disconnected, and spatial distance-weighted network. It includes two major regional maritime markets, the Aegean and the Ionian Sea, and several other local markets. In the network layout, nodes are geo-referenced at the coordinates (WGS Web Mercator) of the port's main mooring platform; and links express annual ferry connection routes (also drawn as linear segments) between ports.

Finally, the fourth layer G_4(41,154) represents the Greek domestic air transport network (Greek Aviation Network – GAN). It is a directed graph constructed in the *L*-space representation. In this layer, the n_2 = 41 nodes represent *active* airports in Greece, and the m_2 = 154 links express direct routes of domestic air transportation between airports, as is shown in Figure 12.1d. The GAN is connected and weighted by spatial distances and flight frequency. In the GAN's layout, nodes are geo-referenced at the coordinates (WGS Web Mercator projection) of the runway centers; and links are their linear segments.

In the layout of the multilayer model, the nodes (railway intersections, ports, and airports) of each layer are converted to the nearest city.

12.2.2 Variable Configuration

To perform the analysis, we configure vector variables, at the interregional level, according to the NUTS III division shown in Figure 12.2. Within this context, each variable includes 51 elements corresponding to prefecture scores of an attribute. On these variables, we further apply empirical analysis.

The available variables participating in the analysis of the GMTN are shown in Table 12.1, organized into the network, network infrastructure, and institutional variable groups: the first group (network variables) includes information about network connectivity (expressed in terms of node degree); the second group (network infrastructure variables) describes in a binary context the prefectures including transportation infrastructures; the third group (institutional variables) includes information relative to the formal and informal institutional configuration in Greece. The time reference of the variables regards the as recent as possible year, according to the data availability shown at the source column of Table 12.1.

12.2.3 Empirical Analysis

The empirical analysis aims to extract information of interrelationship amongst the spheres of economy, institutions, and territories; described in the context of the multimodal interregional transportation Greek market. In technical terms, the analysis consists of *independent-samples t-tests for the comparison of means*, along with *parametric and non-parametric correlation analysis* (Norusis, 2011; Walpole et al., 2012). In particular, the *independent-samples t-test* compares the mean values μ_α and μ_β between two discrete

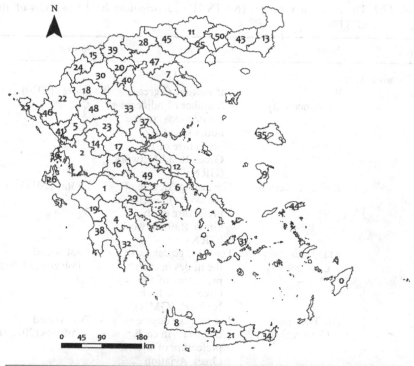

Label	Prefecture	Label	Prefecture	Label	Prefecture	Label	Prefecture
1	Achaea	14	Evrytania	27	Corfu (Kerkyra)	40	Pieria
2	Aetolia-Acarnania	15	Florina	28	Kilkis	41	Preveza
3	Argolis	16	Phocis	29	Korinthia	42	Rethymno
4	Arcadia	17	Phthiotis	30	Kozani	43	Rhodope
5	Arta	18	Grevena	31	Cyclades	44	Samos
6	Attica	19	Elis	32	Laconia	45	Serres
7	Chalkidiki	20	Imathia	33	Larissa	46	Thesprotia
8	Chania	21	Heraklion	34	Lasithi	47	Thessaloniki
9	Chios	22	Ioannina	35	Lesbos	48	Trikala
10	Dodecanese	23	Karditsa	36	Lefkada	49	Boeotia
11	Drama	24	Kastoria	37	Magnesia	50	Xanthi
12	Euboea	25	Kavala	38	Messenia	51	Zakynthos
13	Evros	26	Cephalonia	39	Pella		

Figure 12.2 The NUTS III administrative division of Greece into 51 prefectures.

Table 12.1 The regional variables (NUTS III) participating in the analysis of the GMTN

Rank	Name	Description	Source
■ *Network Variables (N)*[*]			
N_1	Road Connectivity	The average degree (number of adjacent connections) of the nodes in each prefecture of the Greek Road Network (GRN)	Tsiotas (2020)
N_2	Rail Connectivity	The average degree of the nodes in each prefecture of the Greek Rail Network (GRN)	Tsiotas (2017)
N_3	Maritime Connectivity	The average degree of the nodes in each prefecture of the Greek Maritime Network (GMN)	Tsiotas and Polyzos (2015a)
N_4	Air Transport Connectivity	The average degree of the nodes in each prefecture of the Greek Aviation Network (GRN)	Tsiotas and Polyzos (2015b)
■ *Network Infrastructure Variables (NI)*			
NI_1	Railway Infrastructures	Dummy (binary) variable returning 1 whether a prefecture has railway network infrastructures and 0 otherwise.	Tsiotas (2017)
NI_2	Port Infrastructures	Dummy variable returning 1 whether a prefecture has maritime (port) network infrastructures and 0 otherwise.	Tsiotas and Polyzos (2015a)
NI_3	Airport Infrastructures	Dummy variable returning 1 whether a prefecture has airport network infrastructures and 0 otherwise.	Tsiotas and Polyzos (2015b)
■ *Institutional Variables (I)*			
I_1	Religious Metropolises	The number of the religious metropolis, in each prefecture.	OWCG (2021)
I_2	TV Stations (Local)	The number of local TV stations (local range), in each prefecture.	NCRTV (2021)

(Continued)

Table 12.1 (Continued)

Rank	Name	Description	Source
I_3	TV Stations (National)	The number of national TV stations (national range), in each prefecture.	NCRTV (2021)
I_4	Prison Capacity	The maximum number of prisoners can be hosted by the prisons of a region.	MOJTHR (2021)
I_5	3-member Families	The number of three-member families (i.e. families with one child), in each prefecture.	ELSTAT (2011)
I_6	2-member Families	The number of two-member families (i.e. families with no children), in each prefecture.	ELSTAT (2011)
I_7	≥4-member Families	The number of four-member families and above (i.e. families with two and more children), in each prefecture.	ELSTAT (2011)
I_8	High Schools	The number of high schools, in each prefecture.	ELSTAT (2021a)
I_9	Secondary Schools	The number of schools of secondary education, in each prefecture.	ELSTAT (2021a)
I_{10}	Primary Schools	The number of schools of primary education, in each prefecture.	ELSTAT (2021a)
I_{11}	Primary Schools (public)	The number of public primary schools, in each prefecture.	ELSTAT (2021a)
I_{12}	Primary Schools (private)	The number of private primary schools, in each prefecture.	ELSTAT (2021a)
I_{13}	Firms	The number of firms (companies) registered in the trade chamber of each prefecture.	ELSTAT (2021b)
I_{14}	Army Camps	The number of army camps in each prefecture.	vrisko.gr (2021)**

(Continued)

Table 12.1 (Continued)

Rank	Name	Description	Source
I_{15}	Air Force Sites	The number of air force camps in each prefecture.	vrisko.gr (2021)[**]

* Symbols of the variables' groups inside the parentheses.
** Data extracted from the telephone numbers registry website vrisko.gr (2021).

groups $X_\alpha \cup X_\beta = X$ of a variable X that are defined by a dichotomous control variable $X_G = \left\{ x_i = \left\{ \begin{matrix} x_\alpha \\ x_\beta \end{matrix} \right| i = 1,...,n \right\}$. In the case of GMTN, the institutional variables $X \in I = \{I_1, I_2,..., I_{17}\}$ are grouped by the network infrastructure variables $X_G \in NI = \{NI_1, NI_2, NI_3\}$ into two groups, the first one including cases (i.e. the prefectures) that are supported by transportation infrastructures and the second one with those that are not. This approach aims to detect statistically significant differences between these two groups, within each variable, which are expected thus to provide insights into the importance of each transportation mode to the institutional configuration of the country. The computational algorithm applies *Levene's* test (Norusis, 2011) to examine the equality of variances of the comparable groups and produces separate results for unpooled and pooled variances that are chosen according to their significance (Norusis, 2011; Walpole et al., 2012). Moreover, for better supervision and interpretation of the results, the independent-samples *t*-tests are visualized by using error bars of 95% confidence intervals (CIs) for the means. The CIs are computed under a *pair-wisely* exclusion rule of missing values; and on the normality assumption (Norusis, 2011). When CIs overlay, the group's means (expressed by each bar) cannot be considered statistically different. When they do not overlay, the group means can be considered with 1–a% certainty statistically different (where *a* is the significance level).

Secondly, a correlation analysis is applied between the network (*N*) and institutional (*I*) groups of variables; to detect relationships in the variability and ranking of the available variables. This part of the analysis builds on a parametric and non-parametric approach applied by computing *Pearson's bivariate coefficients of correlation* and *Spearman's rank correlation coefficients*, respectively (Devore and Berk, 2012; Norusis, 2011; Walpole et al., 2012). In the parametric part of the analysis, the *Pearson's coefficients of correlation* are computed according to the formula:

$$r_{XY} = \frac{\text{cov}(X,Y)}{\sqrt{\text{var}(X)} \cdot \sqrt{\text{var}(Y)}} \qquad (12.1)$$

where $\text{cov}(X,Y)$ is the *covariance* of variables X, Y, and $\sqrt{\text{var}(\cdot)}$ is the sample standard deviations. The Pearson's coefficient of correlation ranges in the interval $[-1,1]$ and detects linear relations in absolute values close to one (Devore and Berk, 2012). In the non-parametric part, Spearman's correlation coefficients r_s are computed on the standard formula of Pearson's correlation coefficient (relation 1), where the rank-variables $rnk(x)$ and $rnk(y)$ are used instead of the numerical ones X, Y. The rank variables are computed by using the rank of each element instead of its numeric value, therefore providing more structural (that is related to the positioning of the cases) than algebraic information. Following the Pearson's coefficient, Spearman's (rho) coefficient of correlation also ranges within the interval $[-1,1]$, describing a perfect linear (positive or negative) relation when equals to one, expressing a perfectly concordant arrangement between the rankings of the two variables. Overall, the correlation analysis is expected to provide insights into the relationship between the transportation network connectivity and the institutional configuration of Greece, evaluating the importance of transportation networks to the functional (and more specifically institutional) aspects of regional development.

12.3 Results and Discussion

12.3.1 Independent Samples t-test for the Comparison of Means

At the first part of the analysis, an *independent-samples t-test* is applied for the comparison of the mean values $\mu(X_\alpha)$ and $\mu(X_\beta)$ between pair groups X_α, X_β of the institutional variables $X_\alpha \cup X_\beta = X \in I$, grouped by the network infrastructure variables NI. The results of the analysis are shown in detail in the appendix and are illustrated in the error bars of Figure 12.3, which are grouped into three panels, each representing a network (railway, port, and airport) infrastructures' variable. To facilitate comparisons, the error bars are computed on normalized values that are converted to the interval $[0,1]$. As it can be observed, the prefectures equipped with railway infrastructures (Figure 12.3a) have statistically significant greater mean values in all cases of institutional variables than the other prefectures. This observation describes that the prefectures enjoying railway transportation tend to have more developed formal and informal institutional configurations. Provided that railways are infrastructure networks of primal emergence, in the era of motorized land mass-transportation (Rodrigue et al., 2013), this observation implies that the institutional development in Greece was also driven by the primal pattern of the country's spatial development. According to this pattern, railway infrastructures were constructed to connect places where urban agglomeration first emerged, and therefore they provide transportation services to the oldest urban structures (Polyzos and Tsiotas, 2020). As can be observed in the map of Figure 12.1b, these places are located in the east middle of mainland Greece, along with the region of Peloponnesus. This spatial distribution is in line with the major S-shaped pattern of the country's spatial development, driven by gravitational and geomorphological dynamics.

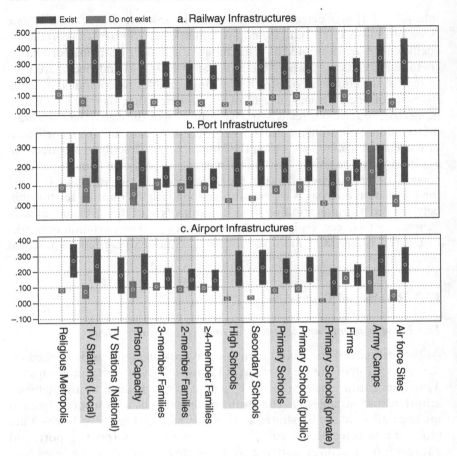

Figure 12.3 Error bars of 95% CIs for the comparison of the mean values $\mu(X_\alpha)$ and $\mu(X_\beta)$ between pair groups X_α, X_β of the institutional variables I that are grouped by the network infrastructure variables of (a) Railway Infrastructures (NI_1), (b) Port Infrastructures (NI_2), and (c) Airport Infrastructures (NI_3). Values of the variables represented by the error bars are normalized within the interval [0,1].

Next, Figure 12.3b shows that the prefectures equipped with ports have statistically significant greater mean values in the variables of religious metropolises (I_1), national stations (I_3), educational structures (I_8–I_{12}), and air force infrastructures (I_{15}) than the other prefectures. This observation illustrates that the prefectures enjoying maritime transportation tend to have more developed religious, media communication, educational, and air force institutional structures than the regions lacking such facilities. Since coastal and insular regions in Greece occupy approximately 88% of the country's population (ELSTAT, 2011), this finding illustrates a gravitational pattern in the configuration of the religious; media; and educational institutional structures in Greece (to the extent that the port infrastructures are highly related to the regions of coastal and insular geomorphology). Especially for air force sites (I_{15}), this finding supports the complementary functionality between maritime and air transportation in Greece, since prefectures with port infrastructures tend to have a higher number of airport infrastructures. Finally, Figure 12.3c describes that the prefectures equipped with airport infrastructures have statistically significant greater mean values in the variables of religious metropolises (I_1), local and national stations (I_2–I_3), educational structures (I_8–I_{12}), and (trivially) air force infrastructures (I_{15}) than the other prefectures. This observation describes that the prefectures enjoying airport facilities also tend to have more developed religious, media communication, educational, and air force institutional structures than the regions lacking such facilities. Since regions with airport infrastructures occupy approximately 72% of the country's population (ELSTAT, 2011), this finding also provides insights into the gravitational configuration of the religious; media; and educational institutional structures of the country. As far as variable I_{15} (air force cites) is concerned, the statistically significant differences detected in the analysis illustrate the complementarity between civil and military aviation. This complementary performance is empirically verified by the hybrid functionality (i.e. supporting both civil and military flights) of a considerable number of airports in Greece.

12.3.2 Correlation Analysis

In the second part, a parametric and non-parametric *correlation analysis* is applied between the groups of network and institutional variables. This analysis aims to detect linear relationships in their variability and ranking. The results are shown in Table 12.2, where rows correspond to institutional (I) and columns to network (N) variables. The parametric correlation analysis (shown in sub-table A) indicates that all cases (except one) of correlation coefficients are significant at the 5% and 1% level. These results imply that the common linear variability captured between the institutional and network variables is not likely (only 1–5%) to be by chance. In particular, all institutional variables, except I_1 (religious metropolises), are significantly correlated with the average GRN degree (N_1), illustrating that prefectures with high road network connectivity tend to have more developed formal and informal institutional configurations. However, religious metropolises are not in line with this result since they are distributed throughout the Greek territory and are sufficiently supported by road connectivity, in their total. Moreover, the correlations captured in this analysis are not high ($r_{XY} \geq 0.7$), indicating that linearity between variables is not strong, which in terms of regional economics can be interpreted as that the functional interrelation between the institutional variables and road network connectivity is basic. This interpretation stresses the primary role of the road network in the service of interregional communication.

Table 12.2 Results* of the correlation analysis

	Variable X eN [a]											
	Av. Degree GRN (N_1)			Av. Degree GRAN (N_2)			Av. Degree GMN (N_3)			Av. Degree GAN (N_4)		
Variable Y el [b]	r_{XY}	Sig. (2-tailed)	N	r_{XY}	Sig.	N	r_{XY}	Sig.	N	r_{XY}	Sig.	N
A. Pearson's Correlation Coefficients												
I_1	.211	.071	74	.530**	.000	69	.911**	.000	58	.889**	.000	71
I_2	.345**	.003	74	.600**	.000	69	.783**	.000	58	.847**	.000	71
I_3	.305**	.008	74	.563**	.000	69	.890**	.000	58	.921**	.000	71
I_4	.331**	.004	74	.627**	.000	69	.807**	.000	58	.839**	.000	71
I_5	.431**	.000	74	.593**	.000	69	.442**	.001	58	.532**	.000	71
I_6	.417**	.000	74	.594**	.000	69	.478**	.000	58	.575**	.000	71
I_7	.438**	.000	74	.599**	.000	69	.433**	.001	58	.538**	.000	71
I_8	.302**	.009	74	.569**	.000	69	.899**	.000	58	.926**	.000	71
I_9	.304**	.009	74	.572**	.000	69	.898**	.000	58	.925**	.000	71
I_{10}	.253*	.029	74	.479**	.000	69	.669**	.000	58	.693**	.000	71
I_{11}	.253*	.029	74	.479**	.000	69	.662**	.000	58	.684**	.000	71
I_{12}	.244*	.036	74	.468**	.000	69	.706**	.000	58	.741**	.000	71
I_{13}	.477**	.000	74	.539**	.000	69	.265*	.044	58	.372**	.001	71
I_{14}	.322**	.005	74	.381**	.001	69	.609**	.000	58	.642**	.000	71
I_{15}	.320**	.006	74	.603**	.000	69	.826**	.000	58	.889**	.000	71
B. Spearman's Correlation Coefficients												
Variable Y	ρ_{XY}	Sig. (2-tailed)	N	ρ_{XY}	Sig.	N	ρ_{XY}	Sig.	N	ρ_{XY}	Sig.	N
I_1	−.037	.754	74	.296*	.014	69	.613**	.000	58	.329**	.005	71
I_2	.319**	.006	74	.437**	.000	69	.330*	.011	58	.366**	.002	71
I_3	.344**	.003	74	.571**	.000	69	.575**	.000	58	.595**	.000	71
I_4	.395**	.000	74	.571**	.000	69	.209	.116	58	.237*	.046	71

	Corr.	Sig.	N	Corr.	Sig.	N	Corr.	Sig.	N	Corr.	Sig.	N
I_5	.601**	.000	74	.691**	.000	69	-.065	.629	58	-.048	.693	71
I_6	.606**	.000	74	**.701**	.000	69	-.036	.791	58	-.017	.886	71
I_7	.607**	.000	74	.697**	.000	69	-.054	.690	58	-.034	.776	71
I_8	.171	.144	74	.358**	.003	69	.613**	.000	58	.473**	.000	71
I_9	.170	.147	74	.384**	.001	69	.622**	.000	58	.473**	.000	71
I_{10}	.161	.172	74	.403**	.001	69	.533**	.000	58	.350**	.003	71
I_{11}	.157	.183	74	.396**	.001	69	.533**	.000	58	.350**	.003	71
I_{12}	.047	.693	74	.259**	.031	69	.461**	.000	58	.389**	.001	71
I_{13}	.607**	.000	74	.626**	.000	69	-.127	.341	58	-.099	.413	71
I_{14}	.219	.061	74	.371**	.002	69	.358**	.006	58	.373**	.001	71
I_{15}	.280*	.016	74	.496**	.000	69	.315*	.016	58	.441**	.000	71

** Correlation is significant at the 0.01 level (2-tailed).
* Correlation is significant at the 0.05 level (2-tailed).
a Network variables.
b Institutional variablesCases shown in **bold** indicate high correlations (≥0.7).

Similarly, all institutional variables (I_1–I_{15}) are significantly correlated with the average degree of the railway (GRAN), maritime (GMN), and air transport (GAN) networks. However, correlations are not high. In the case of GRAN, no high correlations can be observed, also implying the primary functionality of the railway network for regional development and formal and informal institutional configuration of Greece and complies with the findings of the previous t-test analysis. However, we can observe high correlations between institutional variables I_1–I_4 (religious metropolises, local and national TV stations, and prison capacity); I_8–I_9 (high and secondary schools); I_{12} (private primary schools); I_{15} (air force sites); and network variables N_3 (GMN average degree) and N_4 (GAN average degree). These correlations illustrate that regions enjoying high maritime and air transport network connectivity tend to have more developed religious, media communication, imprisonment, medium-level education, and military aviation institutional configuration. Within the context of the gravitational configuration of the maritime and air transport network connectivity, this observation allows developing similar conceptual linkages for their concordant institutional variables.

The results of the non-parametric correlation analysis (shown in sub-table B) shape a similar but smoother picture with the previous approach, indicating that the majority of correlation coefficients are significant at the 5% and 1% levels. This time, the correlation implies structural relevance due to the computation of Spearman's rho coefficient of correlation on ranks (positioning) instead of algebraic values. In particular, institutional variables I_2–I_7 (local and national TV stations, prison capacity, 2, 3, ≥4-member families), I_{13} (firms), I_{15} (air force sites) are correlated (but not highly) with the average GRN degree. This observation indicates a structural relevance between these institutional variables and road network connectivity, within the primary developmental context implied by the road network infrastructure development. Next, all institutional variables are correlated with the average GRAN connectivity, where correlation with I_6 (2-member families) is high. This observation complies with the findings of the previous analyses and also highlights a high structural relevance between railway network connectivity and the spatial configuration of the core (2-member) family structure in Greece. Finally, the institutional variables I_1–I_3 (religious metropolises, local and national TV stations), I_8–I_{12} (high, secondary, primary-public, and private schools), I_{14}–I_{15} (army camps and air force sites) are correlated (but not highly) with the average GMN and GAN connectivity. This result underlines a basic structural relevance between these institutional variables with non-land transportation, though further supporting the gravitational forces in the links of these variables.

12.4 Conclusions

This chapter studied the multilayer network model of interregional multi-modal transport accessibility, which is composed of the layers of the road (GRN), railway (GRAN), maritime (GMN), and air transport (GAN) accessibility, in conjunction with its formal and informal institutional configuration. Different aspects of institutional configuration (family size, religion, prison capacity, media infrastructures, education, army sites, and industry infrastructures were considered) to examine the degree to which the network structure of multimodal transportation is related to the development of formal and informal institutions in Greece. The analysis first revealed significant but not high links between road and railway network connectivity and formal and informal institutional configuration, which were in line with the acknowledged primary role of these transportation networks in regional development. It also revealed significant links between maritime and air transport network connectivity and formal and informal institutional configuration. Correlations with institutional variables related to religion, media communication, imprisonment, medium-level education, and military aviation institutional configuration were high. Finally, the analysis revealed a common gravitational and geomorphological conceptual background in the development of transportation networks and the institutional configuration in Greece, underlining the importance of spatial-economic assessment in the study of social systems and addressing avenues of further research. Within the conceptual context expressed by the triplet relationship "institutions → economic and regional development ← transport infrastructures", the study uncovered a linkage of positive analogy expressing that regions with developed transport infrastructures are also more likely to have highly developed institutional structures. Especially for the more technologically equipped network infrastructures (as maritime and air transport networks are), this analogy becomes higher in cases of more well-established formal institutions. In a cause–effect context, this uncovered linkage illustrates relevance between institutions and transport infrastructures; through their concordant effect on regional and economic development. Overall, this chapter is submitted to promote interdisciplinary research and highlight the effect of space and geography in the configuration of the interregional economy.

A Appendix

A.1 Appendix 12.1   Results of the Independent Samples *t*-tests for the mean differences $\mu_\alpha - \mu_\beta$, between the groups X_α and X_β that are defined by the N_1, and $NI_1 - NI_3$ variables

| | | N_1 | | NI_1 | | NI_2 | | NI_3 | |
		$\mu(N_1 \geq 1.19)$ vs. $\mu(N_1 < 1.19)$		$\mu(NI_1 = 0)$ vs. $\mu(NI_1 = 1)$		$\mu(NI_1 = 0)$ vs. $\mu(NI_1 = 1)$		$\mu(NI_3 = 0)$ vs. $\mu(NI_3 = 1)$	
		Equal variances assumed	Equal variances not assumed	Equal variances assumed	Equal variances not assumed	Equal variances assumed	Equal variances not assumed	Equal variances assumed	Equal variances not assumed
I_1									
Levene's test	F	28.518		56.696		10.102		26.154	
	Sig.	.000		.000		.002		.000	
t-test for equality of means	t	2.079	**2.364**	**−3.313**	**−3.071**	−1.814	**−3.413**	**−2.841**	**−3.631**
	df	72	**44.643**	72	**35.343**	72	**60.957**	72	**46.562**
	Sig. (2-tailed)	.041	**.022**	**.001**	**.004**	.074	**.001**	**.006**	**.001**
	Mean difference	1.917	**1.917**	**−2.912**	**−2.912**	−2.026	**−2.026**	**−2.612**	**−2.612**
	Std. error difference	.922	**.811**	**.879**	**.948**	1.117	**.594**	**.919**	**.719**
	95% CI of the diff. Lower	.079	**.283**	**−4.664**	**−4.836**	−4.253	**−3.213**	**−4.445**	**−4.059**
	Upper	3.754	**3.550**	**−1.160**	**−.987**	.201	**−.839**	**−.779**	**−1.164**
I_2									
Levene's test	F	26.725		64.562		6.754		20.405	
	Sig.	.000		.000		.011		.000	
t-test for equality of means	t	2.907	**3.257**	**−4.127**	**−3.826**	−1.521	**−2.446**	**−2.397**	**−2.950**
	df	72	**50.764**	72	**35.446**	72	**67.262**	72	**57.473**
	Sig. (2-tailed)	.005	**.002**	**.000**	**.001**	.133	**.017**	**.019**	**.005**
	Mean difference	2.307	**2.307**	**−3.094**	**−3.094**	−1.511	**−1.511**	**−1.977**	**−1.977**
	Std. error difference	.793	**.708**	**.750**	**.809**	.994	**.618**	**.824**	**.670**
	95% CI of the diff. Lower	.725	**.885**	**−4.589**	**−4.735**	−3.491	**−2.743**	**−3.620**	**−3.318**
	Upper	3.888	**3.728**	**−1.599**	**−1.453**	.470	**−.278**	**−.333**	**−.635**

I₃	Levene's test	F	49.465	*3.186*	98.096		14.314		36.490	
		Sig.	.000	*41.000*	.000		.000		.000	
	t-test for equality of means	t	2.776	*3.186*	−3.555	*−3.273*	−1.616	*−3.093*	−2.455	*−3.155*
		df	72	*41.000*	72	*33.000*	72	*57.000*	72	*45.000*
		Sig. (2-tailed)	.007	*.003*	.001	*.002*	.110	*.003*	.017	*.003*
		Mean difference	2.143	*2.143*	−2.647	*−2.647*	−1.552	*−1.552*	−1.957	*−1.957*
		Std. error difference	.772	*.673*	.745	*.809*	.960	*.502*	.797	*.620*
		95% CI of the diff. Lower	.604	*.784*	−4.132	*−4.292*	−3.466	*−2.556*	−3.545	*−3.205*
		Upper	3.682	*3.501*	−1.163	*−1.002*	.362	*−.547*	−.368	*−.708*
I₄	Levene's test	F	30.467	*3.534*	57.033		7.794		13.288	
		Sig.	.000	*46.512*	.000		.007		.001	
	t-test for equality of means	t	3.122	*3.534*	−4.278	*−3.963*	−1.510	*−2.515*	−1.531	*−1.866*
		df	72	*46.512*	72	*35.081*	72	*70.815*	72	*59.957*
		Sig. (2-tailed)	.003	*.001*	.000	*.000*	.136	*.014*	.130	*.067*
		Mean difference	452.0	*452.0*	−585.8	*−585.8*	−275.9	*−275.9*	−237.4	*−237.4*
		Std. error difference	144.8	*127.9*	136.9	*147.8*	182.8	*109.7*	155.1	*127.2*
		95% CI of the diff. Lower	163.4	*194.6*	−858.8	*−885.9*	−640.3	*−494.7*	−546.5	*−491.9*
		Upper	740.6	*709.4*	−312.8	*−285.7*	88.4	*−57.2*	71.7	*17.1*
I₅	Levene's test	F	12.229	*4.269*	17.616		5.025		11.789	
		Sig.	.001	*47.160*	.000		.028		.001	
	t-test for equality of means	t	3.778	*4.269*	−4.718	*−4.391*	−.717	*−1.245*	−1.073	*−1.338*
		df	72	*47.160*	72	*36.889*	72	*71.879*	72	*53.653*
		Sig. (2-tailed)	.000	*.000*	.000	*.000*	.475	*.217*	.287	*.186*
		Mean difference	18716.6	*18716.6*	−22232.0	*−22232.0*	−4665.7	*−4665.7*	−5897.3	*−5897.3*
		Std. error difference	4954.6	*4384.3*	4712.0	*5062.5*	6503.6	*3748.2*	5496.3	*4407.0*
		95% CI of the diff. Lower	8839.8	*9897.4*	−31625.2	*−32490.7*	−17630.3	*−12137.7*	−16854.0	*−14734.2*
		Upper	28593.5	*27535.9*	−12838.8	*−11973.3*	8299.0	*2806.4*	5059.3	*2939.5*

(Continued)

Appendix 12.1 (Continued)

		N_1		NI_1		NI_2		NI_3	
		$\mu(N_1 \geq 1.19)$ vs. $\mu(N_1 < 1.19)$		$\mu(NI_1 = 0)$ vs. $\mu(NI_1 = 1)$		$\mu(NI_1 = 0)$ vs. $\mu(NI_1 = 1)$		$\mu(NI_3 = 0)$ vs. $\mu(NI_3 = 1)$	
		Equal variances assumed	Equal variances not assumed	Equal variances assumed	Equal variances not assumed	Equal variances assumed	Equal variances not assumed	Equal variances assumed	Equal variances not assumed
I_6	Levene's test F	16.147		24.527		5.841		13.507	
	Sig.	.000		.000		.018		.000	
	t-test for equality of means t	3.697	*4.195*	-4.571	*-4.244*	-.985	*-1.762*	-1.369	*-1.717*
	df	72	*45.569*	72	*35.984*	72	*69.722*	72	*52.044*
	Sig. (2-tailed)	.000	*.000*	.000	*.000*	.328	*.082*	.175	*.092*
	Mean difference	12780.3	*12780.3*	-15083.0	*-15083.0*	-4438.3	*-4438.3*	-5206.0	*-5206.0*
	Std. error difference	3456.6	*3046.9*	3300.0	*3553.9*	4507.3	*2518.4*	3802.4	*3031.8*
	95% CI of the diff. Lower	5889.7	*6645.6*	-21661.3	*-22290.7*	-13423.5	*-9461.4*	-12786.0	*-11289.7*
	Upper	19670.9	*18915.0*	-8504.7	*-7875.3*	4546.9	*584.8*	2373.9	*877.7*
I_7	Levene's test F	16.874		22.321		6.358		12.890	
	Sig.	.000		.000		.014		.001	
	t-test for equality of means t	3.964	*4.490*	-4.688	*-4.370*	-.996	*-1.756*	-1.179	*-1.461*
	df	72	*46.191*	72	*37.442*	72	*71.102*	72	*55.328*
	Sig. (2-tailed)	.000	*.000*	.000	*.000*	.323	*.083*	.242	*.150*
	Mean difference	13558.3	*13558.3*	-15400.0	*-15400.0*	-4494.5	*-4494.5*	-4502.2	*-4502.2*
	Std. error difference	3420.5	*3019.6*	3285.0	*3524.4*	4512.3	*2559.1*	3819.8	*3080.9*
	95% CI of the diff. Lower	6739.7	*7480.8*	-21948.5	*-22538.2*	-13489.7	*-9597.2*	-12116.8	*-10675.5*
	Upper	20376.8	*19635.8*	-8851.5	*-8261.8*	4500.7	*608.1*	3112.5	*1671.2*

I_8								
Levene's test	F	43.531		83.352		13.344		33.489
	Sig.	.000		.000		.000		.000
t-test for equality of means	t	2.941	*3.371*	−3.582	*−3.301*	−1.908	*−3.640*	−2.750
	df	72	*41.460*	72	*33.295*	72	*57.720*	72
	Sig. (2-tailed)	.004	*.002*	.001	*.002*	.060	*.001*	.008
	Mean difference	75.3	*75.3*	−88.9	*−88.9*	−60.7	*−60.7*	−72.5
	Std. ERROR DIFFErence	25.6	*22.3*	24.8	*26.9*	31.8	*16.7*	26.3
	95% CI of the diff. Lower	24.3	*30.2*	−138.4	*−143.7*	−124.2	*−94.1*	−125.0
	Upper	126.4	*120.5*	−39.4	*−34.2*	2.7	*−27.3*	−19.9
								−3.528
								45.509
								.001
								−72.5
								20.5
								−113.8
								−31.1

I_9								
Levene's test	F	44.053		83.482		12.925		33.302
	Sig.	.000		.000		.001		.000
t-test for equality of means	t	2.949	*3.380*	−3.633	*−3.348*	−1.852	*−3.525*	−2.772
	df	72	*41.433*	72	*33.298*	72	*58.330*	72
	Sig. (2-tailed)	.004	*.002*	.001	*.002*	.068	*.001*	.007
	Mean difference	94.5	*94.5*	−112.7	*−112.7*	−73.9	*−73.9*	−91.3
	Std. error difference	32.1	*28.0*	31.0	*33.6*	39.9	*21.0*	32.9
	95% CI of the diff. Lower	30.6	*38.1*	−174.5	*−181.1*	−153.4	*−115.9*	−157.0
	Upper	158.4	*151.0*	−50.8	*−44.2*	5.6	*−31.9*	−25.6
								−3.556
								45.541
								.001
								−91.3
								25.7
								−143.0
								−39.6

I_{10}								
Levene's test	F	12.766		19.325		4.198		9.759
	Sig.	.001		.000		.044		.003
t-test for equality of Means	t	2.527	*2.878*	−3.209	*−2.972*	−1.579	*−2.855*	−2.353
	df	72	*44.030*	72	*35.024*	72	*68.456*	72
	Sig. (2-tailed)	.014	*.006*	.002	*.005*	.119	*.006*	.021
	Mean difference	75.1	*75.1*	−92.5	*−92.5*	−57.9	*−57.9*	−71.8
	Std. error difference	29.7	*26.1*	28.8	*31.1*	36.7	*20.3*	30.5
	95% CI of the diff. Lower	15.9	*22.5*	−149.9	*−155.7*	−131.0	*−98.4*	−132.6
	Upper	134.3	*127.7*	−35.0	*−29.3*	15.2	*−17.4*	−11.0
								−2.967
								50.428
								.005
								−71.8
								24.2
								−120.4
								−23.2

(Continued)

Appendix 12.1 (Continued)

			N1 μ(N1≥1.19) vs. μ(N1<1.19)		NI1 μ(NI1=0) vs. μ(NI1=1)		NI2 μ(NI1=0) vs. μ(NI1=1)		NI3 μ(NI3=0) vs. μ(NI3=1)	
			Equal variances assumed	Equal variances not assumed	Equal variances assumed	Equal variances not assumed	Equal variances assumed	Equal variances not assumed	Equal variances assumed	Equal variances not assumed
I11	Levene's test	F	12.147		17.628		3.950		9.088	
		Sig.	.001		.000		.051		.004	
	t-test for equality of means	t	2.538	*2.886*	−3.238	*−3.003*	−1.581	*−2.823*	−2.367	*−2.974*
		df	72	*44.627*	72	*35.439*	72	*69.981*	72	*51.545*
		Sig. (2-tailed)	.013	*.006*	.002	*.005*	.118	*.006*	.021	*.004*
		Mean difference	67.3	*67.3*	−83.3	*−83.3*	−51.8	*−51.8*	−64.5	*−64.5*
		Std. error difference	26.5	*23.3*	25.7	*27.7*	32.8	*18.3*	27.2	*21.7*
		95% CI of the diff. Lower	14.4	*20.3*	−134.5	*−139.5*	−117.1	*−88.4*	−118.8	*−108.0*
		Upper	120.2	*114.3*	−32.0	*−27.0*	13.5	*−15.2*	−10.2	*−21.0*
I12	Levene's test	F	20.107		32.560		6.896		15.425	
		Sig.	.000		.000		.011		.000	
	t-test for equality of means	t	2.338	*2.682*	−2.841	*−2.618*	−1.505	*−2.860*	−2.144	*−2.746*
		df	72	*41.271*	72	*33.170*	72	*58.611*	72	*45.886*
		Sig. (2-tailed)	.022	*.010*	.006	*.013*	.137	*.006*	.035	*.009*
		Mean difference	7.77	*7.77*	−9.23	*−9.23*	−6.14	*−6.14*	−7.31	*−7.31*
		Std. error difference	3.32	*2.90*	3.25	*3.52*	4.08	*2.15*	3.41	*2.66*
		95% CI of the diff. Lower	1.15	*1.92*	−15.70	*−16.40*	−14.28	*−10.44*	−14.12	*−12.67*
		Upper	14.39	*13.61*	−2.75	*−2.06*	2.00	*−1.84*	−0.51	*−1.95*
I13	Levene's test	F	8.851		4.018		3.966		6.560	
		Sig.	.004		.049		.050		.013	
	t-test for equality of means	t	4.254	*4.753*	−4.221	*−4.029*	−.796	*−1.267*	−.330	*−.392*
		df	72	*51.975*	72	*48.157*	72	*65.867*	72	*65.695*
		Sig. (2-tailed)	.000	*.000*	.000	*.000*	.429	*.210*	.743	*.696*
		Mean difference	18456.3	*18456.3*	−18230.6	*−18230.6*	−4627.4	*−4627.4*	−1632.6	*−1632.6*
		Std. error difference	4338.2	*3883.2*	4319.5	*4525.2*	5814.8	*3653.4*	4953.6	*4163.1*
		95% CI of the diff. Lower	9808.2	*10663.9*	−26841.4	*−27328.3*	−16218.9	*−11921.8*	−11507.4	*−9945.1*
		Upper	27104.4	*26248.7*	−9619.8	*−9132.8*	6964.2	*2667.1*	8242.3	*6680.0*

		Levene's test[1] F	Sig.	t	df	Sig. (2-tailed)	Mean difference	Std. error difference	95% CI of the diff. Lower	Upper
I₁₄	Equal variances assumed	36.369	.000	2.862	72	.006	1.274	.445	.387	2.161
	Equal variances not assumed			*3.153*	*57.815*	*.003*	*1.274*	*.404*	*.465*	*2.083*
	Equal variances assumed	20.199	.000	-3.523	72	.001	-1.519	.431	-2.379	-.660
	Equal variances not assumed			*-3.392*	*52.657*	*.001*	*-1.519*	*.448*	*-2.417*	*-.621*
	Equal variances assumed	1.742	.191	-.646	72	.520	-.364	.564	-1.488	.759
	Equal variances not assumed			-.727	28.835	.473	-.364	.501	-1.389	.660
	Equal variances assumed	16.671	.000	-1.998	72	.049	-.933	.467	-1.864	-.002
	Equal variances not assumed			*-2.255*	*71.937*	*.027*	*-.933*	*.414*	*-1.758*	*-.108*
I₁₅	Equal variances assumed	31.779	.000	3.121	72	.003	1.074	.344	.388	1.761
	Equal variances not assumed			*3.540*	*45.561*	*.001*	*1.074*	*.303*	*.463*	*1.685*
	Equal variances assumed	60.711	.000	-3.929	72	.000	-1.300	.331	-1.960	-.640
	Equal variances not assumed			*-3.640*	*35.197*	*.001*	*-1.300*	*.357*	*-2.025*	*-.575*
	Equal variances assumed	11.829	.001	-2.193	72	.032	-.938	.427	-1.790	-.085
	Equal variances not assumed			-4.053	64.803	.000	-.938	.231	-1.400	-.475
	Equal variances assumed	20.091	.000	-2.797	72	.007	-.995	.356	-1.705	-.286
	Equal variances not assumed			*-3.480*	*54.315*	*.001*	*-.995*	*.286*	*-1.569*	*-.422*

1 for equality of variances.
Cases in *italics* indicate the proper test according to the Levene's testing.
Cases in **bold** indicate significant tests.

References

Amin, A., and Thrift, N. 1995. *Globalization, institutions, and regional development in Europe*. Oxford: Oxford University Press.

Artavani, M. A., Christoforos-Stefanos, D., and Athanasios, D. 2015. The role of military camps on the process of urban development. *Journal of Regional & Socio-Economic Issues*, 5(2): 71–82.

Barabasi, A.-L. 2016. *Network science*. Cambridge: Cambridge University Press.

Barthelemy, M. 2011. Spatial networks. *Physics Reports*, 499(1–3): 1–101.

Basco, R. 2015. Family business and regional development - A theoretical model of regional familiness. *Journal of Family Business Strategy*, 6(4): 259–271.

Bastian, M., Heymann, S., Jacomy, M. 2009. Gephi: An open source software for exploring and manipulating networks. In *Proceedings of the Third International ICWSM Conference*, San Jose, CA, USA, 17–20 May 2009; The AAAI Press: Menlo Park, CA, USA, 2009; pp. 361–362.

Bille, T., and Schulze, G. G. 2006. Culture in urban and regional development. *Handbook of the Economics of Art and Culture*, 1: 1051–1099.

Boccaletti, S., Bianconi, G., Criado, R., del Genio, C. I., Gomez-Gardenes, J., Romance, M., Sendina-Nadal, I., Wang, Z., and Zanin, M. 2014. The structure and dynamics of multilayer networks. *Physics Reports*, 544: 1–122.

Cohen, B. J. (Ed.) 2017. *International political economy*. New York: Routledge.

Cooke, P., Clifton, N., and Oleaga, M. 2005. Social capital, firm embeddedness and regional development. *Regional Studies*, 39(8): 1065–1077.

Danson, M., and De Souza, P. (Eds.). 2012. *Regional development in Northern Europe: Peripherality, marginality and border issues*. New York: Routledge.

Devore, J., and Berk, K. 2012. *Modern mathematical statistics with applications*. London: Springer-Verlag Publications.

Ducruet, C., and Beauguitte, L. 2014. Spatial science and network science: Review and outcomes of a complex relationship. *Networks and Spatial Economics*, 14: 297–316.

Fischer, M. M., and Nijkamp, P. (Eds.) 2014. *Handbook of regional science* (Vol. 3). Heidelberg: Springer.

Glasmeier, A. K., and Farrigan, T. 2007. The economic impacts of the prison development boom on persistently poor rural places. *International Regional Science Review*, 30(3): 274–299.

Goodin, R. E. 1996. Institutions and their design. *The Theory of Institutional Design*, 1(53), 9–53.

Google Maps. 2021. *Google Mapping Services*, www.google.gr/maps?hl=el [accessed: 15/5/2021].

Haynes, J. 2007. *Religion and development: Conflict or cooperation?*. New York: Springer.

Hellenic Statistical Service – ELSTAT. 2011. Results of the census of population-habitat 2011 referring to the permanent population of Greece. Newspaper of the (Greek) Government (ΦΕΚ), Second Issue (T-B), Number 3465, 28/12.

Hellenic Statistical Service – ELSTAT. 2021a. *Statistics: Education*, www.statistics.gr/el/statistics/pop [accessed: 09/05/2021].

Hellenic Statistical Service – ELSTAT. 2021b. *Statistics: Economy, Indicators*, www.statistics.gr/el/statistics/eco [accessed: 09/05/2021].

Hodgson, G. M. 2006. What are institutions? *Journal of Economic Issues*, 40(1), 1–25.

Hodgson, G. M. 2015. On defining institutions: Rules versus equilibria. *Journal of Institutional Economics*, *11*(3), 497–505.

Janowitz, M. 1976. Military institutions and citizenship in western societies. *Armed Forces & Society*, *2*(2), 185–204.

Kivela, M., Arenas, A., Barthelemy, M., Gleeson, J., Moreno, Y., and Porter, M.A. 2014. Multilayer networks. *Journal of Complex Networks*, *2*: 203–271.

Lavrinenko, P. A., Romashina, A. A., Stepanov, P. S., and Chistyakov, P. A. 2019. Transport accessibility as an indicator of regional development. *Studies on Russian Economic Development*, *30*(6): 694–701.

MacKinnon, D., Cumbers, A., Pike, A., Birch, K., and McMaster, R. 2009. Evolution in economic geography: Institutions, political economy, and adaptation. *Economic Geography*, *85*(2): 129–150.

Marshall, S., Gil, J., Kropf, K., Tomko, M., and Figueiredo, L. 2018. Street-network studies: From networks to models and their representations. *Networks and Spatial Economics*, *18*(3): 735–749.

Napoli, P. M. 2003. *Audience economics: Media institutions and the audience marketplace*. New York: Columbia University Press.

Newman, M. E. J. 2010. *Networks: An introduction*. Oxford: Oxford University Press.

Norusis, M. 2011. *IBM SPSS statistics 19.0 guide to data analysis*, New Jersey: Prentice Hall.

O'Sullivan, A. 2007. *Urban economics* (pp. 225–226). Boston, MA: McGraw-Hill/Irwin.

Official Website of the Church of Greece – OWCG. 2021. *Religious Metropolises in the Greek Territory*, www.ecclesia.gr/Dioceses/Dioceses.asp [accessed: 09/05/2021].

Picard, R. G. 2009. Media clusters and regional development. In *Uddevalla Symposium 2009 – The Geography of Innovation and Entrepreneurship*.

Polyzos, S., and Tsiotas, D. 2020. The contribution of transport infrastructures to the economic and regional development: A review of the conceptual framework. *Theoretical and Empirical Researches in Urban Management*, *15*(1): 5–23.

Rodrigue, J. P., Comtois, C., and Slack, B. 2013. *The geography of transport systems*, New York: Routledge Publications.

Rodriguez-Pose, A. 2013. Do institutions matter for regional development? *Regional Studies*, *47*(7): 1034–1047.

Rodriguez-Pose, A. 2020. Institutions and the fortunes of territories. *Regional Science Policy & Practice*, *12*(3): 371–386.

Schudson, M. (2002). The news media as political institutions. *Annual Review of Political Science*, *5*(1), 249–269.

Skowronski, B., & Talik, E. (2021). Quality of life and its correlates in people serving prison sentences in penitentiary institutions. *International Journal of Environmental Research and Public Health*, *18*(4), 1655.

Spiekermann, K., & Neubauer, J. (2002). *European accessibility and peripherality: Concepts, models and indicators*. Stockholm: Nordregio.

Spierenburg, P. (2007). *The prison experience: Disciplinary institutions and their inmates in early modern Europe*. Amsterdam: Amsterdam University Press.

The Ministry of Justice, Transparency, and Human Rights – MOJTHR. 2021. *Penalty System, Prisoner Statistics*, www.justice.gr/ΣΩΦΡΟΝΙΣΤΙΚΟΣΥΣΤΗΜΑ/Στατιστικάστοιχείακρεουμένων.aspx [accessed: 09/05/2021].

The National Council for Radio and Television – NCRTV. 2021. *Operational TV Stations of National and Regional Range*, www.esr.gr/τηλεόραση [accessed: 09/05/2021].

Tsiotas, D. 2017. Links between network topology and socioeconomic framework of railway transport: Evidence from Greece. *Journal of Engineering Science and Technology Review*, *10*(3): 175–187.

Tsiotas, D. 2020. Drawing indicators of economic performance from network topology: The case of the interregional road transportation in Greece. *Research in Transportation Economics* [10.1016/j.retrec.2020.101004]

Tsiotas, D., and Polyzos, S. 2015a. Analyzing the maritime transportation system in Greece: A complex network approach. *Networks and Spatial Economics*, *15*(4): 981–1010.

Tsiotas, D., and Polyzos, S. 2015b. Decomposing multilayer transportation networks using complex network analysis: A case study for the Greek aviation network. *Journal of Complex Networks*, *3*(4): 642–670.

Tsiotas, D., and Tselios, V. 2021. Understanding the uneven spread of COVID-19 in the context of the global interconnected economy. arXiV: 2101.11036.

Vrisko.gr. 2021. *Telephone Number of Business Directory*, www.vrisko.gr [accessed: 09/05/2021].

Walpole, R.E., Myers, R.H., Myers, S.L., and Ye, K. 2012. *Probability & Statistics for Engineers & Scientists*, 9th ed. New York: Prentice Hall Publications.

Xiaoming, Z. 2006. From institution to industry: Reforms in cultural institutions in China. *International Journal of Cultural Studies*, *9*(3), 297–306.

13 Marginal Returns?

Institutional Dynamics, Peripherality, and Place-Based Development in Canada's and Australia's Natural Resource-Dependent Regions

Laura Ryser, Neil Argent, Greg Halseth, Fiona Haslam-McKenzie, and Sean Markey

13.1 Introduction

Globalization has seen researchers turn their attention to the rise of virtual economies spanning the globe, and the associated sophisticated technologies and techniques by which these networks have developed, along with new and reconfigured urban spaces the phenomenon has generated (Beer and Clower 2019; Peterson 2003). While usually the subject of less research interest, the production and distribution complexes associated with natural resource extraction, usually anchored in a material sense in peripheral regions, are also fundamental to these globalization stories (Argent 2013; Markey et al. 2019). The international trade in mineral and energy (e.g. liquefied natural gas [LNG]) commodities has been and remains crucial to the economic strategies of rapidly industrializing nations such as India and China, as well as to the more-developed world's access to relatively cheap industrial inputs and consumer goods. An understanding of the operation of the space economies that produce these mineral and energy staples is, therefore, vital to understanding the institutional and more place-based factors that influence the long-term viability of such regions of supply. Thus, although research into natural resource extraction has recently been relegated to the peripheries of economic geography (Hayter et al. 2003), and the actual sites of production themselves are usually located in peripheral regions of nation-states, we argue that the issues that we focus on in this chapter are core to appreciating the full set of economic, politico-social and spatial relations of production, distribution and circulation in this fast-evolving sector. We also maintain that this analysis is crucial to an understanding of the broader challenges faced by resource-producing regions caught up in global production chains.

In the context of the rapid expansion of the natural gas economy around the globe, and the equally dramatic neoliberalization of governmental policy over the past four decades, in this chapter we examine the impact of the unconventional Liquid Natural Gas/Coal Seam Gas (LNG/CSG) boom of the 2010s on institutional relationships within the rural communities and economies of the Peace River region of British Columbia, Canada, and the Surat Basin, Queensland, Australia, both of which are somewhat remote

DOI: 10.4324/9781003191049-17

from major metropolitan centres. Conceptually influenced by staples theory and evolutionary economic geography, and drawing on intensive fieldwork conducted by the authors in both case study regions, we chart the socio-spatially uneven distribution of benefits, harms, and responsibilities associated with LNG/CSG expansion in both regions, focusing on the extent to which formal and informal government and governance institutions were able to overcome conditions of lock-in and inertia to capitalize on, and locally embed, the benefits of the boom. While the sheer magnitude and pace of the boom in both jurisdictions initially caught formal and informal institutions off guard, over time some regional and local institutions generated multiscalar and horizontal relationships that enabled them to adapt and capture important and long-term economic benefits. Overall, this chapter reflects on the vital role of institutional relations and dynamics in shaping the evolving uneven geography of resource development.

The chapter will begin by exploring the general challenges for natural resource-dependent economies, and grounds these challenges within the theoretical frameworks provided by staples theory and evolutionary economic geography. Our two case study regions are then described in order to situate our discussion of the uneven distribution of benefits, and the impacts associated with LNG/CSG expansion, focusing on the extent to which government and governance institutions were able to overcome challenges and embed benefits.

Rapid growth in the unconventional oil and gas (UOGD) sector is challenging for communities because much of the exploration activity and pipeline construction takes place outside of municipal boundaries. Once operational, the sector employs relatively few workers locally, making it more difficult for communities to reap benefits from local workforce expenditure or obtain the fiscal revenues through property taxes to respond to the pressures associated with growth. The 'unconventional' industry has moved away from traditional well heads to fracking and coal seam gas development that has more extensive impacts and a much faster pace of development (Buse et al. 2019). This contemporary expansion of resource development has tested state and local actors as often they were not ready for the scale and scope of impacts stemming from the influx of large mobile workforces, nor the potential opportunities to be gained from strategic legacy investments or benefits from supply chains.

A recent resurgence of research focused on the unconventional oil and gas sector's growth, its social impacts on and dynamics in boomtowns, and the shifting patterns of boom-and-bust cycles, has shed much valuable light on these challenges (Benham 2016; Ennis et al. 2013; Measham and Fleming 2014). However, other important research gaps remain concerning the long-term development implications associated with these more unconventional forms of resource development (Luke and Emmanouil 2019). Gaps include the need to better understand the impact of more rapid industry expansion and contraction patterns, the need for better local readiness in managing these impacts, including how to better position small communities to seize potential benefits. There are also gaps in understanding the neoliberal state context in terms of how regulation, capacity, and jurisdiction are shaping these unconventional boomtown environments. Lastly, there is a lack of knowledge around the post-boomtown cumulative impacts and benefits in terms of social, economic, cultural, and environmental issues (Buse et al. 2019; Gillingham et al. 2016). In this chapter, we focus on the institutional dynamics that shape the capacity of rural regions to capitalize on the benefits of the boom and chart a more resilient future.

13.2 General Challenges of Natural Resource Dependent Economies

Political processes shape the redistribution of resource revenues in rural regions. Staples theory and evolutionary economic geography provide a foundation for understanding how these political processes unfold in extraction driven economies (Taylor et al. 2011). Staples theory explains patterns of uneven development and the peripheral role that extraction economies assume within the global economy, based on their dependence on the export of staples commodities to more advanced manufacturing regions (Innis 1933; Nelsen et al. 2010). In this scenario, staples-dependent economies assume the role of 'price takers' driven by the demands of fluctuating global markets. Dependence on the export of staples commodities then 'truncates development' as industry consolidates control over a stable and predictable supply of raw resources and takes advantage of more fluid flows of capital and labour (Ryser et al. 2017). Broader policies are often designed to attract industrial capital rather than diversify rural economies, with those industrial capital investment strategies often predicated on the need to continue raw staples commodity exports. If resource-dependent communities are stuck in the 'staples trap' and unable to diversify via other economic opportunities, they become bound to fluctuating, and often declining, resource industry benefits through the loss of jobs, taxes, and related investments, and have difficulty developing mechanisms to respond to local and regional change (Carson 2011).

The capacity to carve out new economic development pathways, however, has been hindered by the fragmentation and erosion of regional development policies, and the gradual shift towards the greater use of markets, networks, and partnerships. The erosion of such supports has left inadequate planning and financial support for communities experiencing pressures from large-scale industrial projects. Some researchers, however, suggest that states are exercising power and authority in new ways through governance networks that operate between different levels of government and civil society; these tend to be difficult to control due to the 'hollowing out of the state' (Bevir and Rhodes 2003). These governance networks draw upon the role of agency and institutional frameworks as stakeholders negotiate the politics and actions associated with changing rules, protocols, and processes (Kjær 2011). In this context, the form and extent of state intervention is always changing as senior governments re-negotiate the roles and boundaries between the state and civil society. The complexity of these negotiated and re-negotiated spaces can be shaped by the degree to which stakeholders have unequal access to information, expertise, control, resources, and understanding of processes, thereby shaping the balance of power and potential for conflict as stakeholders strive to understand institutional and policy changes (Beer et al. 2005; Howell 2018). These new governance spaces are also shaped by the unequal legitimacy and authority assumed by different stakeholders (Lockie et al. 2006). The shift from government to governance connects the hollowing out of the state to the deterioration of institutional structures, which in turn, exacerbates pressures on local stakeholders (Bevir and Rhodes 2003; Morrison et al. 2012). Nonetheless, the state retains its power and jurisdiction through regulatory and power frameworks (Beer et al. 2007). In this context, Larner (2003, 509) argues, "little attention is paid to the *different variants* of neoliberalism, to the *hybrid nature* of contemporary policies and programmes, or the *multiple and contradictory aspects* of neoliberal spaces, techniques, and subjects" (italics in original).

Evolutionary Economic Geography (EEG) explicitly recognizes the role of competition and adaptation in shaping the relative robustness and resilience of localities and regions to economic 'shocks', regardless of where such perturbations originate. While frameworks such as staples theory articulate and explain the pressures that keep resource-dependent regions focused upon their historical development paths, changes can and do occur. The question is, what types of change, at what pace and scale, and do such changes reinforce path dependence or build foundations towards more diverse and resilient economies? As such, EEG provides a set of tools for evaluating change and recognizing the role of place.

The path dependency of staples-reliant communities is reinforced through historical capital and technology investment strategies that fail to evolve, and which remain fixated on maximizing returns from long-established trading patterns. Often, these developmental trajectories are supported by public policies and institutional structures that insist on providing predictability for those industries (Martin and Sunley 2006; Tonts et al. 2014). Senior governments in many OECD countries have even rolled back regulatory strategies to incentivize jurisdictional environments for resource development, while reducing financial support for communities experiencing social and physical infrastructure pressures from large-scale industrial projects (Ryser et al. 2019). Path dependency is also reinforced when rural regions become 'addicted' to natural resource extraction processes due to limited capacity and access to resources, restricted power and authority, and related structural underpinnings that impede the ability of rural stakeholders to pursue more diverse economic pathways (Smith and Haggerty 2020). These physical, formal, and informal institutional behaviours combine to produce lock-in effects that themselves reinforce inertia around resource regions' path dependence (Markey et al. 2019). When community capital assets are re-combined to strengthen a community's comparative advantage, however, new pathways can emerge through strategic re-arrangements of assets (Mitchell and Shannon 2018). Neoliberal governments, however, often see their primary role as facilitating capital investment at the macro-scale of the nation, with an expectation that benefits will trickle down to rural regions, rather than pursuing a more proactive and purposeful approach to support new pathways.

13.3 Methods

Informed by key EEG and staples theory concepts, and drawing upon two case studies in Canada and Australia, the remainder of this chapter focuses on the extent to which formal and informal government and governance arrangements were mobilized to overcome conditions of lock-in and inertia to capitalize on, and locally embed, the benefits of the LNG/CSG boom to support more diversified development pathways in rural regions. Our selection of the two case study regions, from two different nations, is justified for a range of reasons. Although located in different hemispheres, Australia and Canada share numerous similarities. Both countries were colonised during the height of the Industrial Revolution by European powers and, due to their abundant land and resources, became firmly bound into the trading networks of the British Empire. The two countries also exhibit remarkably similar settlement patterns, including high levels of 'metropolitan primacy'. Australia and Canada are also home to significant Indigenous populations who occupy a largely peripheral position (in a geographical as well as in a social and economic sense) within both societies (Holmes 1987; McCann and Gunn 1998). Although both countries have transitioned towards more service-oriented economies, they are still substantially dependent on the export of staple commodities.

The first case study is the Peace River Region, located in the northeast corner of British Columbia (BC), Canada. It has a regional population of almost 63,000, including seven municipalities that are divided between the North Peace and the South Peace areas. The region is also home to the communities and traditional lands of the Treaty 8 Tribal Association. The Peace River region has extensive experience with natural resource activity in coal, oil and gas, forestry, hydro-electricity, and wind farm developments. Specific to the oil and gas sector, northeastern BC is part of the Western Canada Sedimentary Basin (WCSB), a geologic formation that contains the majority of oil, gas, and crude bitumen in Canada. The Peace River region is the second leading producer of natural gas in Canada. The region was selected because of municipal challenges associated with limited access to an industrial tax base in the surrounding rural areas that produced pressures on municipal budgets, operations, infrastructure, and services. As such, key informant interviews were conducted with 22 stakeholders representing local government, consultants, and provincial government staff who were directly involved with negotiations for two significant agreements to address the limited access to the surrounding industrial tax base: the Fair Share Agreement and the Peace River Agreement.

The Surat Basin, our second case study, is largely located in Queensland, Australia, although it does extend into New South Wales. It has a population of approximately 218,000. The region is situated in the traditional territory of the Mandandanji people and comprises several local government areas, which themselves incorporate the major township of Toowoomba, and the smaller townships of Dalby, Chinchilla, Miles, and Roma. Similar to the Peace River Region, the Surat Basin was selected due to its exposure to the rapid development of unconventional CSG deposits. Fruit, vegetable and livestock production and broad acre agriculture are also key economic drivers of the economy. Twelve interviews were conducted with 16 representatives from local governments, an agricultural lobbyist organization, Queensland State Government agencies, Chambers of Commerce, representatives of the Mandandanji people, a regional development organization, and a regional advocacy and support organization. Key informant participants were selected based on our concern to gain informed perspectives from community leaders and relevant state government representatives about the benefits and investments stemming from large-scale CSG projects in the region. The information and data gathered from both case study regions were analysed with a sensitivity to key concepts within EEG and staples theory (such as lock-in, path dependence and the 'staples trap') as well as emergent themes.

13.4 Findings: Processes of Disruption and Response

We have organized our findings to highlight the pace of disruption from contemporary natural resource development and the late and inadequate responses from community and policy stakeholders in the Peace River Region and the Surat Basin region. This discussion focuses on four key topics: (1) the tensions between impacts and benefits; (2) the lack of readiness; (3) the reactionary state responses; and (4) lessons for staples-dependent economies experiencing growth from LNG/CSG projects in rural regions.

13.4.1 Tensions between Impacts and Benefits

Stakeholders described complex tensions between the impacts and benefits unfolding in these places (Table 13.1). These include, for example, the enormous pressure that industrial traffic has on roads that, hitherto, saw only relatively light local use. Some communities experienced a rapid increase in rental and home ownership costs that led to 'renovictions'[1] of vulnerable populations, as well as the decline of other economic sectors that struggle to recruit workers due to housing costs and difficulty competing with industry wages. Indigenous groups were marginalized due to their limited capacity to participate, and benefit from, these projects. Furthermore, there were concerns about the cumulative impacts from social, economic, cultural, and environmental issues unfolding from these projects.

These impacts, however, are tempered by the 'faint praise' of limited and frequently insufficient benefits. There were concerns that rural businesses and Indigenous stakeholders did not have the capacity to engage in, or benefit from, industry supply chains that seemed to stop at the resource region

Table 13.1 Complex tensions between impacts and benefits

Impacts	Benefits
Physical infrastructure	Infrastructure improvements
Water (ground and surface)	Development of monitoring data
Housing affordability	Asset increase
Other economic sector decline	Jobs, income
Quality of life decline	Youth retention
Identity challenged	Rural and small town revival
Transience	Camps (and economic benefits)
Indigenous marginalization	Revenue sharing
Cumulative impacts	

boundaries. Some regional organizations worked to build the capacity of local businesses to not only compete in supply chains, but also to increase the diversity of businesses that would not be reliant on LNG/CSG activities. In the Surat Basin, for example, the Toowoomba and Surat Basin Enterprise (TSBE) worked with industry to address this goal through the development of two key programs: the Emergent Exporters Program and the Business Navigators Program. These programs provide strategic support to help start-up businesses in agriculture, aquaculture, and other sectors to identify new markets, to gain expertise in exporting products to those markets, to obtain intellectual property protection, and more.

The investments in infrastructure, housing, training, business capacity, and other community assets were often delayed for some time after the construction associated with these large-scale industry projects had already started. This left the community ill-prepared to cope with the pressures or to maximize the benefits of being engaged with these developments in the region. Many stakeholders also acknowledged that gas companies tried to mobilize corporate social responsibility strategies through corporate donations to non-profits, education and training, health centres, or cultural events, but that these industry contributions were often too small or sporadic to support any significant or transformative change.

13.4.2 Lack of Readiness

The lack of community and regional readiness stemmed from a number of factors. First, there was a poor understanding of unconventional natural/coal seam gas development at the local community scale. Inadequate access to data impeded the ability of local and senior governments to understand the speed of development, the scale and scope of regional and local disruptions from these major projects. Responses to these compressed boom periods seemed almost an afterthought as competition amongst local governments to capitalize on the sudden pulse of economic activity led to a 'booster' mentality that placed a premium on development but failed to adequately consider costs during the planning stages. These challenges were exacerbated by the limited capacity of and support for local governance structures and long-term planning. Coherent local responses were also undermined by the limited jurisdiction that local governments had over key policy areas that were instrumental in the shifting boom-and-bust patterns that were unfolding. In the case of the Surat Basin, coordination at the local level was further disrupted by top-down local government amalgamation processes that occurred just prior to the CSG boom and that mostly benefited the larger centres to the detriment of more marginalized small towns.

13.4.3 Reactionary State Responses

Each of our case study regions revealed that they were left with reactionary state responses provided through 'oil and gas commissions' and a set of limited fiscal mechanisms that we call 'royaltyish' programs. As described by these stakeholders:

> We saw the activity in the outlying region could never be incorporated into our boundaries. It was very nimble, mobile, moves around all the time, but it had a huge wealth that wasn't being tapped ... The Province missed the mark.
>
> (RR#6 2017)

> Remember, we had three different companies with three different projects investing $70 billion in about a five-year plan. It was chaos. Government was way behind the 8 ball. Government didn't have a clear plan and nobody anywhere really had any idea of the consequences.
>
> (AA#7 2018)

The construction of institutions and mechanisms to cope with the externalities associated with LNG/CSG projects were not developed early enough to provide effective institutional responses. This reactionary state of senior government responses is further explored below for each of the Peace River Region and the Surat Basin.

13.4.4 Peace River Region

In the Peace River Region, the provincial government and regional communities negotiated a series of fiscal mechanisms designed to mitigate industrial impacts and support quality of life improvements. The Fair Share and Peace River Agreements would provide municipalities collectively with roughly $50 million each year.[2] These are not resource royalty sharing agreements, however, but rather a mechanism to provide local governments in the Peace River Region with access to a tax base that other local governments already have (framed in the region as a 'grant in lieu of taxes'). While most local governments have direct access to an industrial tax base through mines and forestry mills that are located within their municipal boundaries, these agreements were intended to provide municipalities in the Peace River Region with access to a disconnected industrial tax base due to the dispersed nature of oil, gas, and pipeline activities in the surrounding rural regions (Ryser et al. 2019).

However, the agreements were not rolled-out by the provincial government as part of a cohesive approach to address the pressures unfolding in these municipalities. Instead, local governments took the lead to address the empty strategic policy space that had not been addressed by the provincial government. Local governments began to assert their rights in order to gain the Province's attention and engage them in negotiations for the initial Fair Share Agreement. They did this by participating in environmental impact assessment processes to make the case that large-scale industrial development exceeded the carrying capacity of local infrastructure and services, which often increased costs to industry through longer timelines for development project approvals (Markey and Heisler 2010). This made it more appealing for the Province to meet local governments at the negotiation table. These small municipalities then seized the agenda by forming the Northeast BC Resource Municipalities Coalition in order to complete research that would support key principles and arguments in the negotiations. The Coalition had conducted impact studies, obtained information about how much money the provincial government collected from industry in all communities and regions throughout the province, identified the timing of oil and gas developments throughout the region, assessed the size and types of resources that were being extracted, and estimated realistic costs associated with the impacts on municipal infrastructure and services.

However, the Province was not as well prepared and lacked the same detail of information as the Coalition. As one provincial stakeholder explained:

> It was very difficult for us to deal with it because they actually had spent a lot of money. They traveled with a lot of people. When we sat down to talk with them, they would have 10–15 people in the room. We had three. They were able to inundate us with paper, analysis, facts, and numbers … and so from our vantage point, it was difficult, challenging …
>
> (RR#21 2017)

The limited capacity and position of the provincial government was not a surprise to some rural stakeholders who felt that the provincial government was more focused on being a facilitator of industrial development. This position was exacerbated by regulatory roll-back schemes and policies that supported industry self-regulation, leaving state-level actors as weak defenders of rural public interests that reinforce path-dependence. As one rural stakeholder argued:

> For many of the governments going back over the last 34 years, that hasn't been a priority and in fact they've listened more to the global industry voice than they have to the community voice. Right? Because they seem to think their role is just to facilitate, you know the investments along and not to get involved and take any direct role in terms of our

socio-economic planning and looking after the things that could vastly improve the benefits to the rural communities.

(RR#5 2017)

Furthermore, the focus of provincial policies and actions at the time were designed to make the Province of British Columbia more competitive to attract industrial investment with the neighbouring Province of Alberta. As one provincial stakeholder explained:

> We had incentive programs in the natural gas industry. To give you an example, on our best year, we probably drilled about 350 wells here in BC. Alberta in the same token in their best year would be in the 20 000 range. So significant difference. So we had to compete with Alberta in order to attract some of the industry to come over here and research the formations and so on. Shale gas was really a game changer for us. But the incentive programs ... the royalty incentive programs were put in place where some of our gas was deeper. As the technologies were changing, to go deeper costs a lot more than to drill a new well. So we said look we are going to put a deep well royalty incentive program so if you ... start producing until you reclaim your capital ... these are just examples ... we are not going to put a royalty on that. So it gave people an incentive to do new things in our formations and so on. We had a number of the royalty incentive programs that worked that way and they worked very well for us.

(RR#19 2017)

Revenues from the agreements allowed communities to address ageing physical infrastructure, much of which dated back to the immediate post-World War Two era, and had never been replaced. Investments were also made to some roads, intersections, sidewalks, sustainable or renewable energy infrastructure, community halls, protection service infrastructure, and recreational facilities. Under these agreements, these funds are restricted to spending on capital improvements and communities must develop and submit capital plans, official community plans, and annual reports. However, these were not met by complementary investments in infrastructure and services by the provincial government. As the NEBC Resource Municipalities Coalition has argued (2015, 126):

> The municipalities were obligated to carry out capital planning and maintenance of official community plans but did not have specific obligation to do advanced planning to deal with the impacts of all of the resource development being promoted and encouraged by the Province ... The Peace River Agreement does not obligate the Province to prepare or develop plans for the expansion of provincial services such as rural roads, policing, health and education that would parallel and synchronize with the planning requirements that they are now imposing on the Peace Region municipalities.

This vertical fiscal imbalance hampered the establishment of the institutional and fiscal arrangements needed to support diversification. Furthermore, no emergency or legacy funds are permitted under these agreements, leaving these communities trapped in staples dependence.

13.4.5 The Surat Basin

In the Surat Basin, the coal seam gas industry emerged in 2005, and underwent rapid growth during construction. However, the camps to accommodate the large mobile workforce were not put in place before construction began. By 2014, most of the construction was completed. The Queensland Government's 'Royalties for the Regions' fund was also not mobilized until 2012. Almost half-a-billion $AUS was distributed across the region's local governments and Native Title-holding Indigenous groups by this programme from 2012/13 to 2015/16, after which it was replaced by the 'Building our Regions' grants scheme (Argent et al. 2021). Funding from royalty payments, however, are not directly allocated to affected communities. Instead, all funding distributed through these programs comes from consolidated revenue. This last-mentioned fund distributed $AUS 375 million across various programs, primarily for infrastructure (e.g. road, railway, and airport improvements) but also for some services and amenities.

The Building Our Regions program involved a formal process by which the state government delivered royalties to the region in response to applications from local government and other organizations. However, there was varied understanding of these funding programs in the region. Some local stakeholders complained of ad hoc decision-making within the program, which lacked any objective mechanism for articulating benefits to the region. Furthermore, there was no formal policy or process in place to guide how the state should re-invest royalties into the resource-producing region. As one stakeholder explained:

> We must acknowledge that there was a lot of ad hoc decision making. There still is. There's no really centralized cohesive point I don't think that we could say we could go and clearly articulate the benefits to regions based on any kind of formula ... Governments don't want to talk to the regions about how much wealth they take out of the regions to subsidize the activities of government. So, our government has never ... I've never seen either government inclined to actually clarify or formalize any kind of arrangement. So, the best the government will do is say well you know we've got this much from gas and it's going into schools and hospitals and this thing and whatever ... We just had our state budget and everybody knows that coal and gas royalties were a windfall for the state government but I think we went through the entire budget presentation without mentioning coal or gas. So, I think government generally has been reluctant to say this much is coming out of there, this much will go

back because it would have to be a relatively small percentage of what comes out.

(AA#7 2018)

The state's strategy has not been successful on a long-term basis for a number of reasons. While the state government required local governments to complete asset management plans for municipal infrastructure, state support remained focused on the planning process, with limited additional funding to address extant ageing infrastructure pressures unfolding during the boom. At the same time, the small municipalities of the Surat Basin had limited capacity to mobilize resources during the early 'boom' stages and to develop strategic plans capable of seizing opportunities for investments that could better position the community for new development pathways. As one stakeholder explained:

> When I look back, we were behind the ball. I even want to get emotional. We had no idea really and you know, the Miles Chamber in the end developed a community investment plan as a way to try and drive some funding, but that was right at the end. We missed … council missed … we all missed huge opportunities in the beginning.
>
> (AA#10 2019)

Second, the royalty programs never reached their full potential. There was a limited flow of royalty revenues to assist the Surat Basin's resource boom towns because resource companies were allowed to write off their capital expenditures for pipelines, infrastructure, LNG plants, etc. (Commonwealth of Australia 2016). During the early years of the boom, the state government accumulated relatively meagre resource royalties and could consequently argue that they had limited program funds. Similar to the Peace River Region, there was a missed opportunity to invest in a stable, predictable legacy fund that could provide long-term support for this rural region. As another stakeholder told us:

> I think there needed to be a much more sophisticated governance arrangement. And I am critical of particularly state government in perhaps providing some leadership and some facilitation around this. I'm absolutely certain that we missed the opportunity to develop perpetual funding through a community foundation. We should have established a community foundation for the Western Downs and Maranoa, or in the CSG areas, and instead of spending money, the companies would have invested significant dough, millions in a community foundation that had a community board, the money's invested. The dividends from that money provide perpetual investment year after year.
>
> (AA#8 2019)

Furthermore, despite a new period of industrial investment that was initiated in 2005, the state government's efforts to ensure that institutional structures were in place to support community processes were significantly delayed. In 2010, the Government of Queensland established the Surat Basin Coal Seam Gas Engagement Group to facilitate relationships between stakeholders with representatives from rural property owners, industry, and government agencies (Scott 2016). It was not until 2012, however, that this institutional structure was formalized as the Gasfields Commission. In addition to mobilizing a clear communications strategy, the Commission's role was to be an effective and efficient conduit between government, industry, and communities as regulatory frameworks and programs unfolded.

Revenue sharing or royalty agreements, however, represent an opportunity for senior governments to essentially offload the broader public interest for articulating a development vision with rural regions, as well as to deflect critiques that the government is not doing enough. Revenue sharing or royalty agreements may also serve a purpose of maintaining weak or limited regulation. Communities, for example, may use such funds to mitigate some of the negative impacts associated with industry operations which might otherwise have been covered under more stringent regulation.

13.4.6 Lessons

The scale and pace of impacts associated with contemporary natural resource development projects can be significant and can quickly overwhelm both state policy and local governance systems. Initial, intense impacts, however, can also rapidly change as exploration and construction activities cease and industry moves into production. The capacity to cope is embedded in the flexible readiness of both the resource-dependent communities and the senior government policy-makers and agencies, in terms of how they leverage policies and investments to strategically maximize opportunities towards future social and economic diversification. Leveraging such opportunities, however, has been a challenge. Moving forward, our findings suggest a number of key lessons for resource-dependent economies experiencing new growth from LNG/CSG projects.

The first set of lessons focuses upon territories and communities. Regions need the fiscal resources and institutional structures to be able to work collectively and to mobilize a more coherent and coordinated policy vision (Boschma 2015; Mitchell and Shannon 2018). Regions that are able to strategically target their investments to support a coordinated approach are better positioned to build a more resilient future. Local and regional coalitions play an important role in 'scaling-up' their voice and capacity to fill voids created from abandoned regulatory, policy, and investment spaces. All of these lessons point to a need to invest in replacing outdated structures with new governance processes that will better position rural communities to gain the knowledge and tools to overcome their peripherality and be ready and agile as they work through the shifting boom and bust periods associated with these unconventional resource developments.

There are a series of principles that should guide the distribution of fiscal resources (Markey et al. 2019). Greater parity between national and Provincial/

State governments, on the one hand, and local governments and communities, on the other, could be secured by mandating access to the industrial tax base as a property tax issue. This is critical to ensure that accountability exists between the taxation authority and rural taxpayers. The regulatory framework for these mechanisms should be responsive to any increases or decreases in the industrial tax base. Municipalities should be provided with the flexibility to allocate such resources according to local priorities and strategic directions. Any fiscal distribution mechanisms should provide communities with predictable resources in order to support short-term and long-term planning.

The second part of these lessons focuses upon institutions. There is a need to reconstruct institutions to cope with the externalities associated with these rapidly shifting boom and bust periods. This reconstruction should focus on policy capacity and access to sophisticated information in order to track and understand the rapid transformations and impacts that can unfold in these environments. Functional processes are also needed to support effective governance structures that are so critical to the successful oversight commissions. These governance structures need to be complemented with appropriate processes to allocate and distribute royalty benefits in order to support the justice needed to address socio-spatial inequity as stakeholders reflect on who gets what, where, and with what effect, with a particular focus on those communities near extraction sites in order to compensate for the disruption and consequences of exploration, construction, and extraction activities.

The third set of lessons focuses upon rural economies. Stakeholders at the local and regional level will need to explore opportunities to extract benefits from exploration and construction processes, including those stemming from local engagement in supply chains related to work camps. The ability to cope with the rapid pace of change, however, is embedded in flexible readiness. There is a need to strategically focus infrastructure and program investments in order to diversify these rural economies and to better position them to be agile and responsive to emerging challenges and opportunities as communities seek to move beyond path-dependent trajectories.

13.5 Conclusion

While the sheer magnitude and pace of the Peace and Surat region booms initially caught formal and informal institutions off guard, over time some regional and local institutions generated multi-scalar and horizontal relationships that enabled them to adapt and capture important and long-term economic benefits. Others, however, were predominantly exposed to the negative consequences of LNG/CSG development, and felt largely abandoned by regional and central government institutions. Local governments remain limited in their ability to meet basic infrastructure and service obligations without getting ahead of the curve to initiate meaningful strategies for diversified community economic development that are instrumental to both breaking the staples trap and pursuing new development trajectories as highlighted through evolutionary economic geography (Ryser et al. 2019). Fiscal imbalances, vertically between different levels of government, can impede opportunities for pursuing new development path trajectories by limiting the level of control, resources, and flexibility needed to make strategic investments in infrastructure, programs, and services (Blackwell et al. 2015; Eccleston and Woolley 2014). This serves to entrench the peripheral position of rural communities as resource banks for urban centres (Halseth et al. 2014). Negative consequences are exacerbated by the limited capacity of small remote places where more investments and support are needed to strengthen a higher level of sophistication and collaboration at a regional level.

Negative impacts are also exacerbated by a lagging preparedness of policies and communities, despite the extensive experience that stakeholders have with resource-based economies. Our case regions identify the problem, in part, as the absence or inadequacy of mediating institutions to support readiness. Instead, there have been incoherent public policies that emanate from senior governments who see their role as a facilitator of industrial development rather than as a protector of the public interest in resource producing regions. The structural impediments that have restricted the use of resource revenues and limited jurisdictional authority to recombine community assets may continue to reinforce and trap these communities in historic path dependencies (Ryser et al. 2019). Overall, this research reflects on the vital role of institutional relations and dynamics in shaping the evolving uneven geography of resource development.

Notes

1 Renovictions are tenancy evictions that are implemented in order to complete extensive repairs or renovations to a unit (Province of BC 2021). In resource boomtowns, however, 'renovictions' may be used to provide an opportunity to evict lower-paying tenants, raise rental rates, and rent to higher paying industry workers (Ryser et al. 2020).
2 The initial Fair Share Agreement (1994) provided $4 million in provincial non-property tax revenues to the Peace River region's municipalities to mitigate infrastructure impacts from industry (Markey and Heisler 2010). The distribution of revenues was based on the industrial assessment, population levels, and then indexed to growth in the rural industrial tax base. Fair Share was modified twice with the third iteration providing $20 million per year, but with an embedded lift mechanism reaching $46 million in 2015 (Heisler and Markey 2013). In 2015, the new Peace River Agreement (PRA) was finalized to provide $50 million per year for a 20-year period.

References

Argent, N. 2013. "Reinterpreting Core and Periphery in Australia's Mineral and Energy Resources Boom: An Innisian Perspective on the Pilbara." *Australian Geographer* 44, no. 3: 323–340.

Argent, N., S. Markey, G. Halseth, L. Ryser, and F. Haslam-McKenzie. 2021. "The Socio-Spatial Politics of Royalties and Their Distribution: A Case Study of the Surat Basin, Queensland." *Environment and Planning A: Economy and Space.* https://doi.org/10.1177/0308518X211026656

Beer, A., and T. Clower. 2019. *Globalization, Planning and Local Economic Development*. London: Routledge.

Beer, A., T. Clower, G. Haughton, and A. Maude. 2005. "Neoliberalism and the Institutions for Regional Development in Australia." *Geographical Research* 43, no. 1: 49–58.

Beer, A., B. Kearins, and H. Pieters. 2007. "Housing Affordability and Planning in Australia: The Challenge of Policy Under Neoliberalism." *Housing Studies* 22, no. 1: 11–24.

Benham, C. 2016. "Change, Opportunity and Grief: Understanding the Complex Social-Ecological Impacts of Liquefied Natural Gas Development in the Australian Coastal Zone." *Energy Research & Social Science* 14: 61–70.

Bevir, M., and W. Rhodes. 2003. "Searching for Civil Society: Changing Patterns of Governance in Britain." *Public Administration* 81, no. 1: 41–62.

Blackwell, B., B. Dollery, and B. Grant. 2015. "Institutional Vehicles for Place-Shaping in Remote Australia." *Space and Polity* 19, no. 2: 150–169.

Boschma, R., 2015. "Towards an Evolutionary Perspective on Regional Resilience." *Regional Studies* 49, no. 5: 733–751.

Buse, C., M. Sax, N. Nowak, J. Jackson, T. Fresco, T. Fyfe, and G. Halseth. 2019. "Locating Community Impacts of Unconventional Natural Gas Across the Supply Chain: A Scoping Review." *The Extractive Industries and Society* 6, no. 2: 620–629.

Carson, D. 2011. "Political Economy, Demography and Development in Australia's Northern Territory." *The Canadian Geographer* 55, no. 2: 226–242.

Commonwealth of Australia. 2016. *Resources and Energy Quarterly—December 2016*. Canberra: Commonwealth Government of Australia. https://industry.gov.au/Office-of-the-Chief-Economist/Publications/Documents/req/REQ-December-2016.pdf

Eccleston, R., and T. Woolley. 2014. "From Calgary to Canberra: Resource Taxation and Fiscal Federalism in Canada and Australia." *Publius: The Journal of Federalism* 45, no. 2: 216–243.

Ennis, G., M. Finlayson, and G. Speering. 2013. "Expecting a Boomtown? Exploring Potential Housing-Related Impacts of Large-Scale Resource Developments in Darwin." *Human Geographies: Journal of Studies and Research in Human Geography* 7, no. 1: 33–42.

Gillingham, M., G. Halseth, C. Johnson, and M. Parkes (Eds). 2016. *Integration Imperative: Cumulative Environmental, Community and Health Effects of Multiple Natural Resource Developments*. New York: Springer.

Halseth, G., L. Ryser, S. Markey, and A. Martin. 2014. "Emergence, Transition, and Continuity: Resource Commodity Production Pathways in Northeastern British Columbia, Canada." *Journal of Rural Studies* 36: 350–361.

Hayter, R., T. Barnes, and M. Bradshaw. 2003. "Relocating Resource Peripheries to the Core of Economic Geography's Theorizing: Rationale and Agenda." *Area* 35: 15–23.

Heisler, K., and S. Markey. 2013. "Scales of Benefit: Political Leverage in the Negotiation of Corporate Social Responsibility in Mineral Exploration and Mining in Rural British Columbia, Canada." *Society & Natural Resources* 26, no. 4: 386–401.

Holmes, J. 1987. "Population." In *Space and Society: Australia—A Geography*, Vol. II, edited by D. Jeans, 24–48. Sydney: Sydney University Press.

Howell, K. 2018. "Housing and the Grassroots: Using Local and Expert Knowledge to Preserve Affordable Housing." *Journal of Planning Education and Research* 38, no. 4: 437–448.

Innis, H. 1933. *Problems of Staple Production in Canada*. Toronto: Ryerson Press.

Kjær, A. 2011. "Rhodes' Contribution to Governance Theory: Praise, Criticism, and the Future Governance Debate." *Public Administration* 89, no. 1: 101–113.

Larner, W. 2003. "Guest Editorial: Neoliberalism?" *Environment and Planning D: Society and Space* 21: 509–512.

Lockie, S., G. Lawrence, and L. Cheshire. 2006. "Reconfiguring Rural Resource Governance: The Legacy of Neoliberalism in Australia." In *Handbook of Rural Studies*, edited by P. Cloke, P. Mooney, and T. Marsden, 29–43. London and Newbury Park: Sage.

Luke, H., and N. Emmanouil. 2019. "All Dressed Up and Nowhere to Go: Navigating the Coal Seam Gas Boom in the Western Downs Region of Queensland." *The Extractive Industries and Society* 6, no. 4: 1350–1361.

Markey, S., G. Halseth, L. Ryser, N. Argent, and J. Boron. 2019. "Bending the Arc of the Staples Trap: Negotiating Rural Resource Revenues in an Age of Policy Incoherence." *Journal of Rural Studies* 67: 25–36.

Markey, S., and K. Heisler. 2010. "Getting a Fair Share: Regional Development in a Rapid Boom-Bust Rural Setting." *Canadian Journal of Regional Science* 33, no. 3: 49–62.

Martin, R., and P. Sunley. 2006. "Path Dependence and Regional Economic Evolution." *Journal of Economic Geography* 6, no. 4: 395–437.

McCann, L., and H. Gunn (eds.). 1998. *Heartland and Hinterland: A Geography of Canada*. Scarborough: Prentice-Hall Canada Inc.

Measham, T., and D. Fleming. 2014. "Impacts of Unconventional Gas Development on Rural Community Decline." *Journal of Rural Studies* 36: 376–385.

Mitchell, C., and M. Shannon. 2018. "Exploring Cultural Heritage Tourism in Rural Newfoundland through the Lens of the Evolutionary Economic Geographer." *Journal of Rural Studies* 59: 21–34.

Morrison, T. H., C. Wilson, and M. Bell. 2012. "The Role of Private Corporations in Regional Planning and Development: Opportunities and Challenges for the Governance of Housing and Land Use." *Journal of Rural Studies* 28, no. 4: 478–489.

Nelsen, J., M. Scoble, and A. Ostry. 2010. "Sustainable Socio-Economic Development in Mining Communities: North-Central British Columbia Perspectives." *International Journal of Mining, Reclamation and Environment* 24, no. 2: 163–179.

Northeast BC Resource Municipalities Coalition. 2015. *Municipalities Role in Oil, Gas, and Resource Development in Northeastern BC*. Fort St. John, BC: Northeast BC Resource Municipalities Coalition.

Peterson, V. 2003. *A Critical Rewriting of Global Political Economy*. London: Routledge.

Province of BC. 2021. Renovictions. Victoria, BC: Province of British Columbia. https://www2.gov.bc.ca/gov/content/housing-tenancy/residential-tenancies/ending-a-tenancy/renovictions

Ryser, L., G. Halseth, and S. Markey. 2020. "Moving from Government to Governance: Addressing Housing Pressures During Rapid Industrial Development in Kitimat, BC, Canada." *Housing Studies*. https://doi.org/10.1080/02673037.2020.1789564

Ryser, L., G. Halseth, S. Markey, C. Gunton, and N. Argent. 2019. "Path Dependency or Investing in Place: Understanding the Changing Conditions for Rural Resource Regions." *The Extractive Industries and Society* 6, no. 1: 29–40.

Ryser, L., S. Markey, G. Halseth, and K. Welch. 2017. "Moving from Mobility to Immobility in the Political Economy of Resource-Dependent Regions." *Applied Mobilities* 1–22. https://doi.org/10.1080/23800127.2017.1421290

Scott, R. 2016. *Independent Review of the Gasfields Commission Queensland and Associated Matters.* Brisbane: Queensland Department of State Development.

Smith, K., and J. Haggerty. 2020. "Exploitable Ambiguities and the Unruliness of Natural Resource Dependence: Public Infrastructure in North Dakota's Bakken Shale Formation." *Journal of Rural Studies* 80: 13–22.

Taylor, A., S. Larson, N. Stoeckl, and D. Carson. 2011. "The Haves and Have Nots in Australia's Tropical North–New Perspectives on a Persisting Problem." *Geographical Research* 49, no. 1: 13–22.

Tonts, M., P. Plummer, and N. Argent. 2014. "Path Dependence, Resilience and the Evolution of New Rural Economies: Perspectives from Rural Western Australia." *Journal of Rural Studies* 36: 362–375.

14 Economies–Institutions–Territories

Old Issues Revisited and New Research Avenues

Luca Storti, Giulia Urso, and Neil Reid

The book consists of a collection of chapters dissecting the complex interplays between economies, institutions, and territories. As stated in the first chapter, accounting for these interconnections is a way to handle such abstract and multi-faceted concepts. It is also an intriguing way to look at the whole social anatomy (Hedstrom, 2005).

Through the chapters, it has been shown that there are several processes that are more likely to emerge nowadays than in the past (e.g., shocking events, climate change, unforeseen economic crisis) and put the interplays between economies, institutions, and territories under stress. The societies of organized capitalism of the Fordist age were, in fact, well-integrated in their sub-systems. By contrast, recent research has pointed out that in the scenario of financial and globalized capitalism societies have been suffering from new "structural imbalances" (Bagnasco, 2003, 2008). In this context, increasing difficulties have emerged in reconciling economic growth with the containment rise of social inequalities at the territorial level, maintaining high levels of social mobility, and safeguarding the effective and efficient functioning of democratic assets (Trigilia, 2020). From this perspective and by taking multiple disciplinary perspectives, looking at different areas of the world, and examining various topics, several chapters in the book show a series of misalignments among economies–institutions–territories.

Beyond these structural imbalances that have become entrenched in recent decades, the interplay between economies–institutions–territories continuously raise new puzzles pertaining to micro dynamics – i.e., how economic activities are rooted at the territorial level and shaped by local institutional arrangements – and macro dynamics – i.e., the economic orders at the intersection of global trends and institutional frameworks. In this respect, several chapters of the book do not point to the social problematic aspects related to these nexuses but dissect some of their relevant analytical elements.

In sum, the book offers one main general insight. The nexuses between economies–institutions–territories must be investigated through a multidisciplinary comparative perspective. The comparison can be synchronic, i.e., concerning different territorial areas, or, vice versa, diachronic, i.e., relating to the dynamic evolution of these links over time. By observing the connections between economies–institutions–territories, one can grasp both change

DOI: 10.4324/9781003191049-18

and persistence in the social world: the reconfiguration of institutional elements, economic processes, and their kinds of embedding at the territorial level. No single discipline holds the key to unpacking and understanding the complex relationships that exist between economies, institutions, and territories – sociologists, economists, geographers, and scholars from other disciplinary perspectives all have valuable insights to offer.

That being said, in these concluding pages, we can pinpoint some promising research avenues that have already been suggested in some of the chapters and deserve further exploration. We draw upon the four sections of the book to illustrate a tentative future research agenda.

The first thematic area is about the *innovations, tensions and dilemmas relating to the blurring between economy and society*. A research path that can be included in this broad thematic area comes from afar and concerns the regulation of the economy. The theme of regulation can take on a dual meaning. First, the regulation of the economy by politics which define the arrangements that discipline the formal allocation of resources. Second, the informal rules, viz., taken for granted social norms and routines that allow economic organizations to assume legitimacy within a given institutional environment (Stark, 2009). In the presence of a highly changing economy and unstable institutional arrangements, it becomes interesting to observe situations of systematic dissonance between formal and social norms (Sciarrone and Storti, 2019). The systematic discrepancy between social and formal norms can become concrete within specific sectors of the economy or – in a more transversal way – within some territorial areas. In such cases, forms of hybridization between legal and illegal spheres can occur (Sciarrone and Storti, 2019), i.e., institutional environments in which formally illicit economic behaviours are socially supported and deemed legitimate. Beyond the understandable concerns about the spread of illicit economic behaviours, which are not relevant here, it is crucial from an empirical perspective to reconstruct the processes through which illicit economic actions tend to become socially legitimized. These processes generate contradictory dynamics. In the short term, territorial contexts marked by a lowering of the thresholds commitment to formal legality may have certain competitive advantages. Unscrupulous economic actors can, in fact, exploit the lever of informality to make their businesses more profitable (Sciarrone and Storti, 2019; Dagnes et al., 2020). The aggregate outcomes in the long run, however, turn out to be negative. These territories tend to be in trouble: they are abandoned by the more dynamic economic operators who are less willing to act illegally. As a consequence, a decline in innovative processes and downward competitions among economic activities take place. Furthermore, when situations characterized by a systemic imbalance between social and formal norms emerge, a reallocation of material resources and power among social groups is also generated. Under these circumstances, new inequalities, and tensions between economic players, as well as competitions and conflicts between territories, are likely to come out. Empirical explorations of such issues are highly welcome.

We now turn our attention to the *emerging problems related to coordination between state and market*. As thoroughly explained in Chapter 1, when exploring the economies-territories-institutions interplay, one of the crucial aspects is the interaction between the following factors: given, fixed and changing context-specific spatial heterogeneities, on one hand, and market-based selection mechanisms, on the other, with the mediating role of institutions in-between. The study of these multiple, deeply intertwined dimensions has become more and more complex over time, starting from the 1970s – with the awareness that "institutional forms and arrangements that had supported the post-war boom in capitalist countries had become incompatible with, and no longer capable of providing a stable environment for, economic accumulation and competition" (Martin, 2000: 89). These processes led to "de-institutionalize" and socially "dis-embed" the economy, opening a "free enterprise" era till the advent of globalization and the post-industrial information economy, which have shaken, and in part discarded, institutional structures at all scales. This "de-localizing" force has partly disconnected economic activities from their local socio-institutional context, and simultaneously exposed local economies to a world-wide competition and a globally mobile capital, and consequent greater uncertainties (Martin, 2000). Furthermore, over the past two decades and more, financial and labour markets have been progressively de-regulated, public sector activities and the welfare system increasingly privatized and/or profoundly re-structured. This has brought about a widening of existing inequalities and the emergence of new ones between and within territories. Today, these phenomena seem to have been further accelerating, with no longer "a 'fixed landscape' of traditions and institutions structuring economic and social relations" (Ivi: 90), but less rigid institutional arrangements which are called to be more and more flexible so as to adapt to a constantly changing economic space, where new sources of knowledge and market opportunities are incessantly produced. All the above makes clear the extent to which institutions are key to the formation of values and preferences. This is even more evident under conditions of austerity or in time of crisis – as the one experienced due to the COVID-19 pandemic –, when institutions "become transmission mechanisms for the priorities of 'global capital', 'bond markets' and other such abstractions" (Tomaney, 2014: 138). We are witnessing an ever faster evolving and more uncertain economy, with territories being economic and political arenas for within and in-between competition over resources, capitals, values and institutions being relentlessly challenged by the pressures to their regulation and mitigation role. Against this backdrop, the permeation of each of the three dimensions into the others makes it almost impossible to disentangle (and manage) them. This implies a greater difficulty to isolate the single effects they induce on one another and, more generally, on the evolution of the economic and social space. This poses critical issues also from the point of view of academic research dealing with these topics. Academic literature has already widely acknowledged some theoretical and empirical questions that are at stake in unravelling the geographically diverse dynamics of economic development and structures of institutionalization as well as a micro-level comprehension of economic action, market functioning, and technological and institutional change (Bathelt and Glückler, 2014). These mainly pertain to the complexity of these relationships, the lack of easily identifiable indicators and adequate data, the endogeneity of some of the phenomena under scrutiny, the need for a comparative perspective and a multidisciplinary effort to yield more general insights into the economies–institutions–territories nexuses, especially looking at the role of markets and their interactions with formal and informal institutions.

More broadly, with respect to data, "big data" will be increasingly important in the future. The volume, velocity (speed of generation, delivery, and processing), and variety of data are increasing. Both businesses and governments can benefit from big data. While there is a myriad of issues (e.g., security, compliance, and privacy) in utilizing big data, governments see its potential to help them address a variety of challenges ranging from health care to natural disasters and terrorism. Businesses can use it to better understand and be more responsive to consumer preferences and behaviour. For both the private and public sectors, big data can facilitate real-time decision making (Kim et al., 2014). Big data brings challenges, however, and a "careful balance between the threats and opportunities" must be navigated (Cumbley and Church, 2013: 608). As noted by Galloway (2017: 89), "big data represents a radical shift in the balance of power between State and citizen", with big data having the potential to enhance government power, "rather than simply facilitating execution of government activities". Increasingly, researchers are using data from social media sites such as Facebook and Twitter to gain insights into a wide variety of human behaviour and preferences. However, there is evidence that such data sets are more likely to reflect the opinions and preferences of the more privileged than the less privileged members of society (Hargittai, 2020). Social media platforms, such as Facebook, have been accused of steering public opinion in particular directions, even using such nefarious tactics such as censoring particular voices (Carson, 2018). A 2020 survey by the respected Pew Research Center found that 72% of Americans believe that social media companies have too much power and influence in politics (Anderson, 2020). This is a belief that cuts across the political spectrum, with such concerns being raised by liberals and conservatives alike. Understanding the opportunities and threats associated with big data, while figuring out how to enhance the former and mitigating the latter, is a key research need. However, both empirical and theory-building efforts may be arduous in fast-moving and fluid economies and networks, where boundaries – both spatial ones and the ones among different spheres (economic, financial, social, cultural, political, governance) – are often blurring and phenomena are to be grasped at the very time when they occur before they take other shapes or generate impacts of a different nature.

The topics relating to *Social Inequalities, Displacement, and Conflicts among the Interests of Social Groups* raise a series of research questions as well. In several places in the book, references have been made to the topic of social inequality. In sociology, there is a well-established tradition of research dealing with the growth of internal heterogeneity within individual social classes, the relative distance between social classes, and the increment of the gap between the upper and lower class. This tradition of studies – which one refers to with the label of "social stratification" – could cross-fertilize more with the territorialized analyses of geographers and with the investigations of economic sociology concerning the interpenetrations between economy and institutions. This integrated approach would be highly beneficial. A first line of research to be further dealt with concerns the (new) meanings assumed by the urban vs. rural cleavage (Urso, 2021). At first glance, it seems that this distinction is suitable to explain several recent and puzzling phenomena, such as the territorial distribution of the vote and some trends regarding the relationship between civil society and political institutions (Huijsmans et al., 2021). In many large cities in North America and Europe, for example, there seems to be a higher percentage of progressive voters than in other areas, and greater openness to issues such as immigration, the demand for income redistribution policies, and the expansion of civil and social rights. On the contrary, conservative positions, stronger resistance to host immigrants, and a mixed attitude of distrust and individualism seem to prevail in rural and small-town areas, making for little support for progressive fiscal policies and increased social spending (van Leeuwen and Halleck Vega, 2021). The topic deserves to be further explored to understand whether the urban vs. rural distinction is sufficient to understand the above phenomena or are there more elaborated spatial patterns. In fact, differentiated trends also emerge among various cities based on their social structure and the territorial area in which they are placed. At the same time, rural areas can be affected by social, agricultural, and productive renewal processes, which, therefore, produce social cohesion and effervescence. In this regard, new research is much needed elaborating accurate typologies of the amalgam between economy and institutions at the local level, with a view to developing more nuanced models beyond the mere urban vs. rural distinction. In general, the complexification of the nexuses between economies–institutions–territories will make dichotomous explanations, such as core-periphery or North–South, weaker. Therefore, the growing internal variance within regional areas and the spotty distribution from a territorial point of view of socio-economic phenomena will need to be studied in-depth (Larsson et al., 2021). In addition, empirical investigation can reveal possible spurious correlations: are we dealing with actual territorial contextual factors, or is there a tendency for specific social classes to be more rooted in urban environments or, vice versa, in rural areas?

Conflict between groups that feel to be the expression of dissimilar cultures often occurs when one or more occupy or lay claim to the same portion of space, whether that be an entire State or a small neighbourhood within a city. In dissecting and unpacking such conflicts, the geographic concept of "sense of place" may have utility. Geographers have long recognized the cultural significance of place and space (Tuan, 1975). As noted by the renowned human Geographer Tuan (1975: 152), "at a high theoretical level, places are points in a spatial system. At the opposite extreme, they are strong visceral feelings". Conflict over place/space occurs when different groups (defined according to ethnicity, social class, religious beliefs, etc.) attribute competing meanings to and have competing visions for places/spaces. As a result, these places/spaces can become contested (Hayden, 1995), and debates over meaning which are reflective of power relationships ensue. Such debates "ostensibly favour one identity over another" and "define who belongs and who is excluded" from a place/space landscape (Ruoso and Plant, 2018). Identifying new ways to navigate territorial conflicts is a pressing need, and one to which social scientists are well positioned to contribute.

This also requires deepening the connections between social geography, on one hand, and the geography of production structures and market dynamics in its various forms, on the other hand. Much has been written about how specific social segments have been progressively expelled from urban contexts due to speculative residential operations aimed at selecting residents among the upper-middle and upper classes (Atkinson, 2000; Sassen, 2014; Semi, 2015; Storti and Dagnes, 2021). This may have favoured the localization in peri-urban and rural areas of population segments that harboured resentment because of the displacement they have suffered before and then funnelled this resentment towards populistic movements. Also, the need for income support measures is growing in many urban areas and within segments of the population once considered well-established middle classes (Mazzuccato, 2021). These dynamics call for new investigations that highlight the social tensions that are ongoing within large cities, and between cities and rural places.

Finally, when talking of the *challenge of peripherality*, on top of the locational disadvantages connected to a physically marginal position, some of the main issues these places have to cope with are related to access to and exploitation of knowledge, on one side, and access to policy-making networks and the expression of their voice and agency, on the other.

Access to knowledge is critical for both entrepreneurs – to stay innovative and competitive in the market – and policy-makers – to produce well informed strategic plans and gain access to external financing opportunities. The challenge for territories nowadays lies in the fact that they are called to deal with more and more complex economic networks and governance frameworks. This also makes academic research in these domains, especially in the latter, which is under-investigated, very complicated. Our understanding of the role and influence of individual agency has not developed much in past years (Gertler, 2018). More specifically, we still need to advance our knowledge on the impact of individual agents in determining economic and policy/political outcomes, on the extent to which and through which mechanisms institutional structures accommodate such decisions and preferences, and who are the more successful, meaning influential, agents in this regard. Furthermore, there is a social dimension of agency, which pertains to the ability of a place to express itself and act as a collective actor, which ultimately makes the community fully play its political role (Urso, 2021). This is not an easy task from an academic research perspective. It implies, in fact, to first explore and be able to isolate the local aggregate agentic responses and delving into their formation; second, to understand how these actions are enabled and constrained by their specific institutional context and how the local economic and institutional context conditions the activation of agency (Rekers and Stihl, 2021). Ultimately, on a methodological note, this means understanding agency as a mechanism of change in regions. As recently advocated by Rekers and Stihl (2021: 97):

> Agency has been a missing link in literature on institutions in regional development, and clearly holds great promise to better understand the 'transmission mechanism' (Rodríguez-Pose, 2020) by which institutions lead to development outcomes. Future research is needed to identify patterns of change agency in different regional contexts: Different types of regional lock-in (where we find different sets of dominant institutions) will require different types (and combinations) of change agency. Similarly, echoing Gertler (2018), we argue that institutions have been a missing link in the literature on agency in regional development.

As far as peripheral areas are concerned, special attention should be devoted to the role of 'transformative agencies' (Kurikka and Grillitsch, 2020), such as institutional entrepreneurs, place-based leadership, and regional policy intelligence (Gong et al., 2020). Especially in small, marginal, low-density territories, which lack the critical mass of competences and political weight, the role of the transformative agencies of only a few economic and/or institutional actors may be positively disrupting for the local context, producing a relevant change and/or stopping or inverting a path-dependent lock-in trajectory. Research-wise, theoretical and empirical efforts should be especially focused on delving into the role of innovative entrepreneurs who, though operating in disadvantaged areas, manage to transform or create new economic activities through the novel combination of knowledge and resources (Kurikka and Grillitsch, 2020). Even more intriguing in these contexts is the influence of institutional entrepreneurs – i.e., actors that challenge and transform existing rules and practices or aim at creating new ones (Kurikka and Grillitsch, 2020) – who successfully manage to co-ordinate and mobilise different actors and resources for the collective pursuit where previously others failed. Peripheral regions are theoretically interesting because their structural preconditions are in most cases less favourable to the emergence of innovations in both the economic and the institutional or political spheres. Empirically, these contexts are also often more difficult to study because of the lack of data or of their lesser accessibility in many respects. Micro-level, qualitative analyses are probably best suited to the task. However, research is still needed to shed light on the kind of agency which is conducive to a positive change in marginal territories, eventually de-lock them out of a declining spiral. This would be particularly helpful in times of shocks, when these transformative agencies might be crucial in leading to new paths emerging out of the crisis. Crises, as we know, have disorienting and, in some ways, unpredictable effects. In this regard, some of the dynamics affecting peripheral areas during the COVID-19 time deserve to be looked at in the years to come. The pandemic, in fact, has partly turned the centre-periphery relationship upside-down. Starting from the crisis of Fordism, countries and territorial areas being central in the "global value chains" (Gereffi, 2018) have abandoned the labour-intensive and environmentally impactful production activities, relocating them to emerging countries. This allowed rich countries to obtain raw materials at a low cost, thus dedicating themselves only to the assembly phases (one case among others is that of the steel production). However, raw material supply was interrupted or slowed down during the pandemic. As a result, rich countries have had to reprogram the assembly and supply of products on the markets, no longer having an autonomous production of raw materials. Under these circumstances, some territories occupying the most marginal positions in the global value chains are attaining a negotiating power unknown in the recent past. It is likely to be a transitory phenomenon, but it may however produce long-lasting effects.

Peripheral areas may also benefit from considering some of the ideas emanating from the Triple Helix and industrial cluster literatures. Popularized by Henry Etzkowitz and Loet Leydersdorff (1996), the Triple-Helix suggests a rethinking of the roles played by universities, industries, and governments, and the ways in which they interact with each other. A reconfiguration of these institutional relationships has the potential to enhance the speed and scope of innovation in knowledge-based societies. Under the Triple Helix model, "universities, firms, and governments each 'take the role of the other' in triple helix interactions even as they maintain their primary roles and distinct identities" (Etzkowitz, 2008: 1). Developed by Michael Porter, industrial clusters are defined as "geographic concentrations of interconnected companies, specialized suppliers, service providers, firms in related industries, and associated industries (e.g., universities, standard agencies, trade associations) in a particular field that compete but also cooperate" (Porter, 2000: 16). Porter's ideas around collaboration may be particularly relevant in helping peripheral areas overcome some of the challenges associated with their geographic isolation (Reid et al., 2021). Collaboration may allow stakeholders in peripheral regions to achieve things collectively that they cannot not achieve working independently, and allow them to leverage what Nadvi (1999) calls "collective efficiency". As noted by Sölvell (2009: 18), "critical resources and capabilities often do not exist within the firm but are accessible through networks inside the cluster". However, achieving collaboration among the different actors in a peripheral region can be challenging. Lack of trust and competing priorities are among the hurdles that need to be overcome (Gatrell et al., 2010; Pechlaner et al., 2009). As noted by Calignano et al. (2018: 1490), "cluster policy is often ineffective in peripheral regions with weak institutions and significant barriers to knowledge production and exchange". Identifying mechanisms and processes that might overcome these and other hurdles to successful collaboration is a potentially fruitful area of future research. Per the Triple Helix model, universities can facilitate the establishment and growth of industrial clusters in peripheral regions (Čábelková et al., 2017; Karlsen et al., 2017). Where institutions are weak, "universities can play a place leadership role due to their neutrality, accumulated knowledge, and expertise in relation to different stakeholder groups" (Thomas et al., 2021: 771).

Operationalizing the ideas underpinning both the Triple Helix and industrial clusters brings the concept of networks to the fore (Sölvell, 2009). Many institutional relationships are hierarchical-based, with ideas, policies, and resources flowing in a top-down fashion, often to the detriment of those lower down the hierarchy (Nagy et al., 2021; Russell et al., 2021; Skrimizea et al., 2021). In the case of industrial clusters, however, relationships are often network-based. As such, relationships are more egalitarian in nature with a greater variety of individuals and entities having a voice at the table (Kotter, 2011; Reid et al., 2008). Indeed, bottom-up driven clusters can be highly successful (Čábelková et al., 2017; Carroll and Reid, 2013). The networks that comprise a cluster evolve over time and serve a variety of functions, including facilitating knowledge flow and innovation, and improving both firm and regional performance (Boschma and Ter Wal, 2007; Brasier et al., 2007; Giuliani and Bell, 2005). As noted by Breda et al. (2006: 67) "the strategic position of peripheral regions can be accomplished through the identification of product clusters" and "the establishment of public-private partnerships and the creation of networks". In recent years, Social Network Analysis (SNA) has been utilized to better understand the functioning and benefits of relational structures (Alberti et al., 2021; Alberti and Belfanti, 2021). SNA can reveal the "hidden" leaders within an organization/network (Lane et al., 2017). These so-called "hidden" leaders can have more influence than those with positional authority. Emerging models conceptualize leadership as "more of a relational process, a shared or distributed phenomenon dependent on social interactions and networks of influence" (Sanders, 2014: 143). Identifying and understanding effective leadership models, particularly around regional development, is a fruitful area of future research.

Defined around specific issues and concerning circumscribed and well-defined geographical areas, the analysis of the relationship between economies–institutions–territories is destined to feed the literature of various disciplines in the years to come.

References

Alberti, F.G. and Belfanti, F. (2021). Do clusters create shared value? A social network analysis of the motor valley case. *Competitiveness Review*, 31 (2): 326–350.

Alberti, F.G., Belfanti, F. and Giust, J.D. (2021). Knowledge exchange and innovation in clusters: A dynamic social network analysis. *Industry and Innovation*, 28 (7): 880–901.

Anderson, M. (2020). Most Americans say social media companies have too much power, influence in politics. *Pew Research Center*. www.pewresearch.org/fact-tank/2020/07/22/most-americans-say-social-media-companies-have-too-much-power-influence-in-politics/

Atkinson, R. (2000). The hidden costs of gentrification: Displacement in Central London. *Journal of Housing and Built Environment*, 15 (4): 307–326.

Bagnasco, A. (2003). *Società fuori squadra. Come cambia l'organizzazione sociale*. Bologna: Il Mulino.

Bagnasco, A. (2008). *Ceto medio. Come e perché occuparsene*. Bologna: Il Mulino

Bathelt, H. and Glückler, J. (2014). Institutional change in economic geography. *Progress in Human Geography*, 38 (3): 340–363.

Boschma, R. and Ter Wal, A.L.J. (2007). Knowledge networks and innovative performance in an industrial district: The case of a footwear fistrict in the south of Italy. *Industry and Innovation*, 14 (2): 177–199.

Brasier, K.J., Goetz, S., Smith, L.A., Ames, M., Green, J., Kelsey, T., Rangarajan, A. and Whitmer, W. (2007). Small farm clusters and pathways to rural community sustainability. *Journal of the Community Development Society*, 38 (3): 8–22.

Breda, Z., Costa, R. and Costa, C. (2006). Do clusters and networks make small places beautiful?: The case of Caramulo (Portugal). In L. Lazzeretti and C.S. Petrillo(Eds.), *Tourism, Local Systems, and Networking*. London: Elsevier, 67–82.

Čábelková, I., Normann, R. and Pinheiro, R. (2017). The role of higher education institutions in fostering industry clusters in peripheral regions: Strategies, actors and outcomes. *Higher Education Policy*, 30: 481–498.

Calignano, G., Fitjar, R.D. and Kogler, D.F. (2018). The core in the periphery? The cluster organization as the central node in the Apulian aerospace district. *Regional Studies*, 52 (11): 1490–1501.

Carroll, M. and Reid, N. (2013). Social capital and the development of industrial clusters: The northwest Ohio greenhouse cluster. In Frank Giarratani, Geoffrey G.D. Hewings and Philip McCann (Eds.), *Handbook of Industry Studies and Economic Geography*. Northampton: Edward Elgar, 341–354.

Carson, M. (2018). Facebook in the news' social media, journalism, and public responsibility following the 2016 trending topics controversy. *Digital Journalism*, 6 (1): 4–20.

Cumbley, R. and Church, P. (2013). Is "Big Data" creepy? *Computer Law & Security Review*, 29: 601–609.

Dagnes, J., Donatiello, D., Moiso, V., Pellegrino, D., Sciarrone, R., & Storti, L. (2020). Mafia infiltration, public administration and local institutions: a comparative study in northern Italy. *European Journal of Criminology*, 17(5): 540–562.

Etzkowitz, H. (2008). *The Triple Helix: University–Industry–Government Innovation in Action*. London: Routledge.

Galloway, K. (2017). Big Data: A case study of disruption and government power. *Alternative Law Journal*, 42 (2): 89–95.

Gatrell, J.D., Thakur, R., Reid, N. and Smith, B.W. (2010). Clusters and "listening": Situating local economic development. *Journal of Applied Research in Economic Development*, 7 (1): 14–25.

Gereffi, G. (2018). *Global Value Chains and Development. Redefining the Contours of 21st Century Capitalism*. Cambridge: Cambridge University Press.

Gertler, M.S. (2018). Institutions, geography, and economic life. In G.L. Clark, M.P. Feldman, M.S. Gertler and D. Wójcik (Eds.), *The New Oxford Handbook of Economic Geography*. Oxford: Oxford University Press, 230–242.

Giuliani, E. and Bell, M. (2005). The micro-determinants of meso-level learning and innovation: Evidence from a Chilean wine cluster. *Research Policy*, 34: 47–68.

Gong, H., Hassink, R., Tan, J. and Huang, D. (2020). Regional resilience in times of a pandemic crisis: The case of Covid-19 in China. *Tijdschrift voor Economische en Sociale Geografie*, 111 (3): 497–512.

Hargittai, E. (2020). Potential biases in big data: Omitted voices on social media. *Social Science Computer Review*, 38 (1): 10–24.

Hayden, D. (1995). *Urban Landscape History: The Sense of Place and the Politics of Space. The Power of Place: Urban Landscapes as Public History*. Cambridge, MA: MIT Press.

Hedstrom, P. (2005). *Dissecting the Social: On the Principles of Analytical Sociology*. Cambridge: Cambridge University Press.

Huijsmans, T., Harteveld, E., van der Berg, W. and Lancee, B. (2021). Are cities ever more cosmopolitan? Studying trends in urban-rural divergence of cultural attitudes. *Political Geography*, 86: 1–15. 10.1016/j.polgeo.2021.102353.

Karlsen, J., Besedab, J., Sima, K. and Zyzak, B. (2017). Outsiders or leaders? The role of higher education institutions in the development of peripheral regions. *Higher Education Policy*, 30: 463–479.

Kim, G-H., Trimi, S. and Chung, J.-H.. (2014). Big data applications in the government sector. *Communications of the ACM*, 57 (3): 78–85. 10.1145/2500873

Kotter, J.P. (2011). Hierarchy and network: Two structures, one organization. *Harvard Business Review*, May 23. https://hbr.org/2011/05/two-structures-one-organizatio.

Kurikka, H. and Grillitsch, M. (2020). *Resilience in the Periphery: What an Agency Perspective Can Bring to the Table*. Lund University, CIRCLE: Papers in Innovation Studies No. 2020/7.

Lane, K., Larmaraud, A. and Yueh, E. (2017). Finding hidden leaders. *McKinsey Quarterly*, 1: 107–115.

Larsson, J.P., Öner, Ö and Sielker, F. (2021). Regional hierarchies of discontent: An accessibility approach. *Cambridge Journal of Regions, Economy and Society*, rsab015. 10.1093/cjres/rsab015.

Leydersdorff, L. and Etzkowitz, H. (1996). Emergence of a Triple Helix of university-industry-government relations. *Science and Public Policy*, 23 (5): 279–286.

Martin, R. (2000). Institutional approaches in economic geography. In S. Sheppard and T. Barnes (Eds.), *A Companion to Economic Geography*. Oxford: Basil Blackwell, 77–94.

Mazzuccato, M. (2021). *Mission Economy: A Moonshot Guide to Changing Capitalism*. London: Allen Lane-Penguin.

Nadvi, K. (1999). Collective efficiency and collective failure: The response of the Sialkot surgical instrument cluster to global quality pressures. *World Development*, 27 (9): 1605–1626.

Nagy, E., Gajzágó, G., Mihály, M. and Molnár, E. (2021). Crisis, institutional change and peripheral industrialization: Municipal-central state relations and changing dependencies in three old industrial towns of Hungary. *Applied Geography*. 10.1016/j.apgeog.2021.102576.

Pechlaner, H., Raich, F. and Fischer, E. (2009). The role of tourism organizations in location management: The case of beer tourism in Bavaria. *Tourism Review*, 64 (2): 28–40.

Porter, M. (2000). Location, competition, and economic development: Local clusters in the global economy. *Economic Development Quarterly*, 14 (1): 5–34.

Reid, N., Pezzi, M.G. and Faggian, A. (2021) Conclusion – Tourism in peripheral regions: Some challenges. In M.G. Pezzi, A. Faggian and N. Reid (Eds.), *Agritourism, Wine Tourism, and Craft Beer Tourism: Local Responses to Peripherality Through Tourism Niches*, New York: Routledge, 243–254.

Reid, N., Smith, B. and Carroll, M. (2008). Cluster regions: A social network perspective. *Economic Development Quarterly*, 22 (4): 345–352.

Rekers, J.V. and Stihl, L. (2021). One crisis, one region, two municipalities: The geography of institutions and change. *Geoforum*, 124: 89–98.

Rodríguez-Pose, A. (2020). Institutions and the fortunes of territories. *Regional Science Policy and Practice*, 12: 371–386.

Ruoso, L-E and Plant, R. (2018). A politics of place framework for unravelling peri-urban conflict: An example of peri-urban Sydney, *Australia. Journal of Urban Management*, 7 (2): 57–69.

Russell, C., Clark, J., Hannah, D. and Sugden, F. (2021). Towards a collaborative governance regime for disaster risk reduction: Exploring scalar narratives of institutional change in Nepal. *Applied Geography*. 10.1016/j.apgeog.2021.102516.

Sanders, C.G. (2014). Why the positional leadership perspective hinders the ability of organizations to deal with complex and dynamic situations. *International Journal of Leadership Studies*, 8 (2): 136–150.

Sassen, S. (2014). *Expulsions*. Cambridge: Harvard University Press.

Sciarrone, R. and Storti, L. (2019). *Le mafie nell'economia legale. Scambi, collusioni, azioni di contrasto*. Bologna: Il Mulino.

Semi, G. (2015). *Gentrification. Tutte le città come Disneyland?* Bologna: Il Mulino.

Skrimizea, E., Bakema, K., McCann, P. and Parra, C. (2021). Disaster governance and institutional dynamics in times of social-ecological change: Insights from New Zealand, Netherlands and Greece. *Applied Geography*. 10.1016/j.apgeog.2021.10257.

Sölvell, Örjan. (2009). *Clusters – Balancing Evolutionary and Constructive Forces*. Stockholm: Ivory Tower Publishers.

Stark, D.C. (2009). *The Sense of Dissonance. Accounts of Worth in Economic Life*. Princeton: Princeton University Press.

Storti, L. and Dagnes, J. (2021). The super-rich: Origin, reproduction, and social acceptance. *Sociologica*, 15 (2): 5–23. 10.6092/issn.1971-8853/13546.

Thomas, E., Faccin, K. and Asheim, B.T. (2021). Universities as orchestrators of the development of regional innovation ecosystems in emerging economies. *Growth and Change*, 52 (4): 770–789.

Tomaney, J. (2014). Region and place I: Institutions. *Progress in Human Geography*, 38 (1): 131–140.

Trigilia, C. (2020). *Capitalismi e Democrazie. Si possono conciliare crescita e uguaglianza?* Bologna: Il Mulino.

Tuan, Y.-F. 1975. Place: An experiential perspective. *Geographical Review*, 65(2):151–165.

Urso, G. (2021). Metropolisation and the challenge of rural-urban dichotomies. *Urban Geography*, 42 (1): 37–57.

van Leeuwen, E.S. and Halleck Vega, S. (2021). Voting and the rise of populism: Spatial perspectives and applications across Europe. *Regional Science Policy & Practice*, 13(2): 209–219.

Index

Pages in *italics* refer figures, pages in **bold** refer tables and pages followed by n refer notes

Printed in the United States
by Baker & Taylor Publisher Services